Springer Proceedings in Earth and Environmental Sciences

Series editor

Natalia Bezaeva, Moscow, Russia

The series Springer Proceedings in Earth and Environmental Sciences publishes proceedings from scholarly meetings and workshops on all topics related to Environmental and Earth Sciences and related sciences. This series constitutes a comprehensive up-to-date source of reference on a field or subfield of relevance in Earth and Environmental Sciences. In addition to an overall evaluation of the interest, scientific quality, and timeliness of each proposal at the hands of the publisher, individual contributions are all refereed to the high quality standards of leading journals in the field. Thus, this series provides the research community with well-edited, authoritative reports on developments in the most exciting areas of environmental sciences, earth sciences and related fields.

More information about this series at http://www.springer.com/series/16067

Danis Nurgaliev · Natalia Khairullina
Editors

Practical and Theoretical Aspects of Geological Interpretation of Gravitational, Magnetic and Electric Fields

Proceedings of the 45th Uspensky International Geophysical Seminar, Kazan, Russia

Editors
Danis Nurgaliev
Kazan Federal University
Kazan, Russia

Natalia Khairullina
Kazan Federal University
Kazan, Russia

ISSN 2524-342X ISSN 2524-3438 (electronic)
Springer Proceedings in Earth and Environmental Sciences
ISBN 978-3-319-97669-3 ISBN 978-3-319-97670-9 (eBook)
https://doi.org/10.1007/978-3-319-97670-9

Library of Congress Control Number: 2018961198

© Springer Nature Switzerland AG 2019
This work is subject to copyright. All rights are reserved by the Publisher, whether the whole or part of the material is concerned, specifically the rights of translation, reprinting, reuse of illustrations, recitation, broadcasting, reproduction on microfilms or in any other physical way, and transmission or information storage and retrieval, electronic adaptation, computer software, or by similar or dissimilar methodology now known or hereafter developed.
The use of general descriptive names, registered names, trademarks, service marks, etc. in this publication does not imply, even in the absence of a specific statement, that such names are exempt from the relevant protective laws and regulations and therefore free for general use.
The publisher, the authors and the editors are safe to assume that the advice and information in this book are believed to be true and accurate at the date of publication. Neither the publisher nor the authors or the editors give a warranty, express or implied, with respect to the material contained herein or for any errors or omissions that may have been made. The publisher remains neutral with regard to jurisdictional claims in published maps and institutional affiliations.

This Springer imprint is published by the registered company Springer Nature Switzerland AG
The registered company address is: Gewerbestrasse 11, 6330 Cham, Switzerland

Contents

Part I Theoretical Issues of Potential Fields Interpretation

1. **Development of the Finite-Element Technologies in Quantitative Interpretation of Geopotential Fields** 3
   P. I. Balk and A. S. Dolgal

2. **Pareto-Optimal Solutions of Inverse Gravimetry Problem with Uncertain a Priori Information** 13
   T. N. Kyshman-Lavanova

3. **Joint Interpretation of Magnetotelluric and CSAMT Data on the Kola Peninsula (Kovdor Area)** 23
   A. N. Shevtsov

4. **Gridded Potential Fields Data Analysis Based on Poisson Wavelet-Transformations** 31
   K. M. Kuznetsov, A. A. Bulychev and I. V. Lygin

5. **Modified Method S-, F- and R-Approximations in Solving Inverse Problems of Geophysics and Geomorphology** 41
   I. E. Stepanova, I. A. Kerimov, D. N. Raevsky and A. V. Shchepetilov

6. **Reflection of the Petrophysical Basement Rocks Models in Geophysical Fields** 49
   O. M. Muravina, V. N. Glaznev, V. I. Zhavoronkin and M. V. Mints

7. **Continuous Method for Solution of Gravity Prospecting Problems** 55
   I. V. Boikov, A. I. Boikova and O. A. Baulina

8. **The Density Model of the Crystalline Crust the Southwestern Part of the Lipetsk Region** 69
   T. A. Voronova, V. N. Glaznev, O. M. Muravina and I. Y. Antonova

Part II Modern Algorithms and Computer Technologies

9 Neural Network Algorithm for Solving 3d Inverse Problem of Geoelectrics ... 77
M. I. Shimelevich, E. A. Obornev, I. E. Obornev, E. A. Rodionov and S. A. Dolenko

10 Optimization of Computations for Modeling and Inversion in NMR T2 Relaxometry ... 87
L. Muravyev, S. Zhakov and D. Byzov

11 Field of Attraction of Polyhedron and Polygonal Plate with Linear Density Distribution ... 95
K. M. Kuznetsov, A. A. Bulychev and I. V. Lygin

12 Allowance for the Earth's Surface Topography in Processing the Magnetic Field Measurements ... 105
A. S. Dolgal

13 Interpretation Algorithms for Hydrocarbon Deposits ... 113
Yuri V. Glasko

14 Features of Localization of the Poles of the Gravity Potential Regarding to the Field Sources and the Practical Implementation of the «Polus» Method ... 127
G. Prostolupov and M. Tarantin

15 Two Approaches to the Solution of Inversion Problem in the Bear Experiment ... 133
A. A. Zhamaletdinov, M. S. Petrishchev and V. Yu. Semenov

16 About the Numerical Decision of Problem Dirihle for Equation Laplas in a Rectangle in Researches Under the Decision of a Return Problem in Geophysics ... 141
Z. Z. Arsanukaev

17 Calculation of Spherical Layer with Variable Density Gravity Field ... 147
K. M. Kuznetsov, A. A. Bulychev and I. V. Lygin

18 Possibility of Identification of Modeling in Complex Analysis Geological and Geophysical Data ... 157
O. M. Muravina, E. I. Davudova and I. A. Ponomarenko

19 "Native" Wavelet Transform for Solving Gravimetry Inverse Problem on the Sphere ... 163
N. Khairullina (Matveeva), E. Utemov and D. Nurgaliev

Part III Deep Structure Studying

20 Earth's Crust Magnetization Model of the Nether-Polar and Polar Urals 173
N. Fedorova, L. Muravyev and A. Roublev

21 Computer Modeling of Lateral Influence of the Ladoga Anomaly (Janisjarvy Fault Zone) on the AMT Sounding Results 181
A. A. Skorokhodov and A. A. Zhamaletdinov

22 Application of Frequency-Resonance Method of Satellite Images Processing for the Oil and Gas Potential Assessment of "Onisiforos West-1" Well Drilling Site in the Mediterranean Sea 187
S. Levashov, N. Yakymchuk, I. Korchagin and D. Bozhezha

23 Evolution of Ideas on the Nature and Structure of Ladoga Anomaly of Electrical Conductivity 197
A. A. Zhamaletdinov, I. I. Rokityansky and E. Yu. Sokolova

24 The Use of Gravimetry for Studying Shelf of the North Barents Basin 207
M. Chadaev, V. Kostitsyn, V. Gershanok, R. Iblaminov, G. Prostolupov and M. Tarantin

25 An Iterative Solution of the 2-D Non-Linear Magnetic Inversion Problem with Particular Attention to the Anisotropy of Magnetic Susceptibility of Rocks 213
A. B. Raevsky, V. V. Balagansky, O. V. Rundkvist and S. V. Mudruk

26 On 2D Inversion of MTS Data in Tobol-Ishim Interfluve of Western Siberia 223
N. V. Baglaenko, V. P. Borisova, Iv. M. Varentsov, T. A. Vasilieva and E. B. Fainberg

27 Non-hydrostatic Stresses Under the Local Structures on Mars 229
A. Batov, T. Gudkova and V. Zharkov

28 Deep Fluid Systems of Fennoscandia Greenstone Belts 239
A. A. Petrova, Yu. A. Kopytenko and M. S. Petrishchev

29 On Deep Electroconductivity of Tobol-Ishim Interfluve 249
V. P. Borisova, T. A. Vasilieva, S. L. Kostuchenko and E. B. Fainberg

Part IV Geological Interpretation

30 Study of the Magnetic Properties of Geological Environment in Super Deep Boreholes by the Magnetometry Method 259
G. V. Igolkina

31 Underground Water Flows Detection and Mapping by Direct-Prospecting Geoelectric Methods 269
S. Levashov, N. Yakymchuk, I. Korchagin and D. Bozhezha

32 Areas of Negative Excess Density of the Earth's Crust as Sources of Energy for Ore Formation 279
M. B. Shtokalenko, S. G. Alekseev, N. P. Senchina and S. Yu. Shatkevich

33 Distribution of Sources of Magnetic Field in the Earth's Core Obtained by Solving Inverse Magnetometry Problem 285
V. Kochnev

34 Efficiency of High-Precision Gravity Prospecting at Discovery of Oil Fields at Late Stage of Development 293
Z. Slepak

35 Geophysics in Archeology 303
Z. Slepak and B. Platov

36 Using of Probabilistic-Statistical Characteristics in the Interpretation of Electrical Survey Monitoring Observations 313
L. A. Khristenko, Ju. I. Stepanov, A. V. Kichigin, E. I. Parshakov, A. A. Tainickiy and K. N. Shiryaev

37 Multi-Electrode Electrical Profiling Results in the Northern Ladoga Area 321
V. E. Kolesnikov, M. Yu. Nilov and A. A. Zhamaletdinov

38 The Indication in the Potential Fields of Structures Controlling Diamondiferous Magmatism 331
S. G. Alekseev, P. A. Bochkov, N. P. Senchina and M. B. Shtokalenko

39 Horizontal Shear Zones and Their Reflection in Gravitational Field 339
V. Philatov, L. Bolotnova and K. Vandysheva

40 Intermediate Conducting Layers in the Continental Earth's Crust—Myths and Reality 349
A. A. Zhamaletdinov

41	**A Map of the Total Longitudinal Electric Conductivity of the Sedimentary Cover of the Voronezh Crystalline Massif and Its Framing** V. I. Zhavoronkin, V. Gruzdev, I. Antonova and Y. Austova	359
42	**Geophysical Monitoring for the Preservation of Architectural Monuments** Z. Slepak	363
43	**Application of Detailed Magnetics in Intensive Industrial Noise Conditions** P. N. Novikova	371
44	**The Results of Numerical Simulation of the Electromagnetic Field Within the Voronezh Crystalline Massif and its Framing** V. Gruzdev and I. Antonova	377
45	**Predural Depression Structures in the Arctic Urals Magnetic Field** V. A. Pyankov and A. L. Rublev	381
46	**Results of the Complex Airborne Geophysical Survey in the Central African Ridge Area** Yu. G. Podmogov, J. Moilanen and V. M. Kertsman	387
47	**The Forecast of the Structural Surfaces Along the Top of the Pre-Jurassic Base on the Gravitational Field and the Evaluation of Productivity in the Poorly Studied Regions of Western Siberia at Various Stages of Work** N. N. Yaitskii, I. I. Khaliulin and M. V. Melnikova	395
48	**Well Logging During the Processes of Field Development of Native Bitumen and Super-Viscous Oil Deposits** S. I. Petrov, R. Z. Mukhametshin, A. S. Borisov and M. Y. Borovsky	401
49	**Preliminary Results of a Satellite Image Frequency-Resonance Processing of the Gas Hydrate Location Area in the South China Sea** S. Levashov, N. Yakymchuk, I. Korchagin and D. Bozhezha	409

Part I
Theoretical Issues of Potential Fields Interpretation

Chapter 1
Development of the Finite-Element Technologies in Quantitative Interpretation of Geopotential Fields

P. I. Balk and A. S. Dolgal

Abstract Brief description of the assembly method for solving the inverse problem of gravimetry and assessing the reliability confidence validity credibility of interpretational constructions based on the guaranteed approach is presented. It is suggested to estimate the probability of detecting the sources of geopotential fields within the studied geological space by analyzing the variety of the probable interpretations and, then, to use this distribution for criterion-based selecting the model carriers of mass. The synthetic examples of modeling the anomalous disturbing objects are presented.

Keywords Gravimetry · Interpretation · Assembly algorithm · Reliability confidence validity credibility

Quantitative interpretation of gravity anomalies largely employs the fitting method aimed at constructing the unique admissible solution of the inverse problem of gravimetry (IPG) which is assumed to be the best solution. However, because of the practical ambiguity of IPG, a single version of the spatial distribution of anomalous masses is insufficient for objective estimating the accuracy of the interpretation constructions. Such an estimate can only be obtained by a joint analysis of fairly extensive (representative) set Q of admissible solutions of IPG (which ensure the required residual ε between the observed and model fields and satisfy the a priori information about the shapes, sizes, and depths of the anomalous disturbing objects). In the case of the ore-type IPG, set Q can be efficiently constructed by the assembly algorithms which were conceptually introduced in (Ovcharenko 1975;

P. I. Balk
Institute of Applied Geodesy, Berlin, Germany
e-mail: tatianabalk@mail.ru

A. S. Dolgal (✉)
Perm Federal Research Center, Ural Branch, Russian Academy of Sciences,
Perm, Russia
e-mail: dolgal@mi-perm.ru

Strakhov and Lapina 1976) and subsequently developed in the works of P. I. Balk and other researchers.

In the core of the assembly method is integrity of finite-element description of a density medium and a special technique for constructing the approximate solution in the class of such models, not related to the nonlinear optimization methods. We recall that regular tessellation of a plane is the tiling of this plane by a set of regular closed polygons (tessellation elements or tiling elements or simply tiles) ω where the adjacent tiles share a full side; a configurational distribution of masses is the distribution of masses with constant density σ across domain Ω which is the combination of a certain number of tiling elements ω. Tiles ω can be the square bars infinite along the strike (in two-dimensional (2D) case) or cubes (in three-dimensional (3D) case); the sizes of these elements are specified by the interpreter.

The main operations in the class of configurational distributions of masses are performed with the use of the notions of a core $\mathcal{A}[\Omega]$, shell $O[\Omega]$, inner core $\mathcal{A}_0[\Omega]$, and boundary $\Gamma[\Omega]$ of the configuration Ω: $\mathcal{A}[\Omega]$ is the set of elements $\omega_\alpha \in \Omega$; $O[\Omega]$ is the set of all elements $\omega_\alpha \notin \mathcal{A}[\Omega]$ that border on the elements of the core $\mathcal{A}[\Omega]$; $\mathcal{A}_0[\Omega]$ is the set of elements $\omega_\alpha \in \Omega$ that only border on the elements of the same core; and $\Gamma[\Omega]$ is the set of the elements $\mathcal{A}[\Omega]$ that are not included in $\mathcal{A}_0[\Omega]$.

In the simplest IPG statement for an isolated body S with the known density σ, the assembly principle of solving the inverse problem consists in the following: proceeding from a given connected configuration Ω^0 (whose role can be played by the single element ω_0—the center of the crystallization), to construct a finite sequence $\Omega^0, \Omega^1, \Omega^2 \ldots$ converging to a certain limiting configuration Ω^* the field of which, with the fitted density $\sigma^* \approx \sigma^T$, agrees with the measurements of the gravitational field Δg. We note that the connectedness in the assembly algorithms helps to separate the domains in the geological medium that are occupied by the masses with different values of a physical parameter. The simple connectedness of the anomalous bodies implies the absence of voids in them.

In the controlled directed crystallization (CDC) modification, a current approximation is obtained by introducing into the core $\mathcal{A}[\Omega^n]$ a certain (one) element of $O[\Omega^{n-1}]$ that provides the minimal root-mean-square fitting error. The conversion from configuration Ω^{n-1} to configuration Ω^n takes into account the main a priori information about the location, shape, and size of the anomalous bodies that is typically available for the interpreter.

By using the simplest logical operations with indices α of tiling elements $\omega_\alpha \in \Gamma[\Omega]$, we can easily and efficiently control the fulfillment of various a priori constraints including those for the domain that surely contains the source of the field Ω^T and for the domain that surely does not contain this source; for the minimal and maximal probable top and bottom depths of the anomalous object Ω^T; for the vertical and horizontal thicknesses of object Ω^T; for the surface smoothness of object Ω^T, etc.

1 Development of the Finite-Element Technologies …

To make the CDC method more transparent, let us illustrate it by a simple model example. As a carrier of masses with density $\sigma = 0.3$ g/cm^3 we consider an infinite long horizontal prism S. The cross section of the prism is the configuration constructed of 88 squared tiling elements ω with a side of 25 m. The "observed" values of the gravitational field Δg are specified at 36 points on the profile with a step of 50 m and are contaminated by noise ξ with the mean value fairly close to zero and rms deviation of about 0.015 mGal (Fig. 1.1). The noise is constructed by averaging a sequence of a Gaussian random quantity in a moving window with a length of 5 points. It is a priori assumed that the sought anomalous body is simply connected, is limited by a sufficiently smooth boundary, its thickness is at most 1 km laterally and at most 0.5 km vertically; and the depth of its bottom is at most 1 km. The behavior of the residual and optimized density with increasing number t of the iteration is illustrated in the graphs of Figs. 1.2 and 1.3 shows several intermediate iterations approximations including the final IPG solution Ω^*.

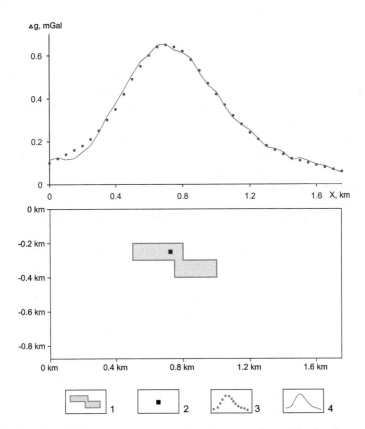

Fig. 1.1 Density model: 1—anomaly-forming prism; 2—center of crystallization; 3—model field Δg; 4—"observed" field

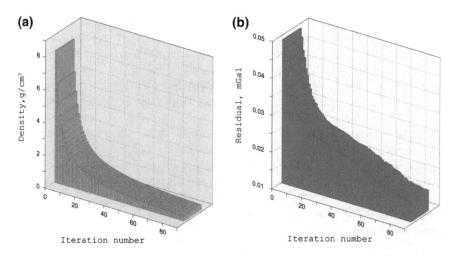

Fig. 1.2 Graphs illustrating the behavior of the residual ε between the observed and model field (a) and density of current configurations Ω^i (b) obtained by CDC solution of the inverse problem of gravimetry

Let us consider the set $Q = Q(\varepsilon)$ of the admissible solutions of the inverse problem for one body in the finite-element class of the models (denote the elements of this class by Ω_i, $i = 1, 2,..., m$) and its two subsets: $D_1 = \bigcup \Omega_m$ containing all the possible IPG solutions (the sum the union of the solutions) and $D_2 = \bigcap \Omega_m$ containing the fragments of the sources that pertain to the sought objects over the entire set of the solutions (the intersection of the solutions). Set Subset D_2 is a fragment that is guaranteed to pertain to the disturbing object S whereas set subset D_1 allows us to delineate the spatial domain that can contain the sought object: $D_2 \subset S \subset D_1$. The pair $\langle D_1, D_2 \rangle$ can be considered as an alternative representation of the results of the quantitative interpretation, and the measure of ε—equivalence can be estimated by metric $\tau(Q) = 1 - \mu(D_2)/\mu(D_1)$ where μ is the Lebesgue measure (Balk 1980).

Let us illustrate the application of the guaranteed approach in the case of joint determination of the physical and geometrical parameters of the object from the noised measurements of gravity. The gravity anomaly Δg is caused by an isolated convex body S with excess density $\sigma = 0.2$ g/cm³; the rms error in the gravity field measurements $\Delta \tilde{g}(x_i)$ is at most 0.1 mGal (Fig. 1.4a). The results of the interpretation with the residual between the observed and model field $\varepsilon \leq 0.2$ mGal in terms of the pair $\langle D_1, D_2 \rangle$ for the case when the exact σ value is specified a priori are shown in Fig. 1.4b and for the case when it is only known that the true value is located in a sufficiently wide interval [0, 1; 0, 3] g/cm³ in Fig. 1.4c. In the second case, as the additional information it is assumed that the maximal vertical thickness of the body is at most 2 km (actually it is 1.75 km). It can be easily seen that in these two variants, the sizes of domain D_1 that is guaranteed to contain the anomaly-forming body are commensurate. Domain D_2 that is surely a part of the unknown body S in the variant with the fuzzy a priori information about density is

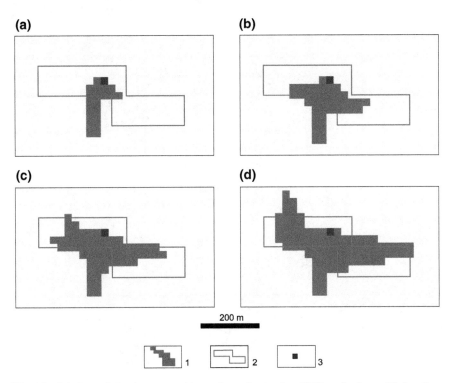

Fig. 1.3 Solution of the inverse problem of gravimetry by CDC method: **a**—20 iterations ($\sigma = 1.248$ g/cm^3; $\varepsilon = 0.026$ mGal); **b**—$\sigma = 40$ iterations (0.656 g/cm^3; $\varepsilon = 0.023$ mGal); **c**—60 iterations ($\sigma = 0.439$ g/cm^3; $\varepsilon = 0.020$ mGal); **d**—result of interpretation Ω^*, 88 iterations ($\sigma = 0.298$ g/cm^3; $\varepsilon = 0.016$ mGal): 1—fitted model; 2—anomalous prism; 3—center of crystallization

only by 15% lower than in the variant with the known density. That is how the principle of interchangeability of the a priori information about the physical and geometrical parameters of the field source (Balk et al. 2012) manifests itself.

The development of the guaranteed approach allows us to estimate the probability of the presence of disturbing masses within the domain D_1/D_2. By directly checking each "elementary" volume ω_i of the density model, we can establish the number m_i of the constructed carriers $\Omega_m \in Q$ for which this elementary volume is a fragment. If the unknown true carrier of masses S is among these carriers, $\omega_i \subset S$. Correspondingly, the frequency with which the IGP solutions containing domain ω_i occur among the entire set of n obtained admissible IGP solutions can be accepted as the estimate of the sought probability $p_i = m_i/n$. Let us refer to the function of the spatial coordinates $\varphi(\omega) = p(x, y, z)$ with the domain of definition [0,1] as the localization function characterizing the structure of domain Q (Dolgal and Sharkhimullin 2011). Also the other characteristics having the similar meaning can be suggested, e.g., guarantee functions, confidence functions, and detection functions (Balk and Dolgal 2016).

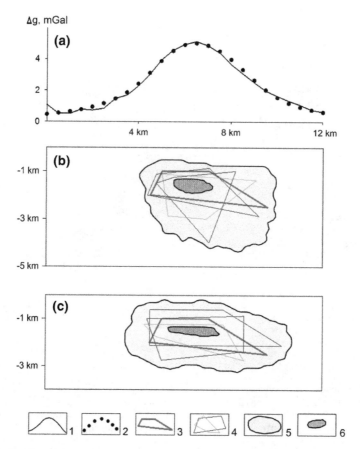

Fig. 1.4 Results of the interpretation of gravity anomaly (**a**) with the density of its source specified by the exact value (**b**) and specified by the interval containing the true value (**c**): 1—"observed" gravity field; 2—fitted gravity field for one of the admissible IGP solutions; 3—anomaly-forming object (source); 4—individual admissible IGP solutions; 5—domain D_1; 6—domain D_2

Let us illustrate the construction principle and the possibilities of the localization function $\varphi(\omega)$ in the 2D IGP statement. Here, the anomalous bodies are associated with their vertical cross sections. Gravity anomaly Δg is caused by three bodies having the effective densities $\sigma_1 = 0.15$, $\sigma_2 = 0.20$, and $\sigma_3 = 0.30 \, \text{g/cm}^3$; the anomaly is measured with a step of 500 m on the profile with a length of 40 km; the measurements are complicated by a weak noise. The model objects can be likened to the certain horizontally extended intrusive bodies located in the graviactive crustal layer with a thickness of about 10 km. The statement of the problem includes the typical constraints: the admissible solutions of the inverse problem are three local (connected) carriers (S^j, $j = 1, 2, 3$) (corresponding to the number of the local maxima of Δg) each of which is simply connected (does not contain voids) and has a sufficiently smooth boundary. The threshold residual ε between the observed and model field is 0.1 mGal.

By selecting different zero approximations for each of the three local bodies, we constructed 1182 variants of density section (3546 admissible carriers S^j). A certain idea of their scatter can be drawn from the following fact: the area (measure μ) of the region covered by these carriers is more than 2.5 times as high as the total area of the real disturbing objects. Figure 1.5 shows the vertical map of the isolines of function $\varphi(\omega)$ constructed from its discrete values within the tiling elements ω—squares with a size of 250 × 250 m. Of course, the interpretation results expressed in terms of the localization function are much more informative than the individual particular IGP solution. For verifying the anomaly by drilling, it is sufficient to highly reliably identify one of the fragments of domain S occupied by the disturbing masses in order to subsequently select the location and depth of the drilling well. However, the results of the quantitative interpretation which do not describe the supposed boundaries of the anomalous bodies are not always applicable. In particular, constructing the expected density boundaries is necessary for solving the problems associated with studying the deep geological structure of the ore regions.

For constructing the particular interpretation models that are preferable over the other ones from the set Q it is reasonable to use various criteria of the decision-making theory (Balk and Dolgal 2015; 2017a). For instance, we may select the IGP solution Ω_1 consisting of m elements for which the average sum of the values of the localization function $\Theta = \sum_1^m \varphi(\omega)/m$ is maximal compared to the other variants of the interpretation. This solution Ω_1 corresponds to the criterion of the maximum of the a posteriori probability (Balk and Dolgal 2017b) and gives the residual of the fields $\varepsilon = 0.030$ mGal (Fig. 1.6a). The traditional method for selecting the "best" IPG solution that corresponds to the minimal residual of the fields $\varepsilon = 0.026$ mGal yields a noticeably worse result (Fig. 1.6b).

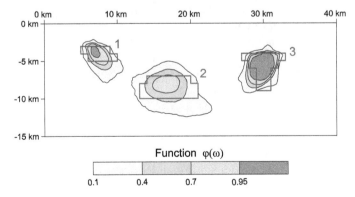

Fig. 1.5 Vertical map of the isolines of the localization function for the model of three of bodies. Field sources and their numbers are shown in red

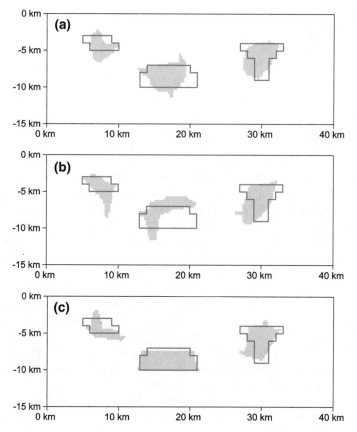

Fig. 1.6 Solutions of the inverse problem of gravimetry: Ω_1 (**a**), Ω_2 (**b**), Ω_3 (**c**). Red polygons show the sources of the field

In addition to the visual estimate let us also quantify this advantage. As a measure of the closeness between the true anomaly-forming object S and its model Ω we consider the ratio of the area (measure μ) of their common fragment to the area of the zone whose points pertain to at least one of them: $\rho(S, \Omega) = \mu(S \cap \Omega)/\mu(S \cup \Omega)$. In the idealized situation when the object and the results of the interpretation completely coincide, the value of $\rho(S, \Omega)$ is maximal and equal to 1; if the object and the results of the interpretation have no common points, $\rho(S, \Omega) = 0$. In our example, $\rho(S, \Omega_1) = 0.55$ (the area of the common fragment of S and Ω_1 is $\sim 71\%$), whereas $\rho(S, \Omega_2) = 0.37$ (here, the area of the common fragment is $\sim 54\%$). However, the maximal value of metrics $\rho(S, \Omega_3) = 0.67$ is provided by the solution Ω_3 (the area of the common fragment of S and Ω_3 is $\sim 73\%$) which has a rather high residual $\varepsilon = 0.052$ mGal (Fig. 1.6c).

Analyzing the entire set of the obtained admissible solutions of IPG, we see that according to the results of the numerical experiment there is a fairly close

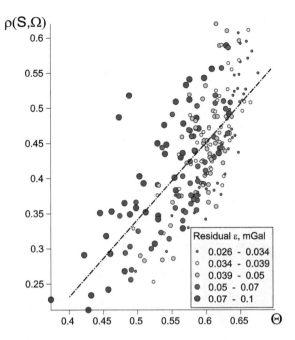

Fig. 1.7 Cross-plot and linear regression dependence $\rho(S,\Omega) = 1.195\Theta - 0.268$ (dash-dotted line). Color lines show the parameters for each tenth IPG solution of set Q

relationship between the quality of carrier ρ overall and the average value Θ of the fragment-based estimates of its quality by the criterion of the maximum a posteriori probability (Balk and Dolgal 2017b). Coefficient of the correlation between ρ and Θ over the set of 1182 IPG solutions is 0.745 (Fig. 1.7). For comparison, the coefficient of the correlation between ρ and residual ε is as low as 0.277.

The described results of our study on estimating the accuracy of quantitative interpretation of gravity anomalies suggests the following conclusions:

(1) As a new form of representation of the results of solving the ore-type IPG, it is suggested to use the information package that includes several distributions of the estimating functions such as $\varphi(\omega)$ and one or a few interpretation models satisfying the optimality criteria (such as Ω_1).

(2) It is obtained that in a number of the experiments, the set Q contains, inter alia, the "best" solutions such as Ω_3 which cannot be revealed without the exact information about the initial model of the field sources. On one hand, this is an additional argument validating the application of the criteria based on estimating the closeness of two different solutions of IPG. On the other hand, this provides potential to increase the accuracy of interpretation constructions by assembling the new partial solutions from elements $\omega \in Q$ with the allowance for the already existing estimates of the probability $\varphi(\omega)$ of these elements to belong to the true carrier of masses S.

References

Ovcharenko, A.V. (1975), Fitting the cross-section of a 2D body based on the gravity field, *Vopr. Neft. Rud. Geofiz.*, Alma-Ata: Kazakh. politekh. inst., 1975, vol. 2, pp. 71–75.

Strakhov, V.N. and Lapina, M.I. (1976), Assembly method for solving the inverse problems of gravimetry, *Proc. Acad. Sci. USSR*, 1976, vol. 227, no. 2, pp. 344–347.

Balk, P.I. (1980), On the reliability of the results of quantitative interpretation of gravity anomalies, *I Proc. Acad. Sci. USSR, Phys. Earth*, 1980. no. 6, pp. 43–57.

Balk, P.I., Dolgal, A.S., and Khristenko, L.A. (2012), Localization of geological objects based on the data of gravity prospecting with incomplete information about the density of rocks, *Dokl. Earth Sci.*, 2012, vol. 442, no. 2, pp. 262–266.

Dolgal, A.S. and Sharkhimullin, A.F. (2011), The increase of the interpretation accuracy for monogenetic gravity anomalies, *Geoinformatika*, 2011, no. 4, pp. 49–56.

Balk, P.I. and Dolgal, A.S. (2016), Additive technologies of quantitative interpretation gravitational anomalies, *Geofizika*, 2016, no. 1, pp. 43–47.

Balk, P.I. and Dolgal, A.S. (2015), A minimax approach to the solution of inverse problems of gravity and magnetic prospecting, *Dokl., Earth Sci.*, 2015, vol. 462, no. 2, pp. 648–652.

Balk, P.I. and Dolgal, A.S. (2017), Inverse problems of gravity prospecting as a decision-making problem under uncertainty and risk, *Izv., Phys. Solid Earth*, 2017, vol. 53, no. 2, pp. 214–229.

Balk, P.I. and Dolgal, A.S. (2017), New possibilities for increasing the informativity of quantitative interpretation of gravity anomalies, *Dokl., Earth Sci.*, 2017, vol. 476, no. 4, pp. 461–465.

Chapter 2
Pareto-Optimal Solutions of Inverse Gravimetry Problem with Uncertain a Priori Information

T. N. Kyshman-Lavanova

Abstract The inverse problem of gravimetry under uncertainty of heterogeneous a priori information is solved. An algorithm using the possibilities of deterministic and probabilistic approaches is developed. In the framework of the probabilistic approach, a priori distribution of model parameters described by fuzzy sets. A deterministic approach is used to calculate fields from a given distribution of model parameters and formalization of a priori information through natural restrictions. Since the establishment of this algorithm is independent, it can be used for solving a wide range of nonlinear geophysical inverse problems.

Keywords Inverse problem · Gravimetry · Uncertain a priori information

Introduction

The consistency of the method for solving the inverse problem is determined by its ability to use both deterministic and statistical approaches to obtain new data on the basis of available information.

In geophysical inverse problems, there are two types of uncertainty: the uncertainty of the observed data is probabilistic, this is the result of random observation errors. The uncertainty of the a priori information about model parameters is often of a non-probabilistic nature, that is, it is associated with a lack of knowledge about explored object. It should be described formally using an non probabilistic measure of uncertainty.

The experience of successful application of non-probabilistic methods of describing uncertainty is well known for the inversion of seismic data (Kozlovskaya 2000; Sambridge 1999; Tarantola 2005; Sambridge and Drijkoningen 1992; Stoffa and Sen 1991). For problems of gravimetry, the paper (Balk et al. 2011), where the

T. N. Kyshman-Lavanova (✉)
Institute of Geophysics, NAS of Ukraine, Kiev, Ukraine
e-mail: kltam@ukr.net

linear inverse problem of gravimetry and magnetometry is solved, is known within the concept of combining deterministic and probability methods.

In the article presented, we show how the uncertainty of a priori information is described by fuzzy sets, and how the non-probabilistic methods, that manipulate uncertain information, can be adapted to solve the inverse nonlinear gravimetric problem. The problem is formulated as a multicriteria optimization problem, a priori information is one of the criteria, and the solution of the problem is a set of interpretational models that satisfy a priori data with a given accuracy.

Description Undefined a Priori Information

The improbability measures to describe the uncertainties have been significantly developed in the 1970s. In particular, some theories that generalize or complement the probability theory were introduced during this decade (Sugeno 1977; Zadeh 1978).

The use of non-probability measures to describe uncertainty can provide a convenient way of representing a priori information in the inverse problems and build an efficient computer algorithm for inversion of geophysical data. It is therefore advisable to elaborate on the possibility theory, based on the definition of fuzzy sets because of its relative computational simplicity in comparison with the theory of probability.

Fuzzy set theory is a well-developed area of mathematics. Its description can be found in Zadeh (1978), Zimmermann (2001). We give the definition of fuzzy sets.

Let U—the so-called universal set, elements of which formed all the other sets under consideration in this class of problems. Fuzzy set A is the set of pairs:

$$A = \{\langle x, \mu_A(x)\rangle | x \in U\},$$

where μ_A is the membership function, i.e. $\mu_A : U \to [0, 1]$. The membership function is an analog of the characteristic binary function in the normal sets.

The uncertainty of information about model parameters of the non-probabilistic nature can be described by means of fuzzy sets. The advantage of this approach is that it provides the possibility of incorporating a wide range of non-probabilistic a priori information in the inversion procedure, and can be applied for the solution of nonlinear problems (Kozlovskaya 2000).

Statement of the Problem

We formulate the problem based on the use of fuzzy sets.

Let geological objects under the earth's surface belong to parameterized region, i.e. described by a set of parameters $\mathbf{m} = [m_1, m_2, \ldots, m_k] \in M$, where M is a parametric space. Each model $\mathbf{m}$ is regarded as a point in a model space M.

In n points of the earth's surface the observed gravity field has a value $\Delta g_{obs} = (g_1, g_2, \ldots, g_n)$. A priori information about the geological object is derived from some experimental observations. Probabilistic approach to inverse problem describes a priori information about the model using the probability density function $p(\mathbf{m})$. The problem is to search a posteriori probability density distribution $p(\mathbf{d})$ of the vector $\mathbf{m}$ on the basis of observed data, the theoretical relationships between model parameters and observed field, and a priori information (Tarantola and Valette 1982):

$$p(\mathbf{d}) = k\, p(\mathbf{m})\, L(\mathbf{m}),$$

where k is the appropriate normalizing constant, $L(\mathbf{m})$—functional, which compares the fit between observed and theoretical data.

In practice in most geophysical experiments it is assumed that both a priori and a posteriori density is Gaussian.

$$p(\mathbf{m}) = \frac{1}{(2\pi)^{L/2}|C_M|^{1/2}} \exp\left\{-\frac{1}{2}[\mathbf{m} - \mathbf{m}_0]^T C_M^{-1}[\mathbf{m} - \mathbf{m}_0]\right\}$$
$$p(\mathbf{d}_{obs}/\mathbf{m}) = \frac{1}{(2\pi)^{L/2}|C_D|^{1/2}} \exp\left\{-\frac{1}{2}[g(\mathbf{m}) - \mathbf{d}_{obs}]^T C_D^{-1}[g(\mathbf{m}) - \mathbf{d}_{obs}]\right\} \quad (2.1)$$

where $\mathbf{m}_0$—a priori model, C_M and C_D—the covariance matrix of the model and observed data, respectively, $|C_M|$ and $|C_D|$—the corresponding determinants.

With these assumptions, the maximization of (2.1) is equivalent to minimize the following objective function:

$$L(\mathbf{m}) = [g(\mathbf{m}) - \mathbf{d}_{obs}]^T C_D^{-1}[g(\mathbf{m}) - \mathbf{d}_{obs}] + (\mathbf{m} - \mathbf{m}_0)^T C_M^{-1}(\mathbf{m} - \mathbf{m}_0) \quad (2.2)$$

The minimization (2.2) can be performed efficiently in the case when $g(\mathbf{m})$ linear and matrix C_M and C_D diagonal. Since the model space is usually multidimensional, then and appropriate a priori distribution is usually quite difficult.

If statistical estimates of a priori information are impossible, then the non-probabilistic a priori information can be described by away more efficient than using the probability density function PDF, namely, by fuzzy sets. An important advantage is that the basic operations of fuzzy sets provide a rather convenient combination of various precise and fuzzy constraints on the model parameters. However, we must remember that the membership function cannot be used instead of the a priori PDF in the classical formulation of the inverse problem, as it not only express two different types of uncertain information, but also correspond to different measures of uncertainty which should satisfy various axioms. To combine different types of uncertainty in inversion scheme, we must change the formulation of the inverse problem.

Transformation of Traditional One-Objective Geophysical Inverse Problem in the Multi-objective Optimization Problem

Let X is a fuzzy set of possible solutions defined in the parametric space with the membership function $\mu_M(\mathbf{m})$ and let $p(\mathbf{d}_{obs}|\mathbf{m})$ be the conditional probability density function of the experimental data (i.e., the probability of obtaining the experimental data for certain values of model parameters).

Then the solution of the inverse problem should:

- to maximize the membership function of a fuzzy set of possible solutions $\mu_M(\mathbf{m})$;
- to maximize the conditional probability density PDF of the observed data $p(\mathbf{d}_{obs}|\mathbf{m})$.

This optimization problem is a multiple objective optimization with two objective functions, that is

$$\mathbf{F}(\mathbf{m}) = (F_1(\mathbf{m}), F_2(\mathbf{m})), \qquad (2.3)$$

where $F_1(\mathbf{m}) = p(\mathbf{d}_{obs}|\mathbf{m})$, $F_2(\mathbf{m}) = \mu_M(\mathbf{m})$.

It is important to note that in the case of one objective optimization problem, the quality of the solution is estimated by the discrepancy function for each solution in the parametric space. In the case of a multipurpose problem, each solution has $L \geq 2$ estimated values, one for each objective function. Solutions with multiobjective values we can to compare with the concept of non-dominance.

For a problem with more than one objective function (i.e., f_j, $j = 1, \ldots, L$ and $L > 1$), two arbitrary solutions x and y can be in the following relations: one dominates the other, or none of the solutions dominates the other.

The solution x dominates the solution y if

1. x is not less than y in all components, that is $f_j(x) \geq f_j(y)$, for all $j = 1, 2, \ldots, L$.
2. x is strictly greater than y, at least for one component, that is $f_j(x) > f_j(y)$, for at least one $j = 1, 3, \ldots, L$.

The solution is Pareto-optimal if it does not dominate in the target space by any other solution, that is $x^* \in R^k$, there is a Pareto-optimal or non-dominant solution of the multipurpose inverse problem with the objective function vector $f(x) = (f_1(x), f_2(x), \ldots, f_L(x))$, if and only if there is no vector $x \in R^k$ such that $f_i(x^*) \leq f_i(x) \quad \forall i \in \{1, 2, \ldots, L\}$ and $f_j(x^*) < f_j(x)$ at least for one $j \in \{1, 3, \ldots, L\}$.

Pareto-set provides a compromise between the criteria for a non-dominant solution: one criterion improves, the other deteriorates.

Algorithm of Global Optimization Using Pareto-Set Definition

For an effective search in a multidimensional parametric space, we use the approach proposed by Sambridge (1999) in his neighborhood algorithm, that is, the approximation of a parametric space by Voronoi diagrams.

Sambridge showed how Voronoi diagrams can be used in global optimization algorithms to reduce the number of direct problem calculations and to increase the resolution of the research area. The calculation of the direct problem is replaced by the search for the nearest element of the Voronoi diagram. In the algorithm proposed in the article, Voronoi's division is used to determine the search area of Pareto-optimal points. The search is performed in a manner similar to that described in Sambridge (1999), that is, the search step is determined by the size and boundaries of the Voronoi cell around the perturbed point.

We use the algorithm of global optimization, proposed by Kozlovskaya (2000), which realized in three stages.

The first stage is the modeling of the initial population from the fuzzy set of feasible solutions. Calculations of the forward problem are not required.

The second stage is the calculation of the initial Pareto-optimal set, which requires the estimation of the misfit function and solution of the forward problem at each point of the initial population.

The third stage is the perturbation of points from the obtained Pareto-optimal set.

The algorithm works until the neighborhoods of all Pareto-optimal points are examined and the distance between the new generated point and its nearest neighborhood becomes less than a certain threshold value. The choice of the final solution from the set P obtained in the last stage can be done using a trade-off analysis between the values of the fuzzy set membership function and misfit function. It is also possible that the finite Pareto set contains only one solution.

Adaptation of the Algorithm to the Inverse Problem of Gravimetry for Three-Dimensional Contact Surfaces

Consider a fairly simple model example to illustrate the action of the proposed algorithm.

Consider the inversion of the nonlinear problem of gravimetry for one contact surface in a limited area of investigation. It belongs to the class of incorrect problems. A detailed statement of the problem is given in (Bulakh and Kyshman 2006), where it is solved by the method of gradient descent.

Here we briefly give the main points of the parameterization of the problem.

The contact surface is determined by the position of the horizontal plane $\zeta = H_0$. At each point $[(\xi_0, \eta_0)_j, j = 1, 2, \ldots, m]$ the surface deviates from this fixed plane by an amount

$$Z = Z(\xi, \eta) = \sum_{j=1}^{m} \frac{Q1_j}{\left[1 + Q2_j(\xi - \xi_{oj})^2 + Q3_j(\eta - \eta_{oj})^2\right]^\alpha} \quad (2.4)$$

The values of the parameters Q1, Q2, Q3 determine the surface configuration. The parameter α is fixed, $\alpha = 2$.

Thus, the contact surface is determined by the function

$$H(\xi, \eta) = H_0 - Z(\xi, \eta) \quad (2.5)$$

Let's write down the parameters that define the geological model

$$P = \{jk; [\sigma; H_0; W1; W2]_t; \quad t = 1, 2, \ldots, jk\} \quad (2.6)$$

$$W1 = [\alpha; m; (\xi_0, \eta_0)_j, j = 1, 2, \ldots, m]; \quad W2 = [(Q1; Q2; Q3)_j, j = 1, 2, \ldots, m].$$

Here *jk* is the number of contact surfaces, σ—the excess density of masses, which are located below the interface, H_0—the horizontal plane, the zero level, relative to which the marks of the relief of the contact surface are counted.

The membership functions for the variable model parameters are written as follows:

$$\mu_{Q1}(Q1) = \begin{cases} 1, & \text{if } 0 < Q1 < 1 \\ 0, & \text{if } Q1 \leq 0 \wedge Q1 \geq 1 \end{cases}$$

$$\mu_{Q2}(Q2) = \begin{cases} 1, & \text{if } 0 < Q2 < 0.5 \\ 0, & \text{if } Q2 \leq 0 \wedge Q2 \geq 0.5 \end{cases}$$

$$\mu_{Q3}(Q3) = \begin{cases} 1, & \text{if } 0 < Q3 < 0.5 \\ 0, & \text{if } Q3 \leq 0 \wedge Q3 \geq 0.5 \end{cases}$$

Let the a priori data on the depth to the contact surface described by the following possible statements:

1. it is possible that H_0 it has values from 1 to 2 km, or less;
2. It is unlikely that H_0 more than 4 km.

In this case, we can describe the parameter by a possible distribution associated with a fuzzy set with the membership function.

$$\mu_H(H_0) = \begin{cases} H_0, & 0 < H_0 < 1 \\ 1, & 1 \leq H_0 \leq 2 \\ 2 - H_0/2, & 2 < H_0 < 4 \end{cases} \quad (2.7)$$

It is obvious that the probability density function of a priori information in this case cannot be considered Gaussian and it is asymmetrical, since $H_0 \geq 0$.

First construct fuzzy sets for different parameters of the models, then combine them into one overall fuzzy set, using the definition of Cartesian product. Ultimately, the membership function of the model will be written

$$\mu_M(m) = \min(\mu_{H0}(H0), \mu_{Q1}(Q1), \mu_{Q2}(Q2), \mu_{Q3}(Q3)) \qquad (2.8)$$

Equation (2.8) can be used to compute values of membership function of the model. The second objective function can be written because it displays the location of the Pareto sets in parametric space:

$$L(m) = [g(m) - t]^T C_D^{-1} [g(m) - t] \qquad (2.9)$$

Equations (2.8) and (2.9) form the vector of objective function multi-purpose optimization problem.

The model for the observed field is the uneven distribution of points on the square 6×6 km, depth $H_0 = 1$ km, the excess density of 0.5 g/cm^3. The other parameters are given in Table 2.1.

The initial population of the models was modeled in the parametric space in accordance with the membership function of a fuzzy set of possible solutions (step 1–3 of the algorithm). It was done 2 test model on regular network 6×6 with a step of 1 km for different values of H_0, Q1, different amounts of points in the initial population and different threshold values ε (Table 2.2).

Table 2.1 The model parameters used to calculate the observed field

N	x_0	y_0	Q1	Q2	Q3
1	−1	2	0.5	0.2	0.2
2	0	−1	0.6	0.2	0.2
3	0	0	0.3	0.2	0.2
4	1	1	0.2	0.2	0.2
5	2	0	0.3	0.2	0.2
6	3	1	0.3	0.2	0.2
7	3	2	0.5	0.2	0.2
8	3	3	0.3	0.2	0.2

Table 2.2 The parameters used to generate the initial population of the model problem

№ test	The number of points in initial population	ε	The parameters of the membership function of a fuzzy set	
			H_0	$Q1_j$
1	60	0.8	1	0.4
2	80	0.7	1.2	0.3

The first test was done for 60 points of the initial population, and a wide region of possible values of depth. Only one Pareto-point was selected from the initial population. The final solution was obtained as the result of a direct search in the neighborhood of this point (Fig. 2.1).

The second test was performed for the same region, but the number of points of the initial population was more. The initial Pareto set, obtained in the second step of the

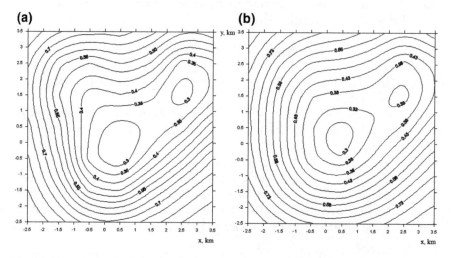

Fig. 2.1 Pareto-optimal solution for the first test. **a** given the topography of the contact surface; **b** obtained surface topography. The function of residuals is 2.44. The membership function is 0.96

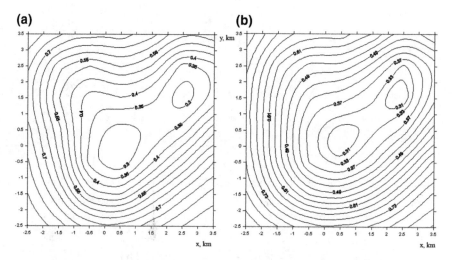

Fig. 2.2 The solution for the second test with two Pareto-optimal points. **a** given the topography of the contact surface; **b** obtained surface topography. The functions of the residual F1 = 2.16, F2 = 3.51. The membership function is 0.97

algorithm contains two different Pareto-optimal solutions, obviously located in the vicinity of two different local minima. Only one Pareto-optimal solution left after performing a direct search in the vicinity of both points and update Pareto set (Fig. 2.2).

The previous two examples show, if the initial search region with equal probability is large, then the number of points in initial population should be large enough in order parameter space during the search was studied in more detail and to avoid the convergence of the algorithm to the same local minimum.

The test results show that the algorithm allows to obtain a satisfactory solution at the stage of search of Pareto-set. Subsequent direct search in the surroundings of Pareto-optimal points greatly reduces the function of the residuals and discards some local minima.

Conclusion

The key idea of this work is to develop an algorithm for the inversion of gravity data in terms of uncertainty of heterogeneous prior information, using the capabilities of deterministic and probabilistic approaches. In the framework of the probabilistic approach, the a priori distribution of model parameters described by fuzzy sets. A deterministic approach is used to calculate fields from a given distribution of model parameters and formalization of a priori information through natural restrictions.

Since the establishment of this algorithm is independent, it can be used for solving a wide range of nonlinear geophysical inverse problems.

References

Balk P.I., Dolgal A.S., Michurin A.V. (2011) Mixed probabilistic and deterministic methods of solving linear inverse problems of gravimetry and magnetometry. Geophysics 2:20–29.

Bulakh E., Kyshman-Lavanova T. (2006) Another approximation approach to solution of inverse problems of gravimetry in the class of three-dimensional contact surfaces. Geophys J 28 (2): 54–62.

Kozlovskaya E. (2000) An algorithm of geophysical data inversion based on non-probalistic presentation of a priori information and definition of Pareto-optimality. Inverse problem 16: 839–861.

Sambridge M. (1999) Geophysical inversion with a neighbourhood algorithm – I. Searching a parameter space. Geophysical journal international 138 (2): 479–494.

Sambridge M., Drijkoningen G. (1992) Genetic algorithms in seismic waveform inversion. Geophys J Int 109: 323–342.

Stoffa P.L., Sen M.K. (1991) Nonlinear multiparameter optimization using genetic algorithms: inversion of plane wave seismograms. Geophysics 56:1749–1810.

Sugeno M. (1977) Fuzzy measures and fuzzy integrals: a survey. *Fuzzy Automata and Decision Processes*. North-Holland, New York, p 89–102.

Tarantola A. (2005) Inverse problem theory and methods for model parameter estimation. SIAM.

Tarantola A., Valette B. (1982) Generalized nonlinear inverse problem using the least squares criterion. Rev. Geophys. Space 20: 219–232.

Zadeh L.A. (1978) Fuzzy sets as a basis of a theory of possibility. Fussy Sets Syst 1: 3–28.

Zimmermann HJ (4th ed.) (2001) Fuzzy set theory—and its applications. Dordrecht, Kluwer.

Chapter 3
Joint Interpretation of Magnetotelluric and CSAMT Data on the Kola Peninsula (Kovdor Area)

A. N. Shevtsov

Abstract Both processes- and with a controlled source and fields of natural origin, contain information about the Earth. The data of the CSAMT research and the results of the AMT-MT measurements should complement each other. To this point of view, in the Geological Institute of RAS has developed a set of programs for data processing and interpretation of the results of joint data of the magnetotelluric research and data of the frequency electromagnetic sounding with controlled sources. Presents the results of the joint inversion of CSAMT data at distances 50 km from two mutually-orthogonal horizontal electric dipoles in the frequency range of 3.82–2185 Hz ("Kovdor-2015") in west sector Kovdor-Jona area and AMT-MT measurements of the 2016 year in point of the disposition of the dipole sources.

Keywords CSAMT · Magnetotelluric · Interpretation · Data processing

Introduction

In frequency electromagnetic sounding method (Frequency Sounding (FS)—in Russia or Controlled Source Audio-Magneto-Telluric CSAMT—foreign jobs) with controlled sources is used idea of harmonic process. As a rule, are assumed known source parameters (geometry and current of the source as a function (harmonic or represented by series harmonics) of time) (Vanyan 1997). In addition, when you use linear grounded sources (electric dipole, a long grounded line) electromagnetic field has a clear elliptical polarization. Processing of MT-AMT soundings requires to abandon presentation of harmonic fields, as real magneto-telluric process is not described by a model of the harmonic fields (Semenov 1985). Instead, it uses the notion of a stationary stochastic process that satisfies a ergodicity condition. Besides, assumes the existence of a linear relationship between the components of the field and of magneto-telluric impedance constancy in a narrow frequency band.

A. N. Shevtsov (✉)
Kola Science Center of RAS, Geological Institute, Apatity, Murmansk region, Russia
e-mail: anshev2009-01@rambler.ru

During the measurements by method of FS (CSAMT) with broadband measurement stations, along with the registration of the electromagnetic field components are recorded and source fields of natural origin in the AMT-MT range that when processing data are often perceived as interference and are filtered out. In AMT measurements—AMT-MT fields often recorded with field of remote sources ELF-VLF range that the standard processing procedures should be filtered as interference of artificial origin (Zhamaletdinov et al. 2012).

But both processes and with a controlled source and fields of natural origin, contain information about the Earth. The data of the CSAMT research and the results of the AMT-MT measurements should complement each other. To this point of view, in the Geological Institute of RAS has developed a set of programs for data processing and interpretation of the results of joint data of the magnetotelluric research and data of the frequency electromagnetic sounding with controlled sources (Zhamaletdinov et al. 2015). In this work are presented the results of the joint inversion of CSAMT data at distances 50 km from two mutually-orthogonal horizontal electric dipoles in the frequency range of 3.82–2185 Hz ("Kovdor-2015") in west sector Kovdor-Jona area (Zhamaletdinov et al. 2017) and AMT-MT measurements of the 2016 year in point of the disposition of the dipole sources.

The values of the components of the impedance tensor $Z_{i,j}$, $i,j = x,y$ are determined from the estimates of spectral densities and the estimate of the errors and the choice of the periods for processing is performed by the coherence values Co_{ij}^2 (Semenov 1985):

Data Treatment

Mathematical Model of the Measurement Data

As a mathematical model of the observed electromagnetic field, we use the generalized spectral representation of the measured components of the electromagnetic field (Semenov 1985):

$$X(t,\kappa) = \int_{-\infty}^{\infty} \exp(i\omega t)\, dZ_{Fx}(\omega,\kappa) + \sum_{k=0}^{\infty} \exp(i\omega_k t)\, X_k$$

here t—is the time, κ—is the parameter called the elementary event, $dZ_{Fx}(\omega,\kappa)$—is the spectral measure, and the amplitude of the random process is the complex value of the spectral measure. The set of all elementary events K is called a probability space.

The first term is the Stieltjes integral in the Cramer representation. It is represented a random process with constraints imposed on it, including stationarity and ergodicity, the continuity of all realizations, the continuous differentiability of the distribution functions and the energy spectrum. In practice, this means the absence of periodic components in the process under investigation. The second term represents just the periodic components of the magnetotelluric field (Semenov 1985). This also includes periodic signals of a controlled field source used for frequency electromagnetic sounding. When processing AMT-MT fields, these components are considered as interference and must be filtered out. When processing CSAMT, random processes and periodic components with harmonics different from the source harmonics are filtered out.

MT-AMT Data

In AMT-MT data treatment by the energy spectrum, the average value of the spectral density in a certain frequency band is determined: $\overline{S_{xx}(\omega)} = (\Delta\omega)^{-1} M\left[dZ_{Fx}^*(\omega,\kappa)dZ_{Fx}(\omega,\kappa)\right]$

Here $M[X(t,\kappa)] = \sum_{i=0}^{N} X(t,\kappa_i)P(\kappa_i)$—is a mathematical expectation for a finite number of realizations (elementary events) κ_i, $i = 1,\ldots,N$, $P(\kappa_i)$—is probability of the event κ_i. In common case we have $M[X(t,\kappa)] = \int X(t,\kappa)P(d\kappa)$.

In magnetotelluric data treatment, in addition to the spectral density for one component, the mutual spectral densities of the two components are used

$$\overline{S_{xy}(\omega)} = (\Delta\omega)^{-1} M\left[dZ_{Fx}^*(\omega,\kappa)dZ_{Fy}(\omega,\kappa)\right]$$

The spectral densities and mutual spectral densities are related by the Wiener-Khinchin relations with the auto- and mutual correlation functions of the random processes corresponding to these field components:

$$S_{xx}(\omega) = 1/2\pi \int_{-\infty}^{+\infty} R_{xx}(\tau)\exp(-i\omega\tau)d\tau; \quad S_{xy}(\omega) = 1/2\pi \int_{-\infty}^{+\infty} R_{xy}(\tau)\exp(-i\omega\tau)d\tau$$

$$R_{xx}(\tau) = 1/2\pi \int_{-\infty}^{+\infty} S_{xx}(\omega)\exp(i\omega\tau)d\omega; \quad R_{xy}(\tau) = 2\pi \int_{-\infty}^{+\infty} S_{xy}(\omega)\exp(i\omega\tau)d\omega$$

$$Arg(Z_{ij}) = arctg(Im(S_{ij})/Re(S_{ij}); \quad |Z_{ij}| = (S_{jj}/S_{ii})^{1/2}; \quad Co_{ij}^2 = |S_{ij}|^2/(S_{ii}S_{jj}); \quad i,j = x,y.$$

Interpretation of the data of magnetotelluric data is carried by the frequency amplitude-phase characteristics of the response functions obtained—either for the components of the impedance tensor or for the apparent resistivity associated with them by the relations.

$$\rho_{T_{ij}}(\omega) = (\omega\mu_0)^{-1}|Z_{ij}|^2, \quad Arg(\rho_{T_{ij}}) = 2Arg(Z_{ij}).$$

CSAMT Data

Frequency probing with a controlled source we be considered in a harmonic mode (Vanyan 1997; Svetov 2008). As a source we use a grounded horizontal electric dipole, in which the current intensity varies in time according to the harmonic law:

$$I = I_0 exp(-i(\omega t - \varphi)).$$

The components of the field at the observation point depend on the time according to the same law.

$$F = \widehat{F_0} exp(-i(\omega t - \varphi F)) = \widehat{F_0} exp(-i(\omega t)).$$

Here the complex amplitude of the measured field component is $\widehat{F_0} = F_0 \exp(i\varphi_F)$.

The field components are normalized by the amplitude of the current in the source, and to the components of the source field at a given observation point over a homogeneous half-space with a unit resistivity value - it is the normal field of the source. The geometric coefficient for measured field component is the value reciprocal of the value of the corresponding component of the normal field at a given point with polar coordinates-distance r from the source center and angle φ, relative to its axis, taking into account the frequency dependence. With the help of geometric coefficients, the values of the apparent resistivity for the field components are calculated.

For horizontal electric field components of a horizontal electric dipole (with length $|AB| \ll r$) in the conditions of the far zone, when the wavelength in the earth $\lambda \ll r$, can be written $\rho_\omega^E = K_E E/I_0$ где $E = E_x$, E_y—are values of the amplitudes of the Cartesian horizontal components of the electric field. Geometric coefficient for the electric field component along the dipole is $K_{Ex} = 2\pi r^3/(|AB|(3\cos^2(\theta) - 2))$, and for the electric field component perpendicular to the dipole is $K_{Ey} = 2\pi r^3/(|AB|(3\cos(\theta)\sin(\theta))$. For horizontal magnetic components $\rho_\omega^H = \omega\mu_0(K_H \cdot H/I_0)^2$, and for a vertical magnetic component $\rho_\omega^{Hz} = \omega\mu_0 K_{Hz} \cdot H_z/I_0$. Here geometric coefficients are: for the horizontal component of the magnetic field along the axis of the source dipole it is $K_{Hx} = 2\pi r^3/(|AB|(3\cos(\theta)\sin(\theta))$ and for the horizontal component of the magnetic field perpendicular to the axis of the dipole it is $K_{Hy} = 2\pi r^3/(|AB|(1-3\sin^2(\theta)))$. For the vertical component of the magnetic field we can write a geometric coefficient as $K_{Hz} = 2\pi r^4/(|AB|3\sin(\theta))$. There are θ—is the angle between dipole momentum $\vec{p} = I\overrightarrow{AB}$ and direction from dipole center to receiver, r is the distance from dipole center to receiver. Besides, one can input apparent resistivity $\rho_{\omega_{ij}}(\omega) = (\omega\mu_0)^{-1}|Z_{ij}|^2$ by horizontal electric components divided by conjugated magnetic components similarly as the magnetotelluric impedance components $Z_{ij} = E_i/H_j$ (Shevtsov et al. 2017).

Measured Data and Inverse Results

To obtain information on the horizontal inhomogeneity (anisotropy) of the underlying half-space in the Geological Institute under the scientific supervision of Zhamaletdinov uses the technique of measuring at one point upon the earth surface the field from two mutually orthogonal ground-based grounded horizontal electric dipoles on the surface of the earth (Zhamaletdinov et al. 2017). For joint interpretation of AMT-MT and CSAMT data, the obtained amplitude-phase frequency curves of the response functions (of impedance or apparent resistivity) must be concordance within the errors of the measured data. It is possible to estimate the effect of galvanic distortions ("statics-shift") at AMT-MT data by using the data of the CSAMT, and the AMT-MT data allow expanding the frequency range of the CSAMT data. Together with obtaining independent, more reliable estimates of the response functions, this allows us to significantly narrow the equivalence domain and increase the stability of the solutions of the inverse problem.

In Fig. 3.1 shows the results of a joint inversion of the CSAMT and AMT-MT measurement data upon the western sector of the Kovdor-Jona area.

The data of the unique experiment "Kovdor 2015" (Zhamaletdinov et al. 2017) with two mutually orthogonal dipoles of length 2 km with azimuths 99° and 179° are presented, in the frequency range from 3.822 to 2185 Hz with spacing 50 km from source and the results of measurements AMT-MT fields of 2016 made at the location of a common electrode of dipoles of the source. Results of AMT-MT and CSAMT measurements in Kovdor-Jona area, for the experiment "Kovdor-2015", the western sector are presented. For points of CSAMT the distances from dipoles of the source are 50 km.

Top panel: *a*—Curves by the AMT-MT data obtained at the location common electrode of the dipoles of the sources of the CSAMT experiment "Kovdor-2015": the effective resistivity $\rho_T = \sqrt{\rho_{xy} \cdot \rho_{yx}}$ (Ohm m) from period T (seconds). The panels *b, c, d*—are presented composite curves CSAMT-AMT-MT of the effective resistivity ρ_T with $\rho_\omega = \sqrt[4]{\rho^\perp_{\omega_{xy}} \cdot \rho^\perp_{\omega_{yx}} \cdot \rho^{II}_{\omega_{xy}} \cdot \rho^{II}_{\omega_{yx}}}$ obtained with the two mutually-orthogonal polarizations of the CSAMT source field for observation points located, respectively, to the north, south and east of the supplying dipoles, supplemented by AMT-MT data; *b*—to the north with azimuths from the sources centers 92° for the equatorial ($\perp$) array and 11° for the axial (II) array; *c*—to the south with azimuths 57° ($\perp$), 158° (II); *d*—to the east with azimuths 158° ($\perp$), 78° (II). ***Low panels***:—*e*—phase curve of effective impedance by AMT-MT data $\varphi_T = 0.5 \cdot (Arg(Z_{xy}) + Arg(Z_{yx}))$; *f, g, h*—composite curves CSAMT-AMT-MT in the same measured points—the phase of the effective impedance φ_T and $\varphi_\omega = 0.25 \cdot \left(Arg\left(Z^\perp_{\omega_{xy}}\right) + Arg\left(Z^\perp_{\omega_{yx}}\right) + Arg\left(Z^{II}_{\omega_{xy}}\right) + Arg\left(Z^{II}_{\omega_{yx}}\right)\right)$

The results of the inversion by the conjugate gradients method presented on Fig. 3.2.

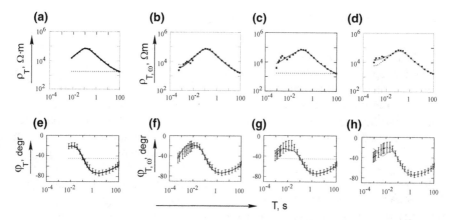

Fig. 3.1 Results of AMT-MT and CSAMT measurements Kovdor-Jona area, the experiment "Kovdor-2015", the western sector Points of CSAMT at the distances from dipoles of the source are 50 km. Top panels: **a**—Curves by the AMT-MT data obtained at the location common electrode of the dipoles of the sources of the CSAMT experiment "Kovdor-2015": the effective resistivity ρ_T.(Ohm m) from period T (seconds). The panels *b, c, d*—are presented composite curves CSAMT-AMT-MT of the effective resistivity ρ_T with ρ_ω obtained with the two mutually-orthogonal polarizations of the CSAMT source field for observation points located, respectively, to the north, south and east of the supplying dipoles, supplemented by AMT-MT data **b**—to the north with azimuths from the sources centers 92° for the equatorial ($\perp$) array and 11° for the axial (II) array; **c**—to the south with azimuths 57° ($\perp$), 158° (II); **d**—to the east with azimuths 158° ($\perp$), 78° (II). The bottom panels: **e**—phase curve (degrees) of effective impedance by AMT-MT data φ_T; *f, g, h*—composite curves CSAMT-AMT-MT in the same measured points—the phase of the effective impedance φ_T and φ_ω (degrees)

On the panels *a, b, c, d* the initial approximation is shown as dashed lines, the solution of the inversion is a solid line. The starting model for the composite curves of CSAMT and AMT-MT data is the solution of the inverse problem for AMT-MT curves.

The misfit estimated for MT-AMT data as

$$S(\sigma,h) = \frac{1}{N+1}\sum_{j=0}^{N}\left[\left(\frac{\log\left(\frac{\rho_{Tj}}{\rho_{T0}(\sigma,h,\omega_j)}\right)}{d\rho_j}\right)^2 + \left(\frac{\varphi_{Tj}-\varphi_{T0}(\sigma,h,\omega_j)}{d\varphi_j}\right)^2\right]$$

And for CSAMT data it is

$$S(\sigma,h,r,\theta) = \frac{1}{N+1}\sum_{j=0}^{N}\left[\left(\frac{\log\left(\frac{\rho_{\omega j}}{\rho_{\omega 0}(\sigma,h,\omega_j,r,\theta)}\right)}{d\rho_j}\right)^2 + \left(\frac{\varphi_{\omega j}-\varphi_{\omega 0}(\sigma,h,\omega_j,r,\theta)}{d\varphi_j}\right)^2\right]$$

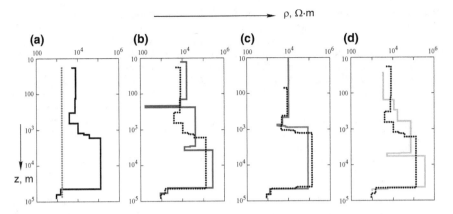

Fig. 3.2 The results of the inversion measured data presented on Fig. 3.1 by the conjugate gradients method. On the panels *a, b, c, d* the initial approximation (uniform half-space for panels a and c) is shown as dashed lines, the solution of the inversion is a solid line. The starting model for the composite curves of CSAMT and AMT-MT data is the solution of the inverse problem for AMT-MT curves

There are σ—array of conductivities, h—array of the thickness of the layered model, apparent resistivity $\rho_{T_0}(\sigma, h, \omega)$, and phase of the impedance $\varphi_{T_0}(\sigma, h, \omega)$—solution of the forward problem of MT-AMT sounding, $\rho_{\omega_0}(\sigma, h, \omega, r, \theta)$—solution of the forward problem of the CSAMT, $j = 0,...,N$ index of the measured frequencies, measured apparent resistivity values $\rho_{T_j}, \rho_{\omega_j}$ and measured phases of the impedance $\varphi_{T_j}, \varphi_{\omega_j}$ for MT-AMT and CSAMT. $d\rho_j$—relative error for apparent resistivity, $d\phi_j$—absolute error for phase of the impedance.

Conclusion

The results of joint inversion AMT-MT and CSAMT data in western sector of the Kovdor-Jona area supported presence high resistivity relatively horizontally uniformed media at this region. Besides, the conductive non-uniformed zone at horizontal is detected at the depths range 0.7–6 km.

References

Semenov V.Yu. (1985) Data treatment of the magnetotelluric sounding. M.: Nedra, 1985. 133 p.
Shevtsov A.N., Zhamaletdinov A.A., Kolobov V.V., Barannik M.B. (2017) Frequency Electromagnetic Sounding with Industrial Power Lines on Karelia-Kola Geotraverse. Zapiski Gornogo instituta. 2017. Vol. 224. P. 178–188. https://doi.org/10.18454/pmi.2017.2.178.
Svetov B.S. (2008) Fundamentals of geoelectrics. Moscow: Publisher LKI, 2008. 656 p.

Vanian L.L. (1997) Electromagnetic soundings. M.: Scientific World, 1997. 219 p.

Zhamaletdinov A.A., A.N. Shevtsov, E.P. Velikhov, A. A. Skorokhodov, V. E. Kolesnikov, T. G. Korotkovaa, P. A. Ryazantsev, B. V. Efimov, V. V. Kolobov, M. B. Barannik, P. I. Prokopchuk, V. N. Selivanov, Yu. A. Kopytenko, E. A. Kopytenko, e, V. S. Ismagilov, M. S. Petrishchev, P. A. Sergushin, P. E. Tereshchenko, f, B. V. Samsonov, M. A. Birulya, M. Yu. Smirnov, T. Korja, Yu. M. Yampolski, A. V. Koloskov, N. A. Baru, S. V. Poljakov, A. V. Shchennikov, G. I. Druzhin, W. Jozwiak, J. Reda, and Yu. G. Shchors. (2015) Study of Interaction of ELF–ULF Range (0.1–200 Hz) Electromagnetic Waves with the Earth's Crust and the Ionosphere in the Field of Industrial Power Transmission Lines (FENICS Experiment)// Proceedings of the Atmospheric and Oceanic Physics, 2015, Vol. 51, No. 8, pp. 826–857.

Zhamaletdinov A.A., M.S. Petrishchev, A.N. Shevtsov, V.V. Kolobov, V.N. Selivanov, O.A. Esipko, E.A. Kopytenko, V.F. Grigor'ev. (2012) Electromagnetic Sensing of the Earth Crust in the area of the Super-Deep Holes, SDH-6 and SDH-7 in fields Natural and powerful controlled Sources. // Doklady Earth Sciences, 2012, Vol. 445, No. 2, p. 205–209.

Zhamaletdinov A.A., Velikhov E.P., Shevtsov A.N., Kolobov V.V., Kolesnikov V.E., Skorokhodov A.A., Korotkova T.G., Ivonin V.V., Ryazantsev P.A., Birulya M.A.(2017) The Kovdor-2015 Experiment: Study of the Parameters of a Conductive Layer of Dilatancy–Diffusion Nature (DD Layer) in the Archaean Crystalline Basement of the Baltic Shield. // Doklady Earth Sciences, 2017, Vol. 474, Part 2, pp. 641–645. © Pleiades Publishing, Ltd., 2017.Original Russian Text © ISSN 1028-334X, https://doi.org/10.1134/s1028334x17060095.

Chapter 4
Gridded Potential Fields Data Analysis Based on Poisson Wavelet-Transformations

K. M. Kuznetsov, A. A. Bulychev and I. V. Lygin

Abstract Current article discusses a possibility of applying wavelets based on the Poisson kernel for processing and interpretation the area potential fields. By analyzing their wavelet-spectrum, it is possible to determine the position of sources. It is also possible to reconstruct or calculate the transformants of the original signal using the inverse wavelet transform of the wavelet-spectrum.

Keywords Continuous wavelet transformation · Poisson wavelet · Potential fields

Bases of Wavelet-Transformation for 2D Data

Nowadays one of the most frequently used approaches in potential fields data processing is based on their representation by trigonometric Fourier series. At the same time, wavelet-transformations of signals become more important in various sciences. In the works of various authors, the possibilities of singular points localization for profile potential fields and their transformations based on wavelets constructed with basis of the Poisson kernel (Utyomov et al. 2010; Obolenskiy and Bulychev 2011; Kuznetsov et al. 2015; Pugin 2004).

Results of the expansion of functions into three-parameter wavelets can be used in data processing with spatial gravity and magnetic fields. Generally, a continuous wavelet-transformation can be described by a convolution in the following form (Yudin et al. 2001):

$$W(a_x, a_y, x, y) = \int_{-\infty}^{\infty} \int_{-\infty}^{\infty} g(\xi, \eta) \psi_{a_x, a_y, x, y}(\xi, \eta) \, d\xi d\eta. \tag{4.1}$$

K. M. Kuznetsov (✉) · A. A. Bulychev · I. V. Lygin
Department of Geophysical Methods of Earth Crust Study, Faculty of Geology,
Lomonosov Moscow State University, Moscow, Russian Federation
e-mail: kirillkuz90@gmail.com

© Springer Nature Switzerland AG 2019
D. Nurgaliev and N. Khairullina (eds.), *Practical and Theoretical Aspects of Geological Interpretation of Gravitational, Magnetic and Electric Fields*, Springer Proceedings in Earth and Environmental Sciences, https://doi.org/10.1007/978-3-319-97670-9_4

Function $W(a_x, a_y, x, y)$ is wavelet-spectrum of function $g(\xi, \eta)$. Wavelet $\psi_{a_x, a_y, x, y}(\xi, \eta)$ deduces from base (mother) wavelet $\psi_0(\xi, \eta)$:

$$\psi_{a_x, a_y, x, y}(\xi, \eta) = \frac{1}{\sqrt{a_x}\sqrt{a_y}} \psi_0\left(\frac{\xi - x}{a_x}, \frac{\eta - y}{a_y}\right), \quad (4.2)$$

where a_x и a_y—scale coefficients, x and y—parameters of wavelet's shift. Function $\psi_0(\xi, \eta)$ must meet certain requirements:

$$\int_{-\infty}^{\infty}\int_{-\infty}^{\infty} \psi_0(\xi, \eta)\,d\xi d\eta = 0, \quad \int_{-\infty}^{\infty}\int_{-\infty}^{\infty} |\psi_0(\xi)|^2 \,d\xi d\eta < \infty. \quad (4.3)$$

For analysis of potential field's signals, we can use wavelets based on the Poisson kernel, as its partial derivatives:

$$K^{nx, my, kz}(x, y, z) = \frac{\partial^{n+m+k}}{\partial^n x \,\partial^m y \,\partial^k z}\left(\frac{z}{\sqrt{(x^2 + y^2 + z^2)^3}}\right). \quad (4.4)$$

In this case, the normalizing coefficients a_x and a_y should be assumed equal and will be denoted as h further. Then, Eq. (4.2) looks like:

$$\psi_{h, x, y}(\xi, \eta) = \frac{1}{h} \psi_0\left(\frac{\xi - x}{h}, \frac{\eta - y}{h}\right) \quad (4.5)$$

Poisson kernel (4.4), corresponding to the calculation of the first vertical derivative at height z is represented by the relation (Gravimetry 1990):

$$K(x, y, z) = \frac{1}{2\pi}\left(\frac{2z^2 - x^2 - y^2}{(\sqrt{x^2 + y^2 + z^2})^5}\right). \quad (4.6)$$

By fixing $z = 1$ and omitting factor $(1/2\pi)$ it is possible to write base wavelet:

$$\psi_0^z(\xi, \eta) = \frac{2 - \xi^2 - \eta^2}{(\sqrt{\xi^2 + \eta^2 + 1})^5} \quad (4.7)$$

Basing on this equation it is possible to construct wavelets of the 1st order corresponding to the second vertical derivative of the potential:

$$\psi_{h, x, y}^z(\xi, \eta) = \frac{1}{h}\psi_0\left(\frac{\xi - x}{h}, \frac{\eta - y}{h}\right) = \frac{1}{h}\frac{2 - \left(\frac{\xi-x}{h}\right)^2 - \left(\frac{\eta-y}{h}\right)^2}{\left(\sqrt{\left(\frac{\xi-x}{h}\right)^2 + \left(\frac{\eta-y}{h}\right)^2 + 1}\right)^5}$$

$$= \frac{1}{h}h^3\frac{2h^2 - (\xi - x)^2 - (\eta - y)^2}{(\sqrt{(\xi - x)^2 + (\eta - y)^2 + h^2})^5}. \quad (4.8)$$

It should be noted, that basing on expression (4.4) it is possible to construct functions analogous to (4.8) corresponding to horizontal derivatives of the potential. Specifically, it can be written:

$$\psi^x_{h,x,y}(\xi, \eta) = \frac{1}{h} h^3 \frac{3h(\xi - x)}{\left(\sqrt{(\xi - x)^2 + (\eta - y)^2 + h^2}\right)^5}. \quad (4.9)$$

Similarly, higher order wavelets based on Poisson kernel derivatives can be constructed.

If initial function's $g(\xi,\eta)$ mean value equal zero and its wavelet-spectrum was received by convolution with the axisymmetric wavelets, which obeying (4.3), than to possible to reconstruct initial signal $g(\xi,\eta)$ by having executed inverse continuous wavelet-transformation to function $W(h,x,y)$ by formula:

$$g(x, y) = \frac{1}{C_\psi} \int_0^\infty \frac{1}{h^3} \left(\frac{1}{h} \int_{-\infty}^\infty \int_{-\infty}^\infty W(h, x, y) \psi_0 \left(\frac{\xi - x}{h}, \frac{\eta - y}{h} \right) dx \, dy \right) dh, \quad (4.10)$$

C_ψ—constant, that calculates by ψ_0:

$$C_\psi = \int_0^\infty \frac{\left|\hat{\psi}_0(\omega)\right|^2}{\omega} d\omega, \quad (4.11)$$

where $\hat{\psi}_0(\omega)$—Fourier-spectrum of function $\psi_0(x,y)$.

It should be noted that by inverse wavelet-transformation with the Poisson wavelets, which corresponds to horizontal derivatives, it is impossible to restore an initial function. But, despite this, received on their basis spectrums allows to locate singular points of the fields anomalies.

Localization of Anomalies Sources

3D wavelet-spectrum allow to fix position of potential field's anomalies singular points. Let's consider model of a point source at depth 50 m and analyze wavelet-spectrum of its magnetic field. In the Fig. 4.1a field ΔT is presented of model. The results of wavelet-transformation with Poisson wavelet corresponding to vertical derivative (4.10) is presented in the Fig. 4.1. Extremum of wavelet-spectrum coordinates, coincide with position of point source.

Now we will consider gravity effect of a point source. In the Fig. 4.2 the vertical slices of wavelet-spectrum passing through a source and calculated by

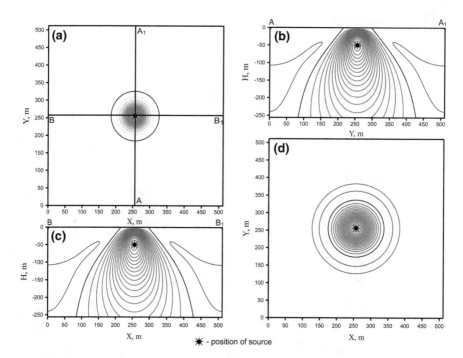

Fig. 4.1 Results of wavelet-transform of point source's field ΔT with Poisson wavelet of first order (4.8). **a**—initial field ΔT, **b**—cross-section of wavelet-spectrum across line AA$_1$, **c**—cross-section of wavelet-spectrum across line BB$_1$, **d**—slice of wavelet-spectrum at level 50 m

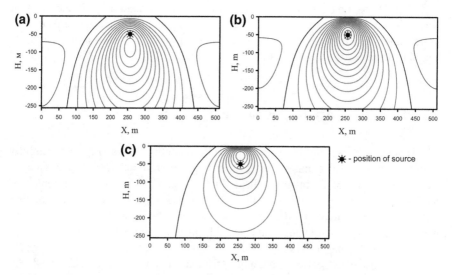

Fig. 4.2 Vertical slices of V_z field wavelet-spectrum for model of point source, calculated by wavelet-transformation (4.8) with scale factors: **a**—$1/h$, **b**—$1/h^{3/2}$, **c**—$1/h^2$

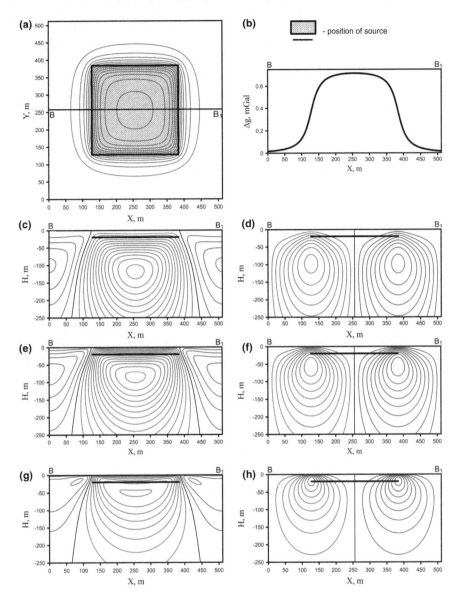

Fig. 4.3 Results of wavelet-transforms of horizontal plate gravity effect with Poisson wavelets of first order (4.8) and (4.9). **a**—initial gravity field, **b**—gravity field plot along line BB_1, **c, e, g**—cross-section of wavelet-spectrum calculated with wavelet (4.8) with scale coefficients $1/h^{3/2}$, $1/h^2$, $1/h^{5/2}$; **d, f, h**—cross-section of wavelet-spectrum calculated with wavelet (4.9) with scale coefficients $1/h^{3/2}$, $1/h^2$, $1/h^{5/2}$

wavelet-transformations of the field V_z by Poisson wavelet of the first order (4.10) with different scale coefficients. At the picture it is possible to note, that extremum of wavelet-spectrum coincides with position of source if to multiply an initial scale factor with $1/h^{1/2}$. Thus for sources' center of mass localization for gravity field it is better to use a scale factor $1/h^{3/2}$.

Let's consider model of a flat horizontal plate at depth of 20 m with width and length—256 m. 3 vertical slices of the wavelet-transformations results along an axis Ox and crossing model center are presented in the figure. Wavelet-spectrums were calculated by wavelets which corresponds vertical (4.8) and horizontal (4.9) derivatives with various scale coefficients. Corners of model can be localized in the best way by wavelet-spectrums extremums in the Fig. 4.3h. It was calculated by wavelet-transformations with wavelets corresponds horizontal derivative along an axis Ox (4.9) with scale coefficients $1/h^{5/2}$. It should be noted that all results of wavelet-transformations with the Poisson wavelets corresponding to horizontal derivatives allow to localize lateral position of model's borders.

Calculation of Equivalent Density and Magnetization Distribution Calculation

Let's function $g(x,y)$ is the gravity anomalies field which is created by density distribution $\sigma(h,\xi,\eta)$. Then it is possible to write down:

$$g(x,y) = G \int_{-\infty}^{\infty} \int_{-\infty}^{\infty} \int_{0}^{h} \sigma(h,\xi,\eta) \frac{h}{\sqrt{(\xi-x)^2 + (\eta-y)^2 + h^2}^3} d\zeta d\xi d\eta, \quad (4.12)$$

where G is the gravitation constant.

Wavelet-spectrums of gridded gravity field can be connected with equivalent density distribution. Let's define function $\rho(h, x, y)$, which is bound to the initial field $g(x, y)$:

$$\rho(h,x,y) = \frac{1}{2\pi} \int_{-\infty}^{\infty} \int_{-\infty}^{\infty} g(,\xi,\eta) \frac{2h^2 - (\xi-x)^2 - (\eta-y)^2}{\sqrt{(\xi-x)^2 + (\eta-y)^2 + h^2}^5} d\xi d\eta. \quad (4.13)$$

This equation coincides with calculation of the first vertical derivative at the level h (Gravimetry 1990). In frequency domain it is possible to write (Bulychev 1985):

$$\hat{\rho}(h,\omega_x,\omega_y) = \hat{g}(\omega_x,\omega_y)\left(|\omega|e^{-|\omega|h}\right), \text{ where } \omega = \sqrt{\omega_x^2 + \omega_y^2}. \quad (4.14)$$

Let's give functions ρ(h, x, y) meaning of density distribution and calculate it's effect f(x,y). In frequency domain it is possible to write:

$$\hat{f}(\omega_x, \omega_y) = G \int_0^h \hat{\rho}(h, \omega_x, \omega_y)(2\pi\, e^{-|\omega|h})dh = \pi G \hat{g}(\omega_x, \omega_y) \quad (4.15)$$

Thus it turns out that function $\delta(h,x,y) = \frac{1}{\pi G}\rho(h,x,y)$ describes density distribution (Kobrunov and Varfolomeev 1981), which matches g(x,y). Function ρ(h,x,y) can be connected with W(h,x,y), calculated with Poisson wavelets (4.8) by equation:

$$\rho(h,x,y) = \frac{1}{2\pi}\frac{h}{h^3} W(h,x,y). \quad (4.16)$$

Then the equivalent density distribution δ(h,x,y), which effect is g(x,y), can be described as follows:

$$\delta(h,x,y) = \frac{1}{\pi G}\rho(h,x,y) = \frac{1}{2\pi}\frac{h}{h^3}\frac{1}{\pi G} W(h,x,y) = \frac{1}{2\pi^2 h^2 G} W(h,x,y). \quad (4.17)$$

Let's consider one more way of transformation of wavelet-spectrum to density. Let function g(x,y) is a vertical derivative of gravitational field V_{zz}, which is effect of density distribution σ(h,ξ,η) (Gravimetry 1990):

$$g(x,y) = G \int_{-\infty}^{\infty}\int_{-\infty}^{\infty}\int_0^h \sigma(h,\xi,\eta)\frac{2h^2 - (\xi-x)^2 - (\eta-y)^2}{\sqrt{(\xi-x)^2 + (\eta-y)^2 + h^2}^{\,5}} d\zeta d\xi d\eta. \quad (4.18)$$

If the initial field was expantioned to wavelet-spectrum by Poisson wavelet of the 1st order, which corresponds vertical derivative, than it can be reconstructed by reverse wavelet-transform (4.10) and it can be written:

$$g(x,y) = G \int_{-\infty}^{\infty}\int_{-\infty}^{\infty}\int_0^h \delta(h,\xi,\eta)\frac{2h^2 - (\xi-x)^2 - (\eta-y)^2}{\sqrt{(\xi-x)^2 + (\eta-y)^2 + h^2}^{\,5}} dhd\xi d\eta, \quad (4.19)$$

where

$$\delta(h,\xi,\eta) = \frac{1}{GC_\psi h} W(h,x,y). \quad (4.20)$$

Thus function δ(h,ξ,η) describes such spatial distribution of density, which effect V_{zz} coincides with initial function g(x,y). It should be noted that both offered ways

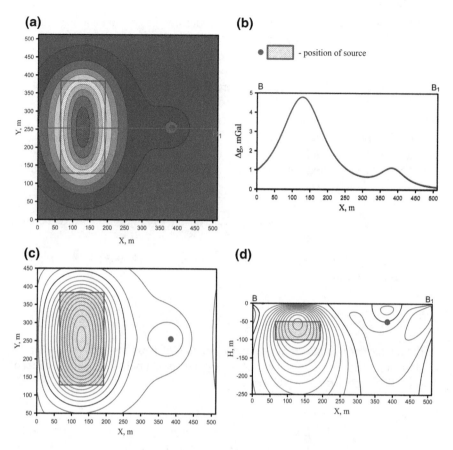

Fig. 4.4 3D cube of equivalent density: **a**—initial gravity field, **b**—initial gravity filed and calculated field V_{zz} plots along line BB_1, **c**—horizontal slice of density cube at level 50 m, **d**—cross-section of density cube along line BB_1

allow to calculate only the equivalent to the true density distributions which do not bear in themselves obvious geological meaning.

In the Fig. 4.4 it is presented results of transformation to density for gravity field of the model which consists of prism and point source at the assumption that it is the V_{zz} field. The imposed schedules of the initial and calculated fields are presented in point b. The standard deviation between the initial field and calculated is less than 1%.

The same approach can be used also for calculation of a cube of the equivalent magnetizations for a magnetic field. Let function $g(x,y)$ corresponds to a vertical component of magnetic field $Z(x,y)$, which is created by distribution of vertical magnetization $I(h,x,y)$. Then it is possible to write down in the SGS system:

$$Z(x,y) = \int_{-\infty}^{\infty} \int_{-\infty}^{\infty} \int_{0}^{h} I(h,\xi,\eta) \frac{2h^2 - (\xi - x)^2 - (\eta - y)^2}{\sqrt{(\xi - x)^2 + (\eta - y)^2 + h^2}^5} \, d\zeta d\xi d\eta \quad (4.21)$$

If to compare this Eq. (4.21) with reverse wavelet-transformation (4.10), then it is possible to write down:

$$I(h,\xi,\eta) = \frac{1}{C_\psi h} W(h,x,y). \quad (4.22)$$

Upward and Downward Continuations

As it shown in previous paragraph it is possible to transform results of wavelet-transformations $W(h,x,y)$ with Poisson wavelets to equivalent density distribution or magnetizations (16, 20, 22) in the lower half-space which creates the initial field. Then it is possible to calculate values of field or its derivatives at the appointed height. Thus, it is possible to realize continuation of the field in the top half-space.

In case of downward continuation between levels of initial field and continuation at such approach there can be sources. By nullified these values of source function it is possible to calculate direct gravity (magnetic) effect at the appointed level. At downward continuation it is more preferable to consider an initial signal as the field corresponding to higher partial derivatives of a gravity potential than the field V_z, for example like field V_{zzzz}. In this case efficient masses will be located at larger depths in comparison with a case of application of wavelet of the 1st order.

Filtering Gridded Potential Field Based on Poisson Wavelet-Transformations

Separation of potential fields into components, which characterize particular studied objects, is one of the major problem at interpretation of potential fields. It is possible to reconstruct initial field $g(x,y)$ by revers wavelet-transformation (4.10) of wavelet-spectrum values $W(h,x,y)$. Thus the problem of filtration of the field (signal) can be solved by various transformations and filtrations of the wavelet-spectrum. It should be noted that at division of fields it is possible to use not only Poisson wavelets, but also other types of wavelets which in some cases can be more.

References

Bulychev A.A. (1985) Methods of potential fields processing and interpretation based on the apparatus of spectral transformations: Doct. Diss. Moscow., 1985. 145 pp. (in Russian)

Gravimetry. Reference book of geophysics (1990)/edited by E.A. Mudrecova, K.E. Veselov. Moscow.: Nedra 1990. 607 pp. (in Russian)

Kobrunov A.I., Varfolomeev V.A. (1981) On one method of ε-equivalent redistributions and its use in the interpretation of gravitational fields. //Izv. AN SSSR. Fizika Zemli. 1981. № 10. pp. 25–44. (in Russian)

Kuznetsov K.M., Obolenskiy I.V., Bulychev A.A. (2015) Potential fields transformations based on continuos wavelet-transformation//Moscow University Bulletin. Series 4: Geology. — 2015. — № 6. — pp. 61–70 (in Russian)

Obolenskiy I.V., Bulychev A.A. (2011) Application of complex continuous Poisson wavelet-transformations for sources of potential field's anomalies determination. // Geophysical research. 2011. vol. 12. №3. pp. 5–21 (in Russian)

Pugin A.V. (2004) Wavelets: a new tool for interpreting potential fields. //Gornoe eho. Vestnik Gornogo inctituta UrO RAN. 2004. № 3. pp. 20–23 (in Russian)

Utyomov E.V., Nurgaliev D.K., Hamidullina G.S. (2010) Technology of gravity data processing and interpretation based on the "natural" wavelet-transform, Uchenye zapiski Kazanskogo universiteta. 2010. vol. 152, book 3. pp. 208–222 (in Russian)

Yudin M.N., Farkov Yu.A., Filatov D.M. (2001) Introduction in wavelet-analysis. Moscow.: MGAA. 2001 (in Russian)

Chapter 5
Modified Method S-, F- and R-Approximations in Solving Inverse Problems of Geophysics and Geomorphology

I. E. Stepanova, I. A. Kerimov, D. N. Raevsky and A. V. Shchepetilov

Abstract The connection of different modifications of the linear integral representation method is studied. Solutions of the related inverse problems based upon a «gibrid version of three approximations» of the topography and geopotential fields enable more refined tuning of the method in solving the inverse problems of geophysics and geomorphology and more complete interpretation for the a priori information about the surface elevation data and elements of anomalous fields. The technique for finding a stable approximate solution for the inverse problem of determining the mass distributions equivalent to the external gravitational (or any other potential) field is presented. The results of the mathematical experiment are discussed.

Keywords Regularization · Integral · Representation · Modified Optimal

Introduction

Over the last few years substantial progress has been achieved in measuring the global gravity field using satellites. The new satellite system GRACE, launched in 2002, increased by almost two orders of magnitude the accuracy of the first 90 harmonics in the spherical expansion model of the global gravity field and for

---

I. E. Stepanova (✉) · I. A. Kerimov · D. N. Raevsky
Schmidt Institute of Physics of the Earth, Russian Academy of Sciences, Moscow, Russia
e-mail: tet@ifz.ru

A. V. Shchepetilov
Faculty of Physics, Moscow State University, Moscow, Russia

the first time permits registration of its temporal variations. The new satellite GOCE will provide high accuracy gravity field models containing 250 spherical harmonics.

As it was stressed in (Stepanova 2008, 2009; Stepanova and Raevsky 2014; Stepanova and Raevsky 2015a, b; Strakhov and Stepanova 2002a, b), the problem of linear approximation of the gravity and magnetic field (potential) may be reduced to the linear algebraic system:

$$Ax = f_\delta = f + \delta f, \qquad (5.1)$$

where, in general, the design matrix A is a full $M \times N$ matrix with real coefficients a_{ij}, $1 \leq i \leq M$, $1 \leq j \leq N$, x is the N-vector to be determined; f is the M-vector describing the signal, f_δ denotes the given vector of the observed gravitational functional that contains errors, δf characterizes the M-vector of the noise, resp. of the errors. The main problem of the gravity field modeling is to obtain stable approximate solutions of linear algebraic equation systems (5.1).

There are various techniques for solving (5.1). Two versions (local and regional) of the method of S-approximations (local and regional), F-approximation (based upon Fourier-transform) and R-approximation are modifications of the method of linear integral representations. The main characteristics of this method are presented in our previous papers (Strakhov and Stepanova 2002a, b).

In the method of S-approximations, the known component of the gravitational field is approximated by a sum of a simple and double layers which are distributed on a certain set of areas (domains). In the local case, these areas are horizontal planes, and in the regional version these are spheres or spheroids.

The solution obtained by the methods of R-, F- and S-approximations allows efficient construction of the linear transforms of the field and can be used as zero approximation for solving the nonlinear inverse problem on localizing the sources (anomalous bodies).

In this work, we present the results of constructing analytical approximations of the anomalous gravitational and magnetic field and surface topography in the local version with the use of rectangular Cartesian coordinates. Here, an important fact is that the S-, F- and R-approximations are closely interrelated with each other. This prompts the idea to use jointly the three or two modifications of the method of linear integral representations which were described in our previous papers in order to construct more advanced models of the geoid and the anomalous potential fields. The R-approximations can be useful due to their ability to «transilluminate» the unknown sources (these sources become «transparent»), to derive the unknown characteristics of the object from the integral information about it. The method of R-approximations is particularly important in solving the problems of seismic tomography, in processing large amounts of seismic profiling data, etc.

Constructing the R-Approximations of the Elements of Anomalous Potential Fields

For understanding why there is a close relationship between different modifications of the method of linear integral representations, we recall the construction procedure of R-approximations.

As was described in detail in (Stepanova 2009) for function $f(x) \in S(R^n)$, where $S(R^n)$ denotes the Schwartz space consisting of all continuously differentiable functions steeply decaying at infinity, the Radon transform exists:

$$\hat{f}(\omega, p) = \int_{(\omega,x)=p} f(x) dm(x), \tag{5.2}$$

where ω is identity vector and $dm(x)$ is the measure on straight line $(\omega, x) = p$.

In the two-dimensional (2D) case, formula (5.2) has the following form:

$$\hat{f}(\omega, p) = \int_{-\infty}^{\infty} f(-t \sin s + x_1 \cos \varphi, t \cos s + x_2 \sin \varphi) ds, \omega = (\cos \varphi, \sin \varphi), x = (x_1, x_2). \tag{5.3}$$

We should stress here the intimate relationship between the Radon transform and n-dimensional Fourier transform:

$$\tilde{f}(u) = \int_{R^n} f(x) e^{-i(x,u)} dx, u \in R^n. \tag{5.4}$$

We will use the integral representation (5.3) for finding the spatial distribution of the elements and localizing the sources of gravitational field. Specifically, we record the main formula of harmonic function theory for the halfspace bounded by the plane $x_3 = 0$ (hereinafter referred to as the Π-plane) (Strakhov and Stepanova 2002a, b):

$$V(M) = \int_{-\infty}^{+\infty}\int_{-\infty}^{+\infty} \frac{p_1(\xi_1, \xi_2) d\xi_1 d\xi_2}{\sqrt{(x_1-\xi_1)^2 + (x_2-\xi_2)^2 + x_3^2}} + \int_{-\infty}^{+\infty}\int_{-\infty}^{+\infty} \frac{p_2(\xi_1, \xi_2) x_3 d\xi_1 d\xi_2}{[\sqrt{(x_1-\xi_1)^2 + (x_2-\xi_2)^2 + x_3^2}]^3},$$
$$M = (x_1, x_2, x_3), \xi = (\xi_1, \xi_2, \xi_3). \tag{5.5}$$

We selected the coordinate system is such a way that the plane of the simple and double layer be specified by equation $x_3 = 0$. Then, the derivative of potential V with respect to x_3 taken with opposite sign has the following form:

$$-\frac{\partial V}{\partial x_3}(M) = \int_{-\infty}^{+\infty}\int_{-\infty}^{+\infty} \frac{\rho_1(\hat{\xi})x_3 d\hat{\xi}}{[\sqrt{(x_1-\xi_1)^2+(x_2-\xi_2)^2+x_3^2}]^3}$$
$$+ \int_{-\infty}^{+\infty}\int_{-\infty}^{+\infty} \frac{\rho_2(\hat{\xi})(2x_3^2-(x_1-\xi_1)^2-(x_2-\xi_2)^2)^2 d\hat{\xi}}{[\sqrt{(x_1-\xi_1)^2+(x_2-\xi_2)^2+x_3^2}]^5}, \quad M=(x_1,x_2,x_3),\ \hat{\xi}=(\xi_1,\xi_2).$$

(5.6)

Functions ρ_1, ρ_2 are not known. Let the components of the field be specified in a finite set of points $M_i = (x_1^{(i)}, x_2^{(i)}, x_3^{(i)})$, $i = 1, 2, \ldots, N$. We denote the integration function in the first term of (5.5) at point M_i by $Q_1^{(i)}$ and in the second term by $Q_2^{(i)}$. Hence, we obtain:

$$-\frac{\partial V(M_i)}{\partial x_3} \equiv f_i = \int_{-\infty}^{+\infty}\int_{-\infty}^{+\infty}(\rho_1(\hat{\xi})Q_1^{(i)}(\hat{\xi})+\rho_2(\hat{\xi})Q_2^{(i)}(\hat{\xi}))d\hat{\xi}, i = 1,2,\ldots,N.$$

(5.7)

It should be noted that formulas (5.4)–(5.7) are the backbone for constructing S-approximations of the sought element of the anomalous potential field.

Let us subject the both sides of (5.5) to the Radon transform. In practice, the components of the field are usually specified with some uncertainty therefore the input information is the values of $f_{i,\delta}$. Using the solution of the variational problem (the variational statement in the general form is described in (Stepanova 2009; Strakhov and Stepanova 2002a, b).

$$\Omega(\rho) = \int_0^{+\infty}\int_{-\infty}^{+\infty} dq \int_0^{2\pi} (\hat{\rho}_1^2(\omega,q)+\hat{\rho}_2^2(\omega,q))dpd\phi = \min_\rho,$$

$$f_{i,\delta} = -\frac{1}{2\pi}\int_0^{2\pi} d\phi \int_{-\infty}^{\infty}\left\{\left[\int_{-\infty}^{\infty}\frac{(\hat{Q}_1^{(i)})'_p(\omega,p-q)dp}{p}\right]\rho_1(\omega,q)\right.$$
$$\left. + \left[\int_{-\infty}^{\infty}\frac{(\hat{Q}_2^{(i)})'_p(\omega,p-q)dp}{p}\right]\rho_2(\omega,q)\right\}dq$$

We obtain that the sought functions should have the form described in (Stepanova 2009). Thus, we come to a system of linear algebraic equations (SLAE):

$$A\lambda = \mathbf{f}_\delta,\ \lambda = (\lambda_1,\ldots,\lambda_N),\ \mathbf{f}_\delta = (f_{1,\delta},\ldots,f_{N,\delta}) \tag{5.8}$$

with easily computable matrix entries.

We should remind a very important fact: integral representations of the anomalous potential fields (i.e. the fields that are harmonic in some domains of the space of sourcewise functions) are fairly closely linked with each other. If we recall formulas for matrix elements in the method of S-approximations (Strakhov and Stepanova 2002a, b):

$$a_{ij} = 2\pi \left\{ \frac{z_i + z_j}{\rho_{i,j}^3} + \frac{(z_i + z_j)(9\rho_{i,j}^2 - 6(z_i + z_j)^2)}{\rho_{i,j}^7} \right\}, \qquad (5.9)$$

$$\rho_{i,j}^2 = (z_i + z_j)^2 + (x_i - x_j)^2 + (y_i - y_j)^2, \ 1 \leq i,j \leq N$$

We can conclude that the Radon transform yields exactly the same SLAE as the local version of S-approximation in which the sought element of the field is represented in the form of the potential of the simple layer [it corresponds to the first term in (5.5)]. What can we gain from the close linkage between the Radon transform and S-approximation of the elements of anomalous fields and functions describing the Earth's surface topography? This interdependence of the different integral transformations allows us to determine, based on the obtained SLAE solutions, the important characteristics of the geological medium under study: we can calculate the ray transform of the observations and reveal the structural pattern of the Earth's crust along the directions of interest for us. We can calculate the mean value of the ray transform of the field element at a given point, etc. Hence, it becomes possible to reconstruct the three-dimensional (3D) medium, i.e. to solve the problem similar to the ones addressed by computer tomography.

Analytical Approximation of the Magnetic Field from the Data of a Hydromagnetic Survey

The hydromagnetic survey's data of the Kineret Lake were used for testing the algorithm of modified S- and R-approximations (Fig. 5.1). The step of a grid is h = 25 m. The total number of valid measurements is 264,442.

The magnetic field was represented with the sum of a simple and double layers distributed on 4 planes at the depths of 250, 600, 1100 and 1100 m beneath the depth of the survey (it was equal to —220 m) respectively. The constants limiting the misfit's square were set to $\delta_{min}^2 = 6000$ nTsl^2, $\delta_{max}^2 = 30{,}000$ nTsl^2.

Let us represent the function describing the magnetic field by the sum of the potentials from the simple and double layers distributed on the plane $x_3 = -H$.

In this section we consider the technologies for constructing analyticl approximation based on the joint S- and R-approximations. In this case, the matrix elements for solving the SLAE can be calculated in the following way:

Fig. 5.1 The difference map of the magnetic field obtained by combined S- and R-approximations and real anomalous magnetic field of Lake Kinneret

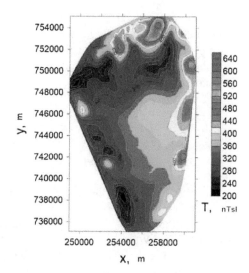

$$a_{ij} = \alpha a_{ij}^S + (1-\alpha) a_{ij}^R,$$

$$a_{ij}^S = 2\pi \sum_{r=1}^{R} \left\{ \frac{z_i + z_j - 2H_r^S}{\rho_{i,j}^3} + \frac{(z_i + z_j - 2H_r^S)\left(9\rho_{i,j}^2 - 6(z_i + z_j - 2H_r^S)^2\right)}{\rho_{ij}^7} \right\},$$

$$\rho_{ij}^2 = (z_i + z_j - 2H_r^S)^2 + (x_i - x_j)^2 + (y_i - y_j)^2,$$

$$a_{ij}^R = 2\pi \sum_{k=1}^{Q} \frac{z_i + z_j - 2H_k^R}{\eta_{i,j}^3}, \quad \eta_{ij} = \sqrt{(z_i + z_j - 2H_r^R)^2 + (x_i - x_j)^2 + (y_i - y_j)^2},$$

(5.10)

where $0 \leq \alpha \leq 1$ is the weighting coefficient, a_{ij}^S are matrix elements for S-approximation, a_{ij}^R are matrix elements for R-approximation, $M = (x_i, y_i, z_i)$, $1 \leq i \leq N$ are the points of surface topography, H_r^S, $1 \leq r \leq R$ is the depth of the r-th carrier of the simple and double layer in the method of S-approximations, H_k^R, $1 \leq k \leq Q$ is the characteristics of the representation subspace of the Fourier integral (i.e. it is assumed that the field is generated by the set of the functions that are harmonic at $z > H_k^R$). It is more reasonable to set $R = K$ and $H_k^R = H_r^S$ in order to represent the field in the form of the functions each of which is harmonic in a certain domain. Hereinafter we assume that $R = K$ and $H_k^F = H_r^S = H_r, H_k^R = H_r^S = H_r$.

For this model example, we carried out the corresponding calculations with the different model parameters (α, R, H_r). The SLAE was solved by the regularized iterative Chebyshev method (CH). All computations were performed on the Intel Core i7-4700HQ 2.4 GHz PC.

The difference map (Fig. 5.1) shows zones of maximal divergence between the real and approximated fields.

Conclusions

The linkage between the S-, F-, and R-approximations is demonstrated. It is shown that the elements of SLAE matrix in the three considered cases are interrelated, and based on the corresponding solution it is possible to find various characteristic of the anomalous geophysical fields.

References

Stepanova, I. (2008). On the S-approximation of the Earth's gravity field, Inverse Problems in Science and Engineering. (16)5,. 547–566.
Stepanova, I. (2009). On the S-approximation of the Earth's gravity field. Regional version, Inverse Problems in Science and Engineering. (17)8, 1095–1111.
Stepanova, I. and D. Raevsky. (2014). On solving reverse problems of geophysics applying the methods of the theory of dynamic systems. Geophysical Journal. (36) 3, 118–131.
Stepanova, I. and D. Raevsky (2015). On the solution of inverse Problems of gravimetry. Izvestiya Physics of the Solid Earth. (51)2, 207–218.
Stepanova, I. and D. Raevsky (2015). The modified method of S-approximation. Regional version. Izvestiya Physics of the Solid Earth. (51)2, 197–206.
Strakhov, V.N and I.E. Stepanova (2002). The S-Approximation Method and Its Application to Gravity Problems. Izvestiya, Physics of the Solid Earth. (38)2, 91–107.
Strakhov, V.N. and I.E. Stepanova (2002). Solution of Gravity Problems by the S-Approximation Method (Regional Version). Izvestiya, Physics of the Solid Earth. (38)7, 535–544.

Chapter 6
Reflection of the Petrophysical Basement Rocks Models in Geophysical Fields

O. M. Muravina, V. N. Glaznev, V. I. Zhavoronkin and M. V. Mints

Abstract The problems of using petrophysical data for studying deep structure of areas covered by sedimentary rocks are discussed. The correlation analysis of the density and gravity anomalies values revealed a complex character of the reflection of crystalline rocks of different density in anomaly of gravity field.

Keywords Petrophysics · Density of rocks · Correlation analysis · Gravity field · Voronezh crystalline massif

Geophysical data and drilling results are the basis for studying the deep structure of areas covered by sedimentary rocks. As a rule, the study by drilling is extremely uneven, which often leads to a simplified interpretation of geophysical fields due to the lack of a reliable actual petrophysical basis. Geological mapping is carried out by analogy in accordance with the images of physical fields without taking into account the structural and geodynamic situation. Numerous studies performed on the Voronezh crystalline massif (VKM) showed that the petrophysical characteristics of rocks depend on the geodynamic conditions of their formation (Afanasiev 2012).

The spatial basis of petrophysical information of the VKM territory was developed using GIS-technologies in the ArcView 3.2. This database in addition to physical properties contains the necessary information about the geological belonging of the core samples to certain types of rocks, indicated in a sufficiently detailed classification. Thus, the database is the most representative petrophysical description of typical crystalline and sedimentary rocks of the VKM territory. It contains information on the results of more than 90,000 petrophysical determinations of various properties of the rocks for 4418 wells (Fig. 6.1).

O. M. Muravina (✉) · V. N. Glaznev · V. I. Zhavoronkin
Voronezh State University, Voronezh, Russia
e-mail: muravina@geol.vsu.ru

M. V. Mints
Geological Institute of the RAS, Moscow, Russia

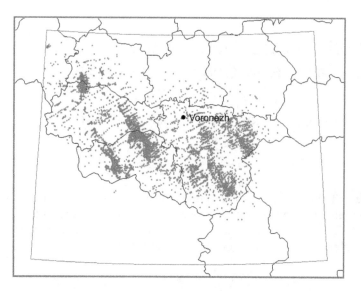

Fig. 6.1 Scheme of well locations

Density models of the sedimentary cover and the upper part of the lithosphere of the VKM region and adjacent territories were created within the framework of the formed petrophysical database on a geological basis at scale 1:500000 (Fig. 6.2) (Muravina et al. 2013, 2014a, b; Muravina and Zhavoronkin 2015; Muravina et al. 2015, 2016). All of these materials were used for the analysis of geophysical fields (Glaznev et al. 2012, 2013; Muravina and Glaznev 2013, 2014) and constructing density models of the upper crust in the VCM region on the basis of new technic of the interpretation approaches (Glaznev et al. 2014, 2015a, b, c; Muravina and Glaznev 2015; Voronova and Muravina, 2017).

At present, work is underway to create similar petromagnetic models of the region. Comparison of developed petrophysical models and corresponding geophysical fields allows to eliminate contradictions in the interpretation of geophysical data. Correlation analysis, in which the values of density and gravitational field were compared, revealed the features of the reflection of crystalline rocks of different density in the gravity field.

Figure 6.3 shows a schematic map of statistical probability density, which demonstrates the complex nature of the relationship between the density of crystalline rocks and the anomaly of gravity field. For example, crystalline rocks with a density of about 2.7 g/cm^3, corresponding to gneisses of different composition, correspond to both lower (down to −28 mGal) and elevated (up to +3 mGal) values of the gravitational field. The average density of crystalline VCM rocks is 2.721 g/cm^3, which is typical for the Archaean crystalline rocks (Galitchanina et al. 1995; Buyanov et al. 1995; Kozlov et al. 2006). The average value of the gravitational field for the area is −10.74 mGal. At the same time, high-density crystalline rocks are reflected in the gravitational field only by an insignificant increase with respect

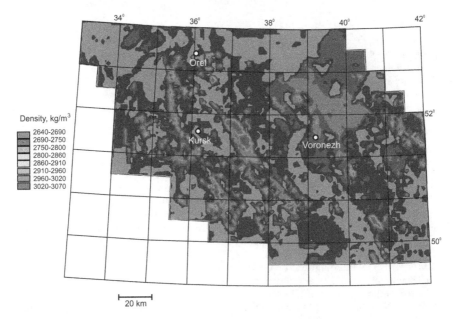

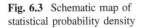

Fig. 6.2 Map of density distribution of crystalline rocks of the Voronezh crystalline massif

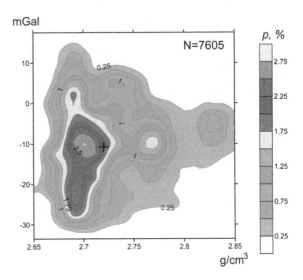

Fig. 6.3 Schematic map of statistical probability density

to the mean values of the field (Muravina et al. 2016; Voronova and Glaznev 2014; Voronova and Muravina 2014).

These results were taken into account when creating a complex model of the VKM lithosphere (Glaznev et al. 2016; Muravina et al. 2016; Muravina 2016).

This work was supported by RFBR, grant No 16-05-00975.

References

Afanasyev, N. S. (2012). Petrophysics of the Earth's crust VKM. Lithosphere of the Voronezh crystalline massif by geophysical and petrophysical data. Ed.: N.M. Chernyshov. Voronezh: Scientific Book, pp. 21–88. (in Russian).

Buyanov, A. F., V. N. Glaznev, F. P. Mitrofanov and A. B. Raevsky. (1995). Three-dimensional modelling of the Lapland Granulite Belt and adjacent structures of the Baltic Shield from geophysical data. Geology of the eastern Finnmark – western Kola peninsula region. Geological Survey of Norway, Special Publication. Eds. D. Roberts, N. Nordgulen. V. 7, pp. 167–178.

Galitchanina, L. D., V. N. Glaznev, F. P. Mitrofanov, O. Olesen and H. Henkel (1995) Surface density characteristics of the Baltic Shield and adjacent territories. Norges Geologiske Undersøkelse. Special Publication «Geology of the Eastern Finnmark - Western Kola Peninsula Region» . Proceedings of the 1st. International Barents Symposium «Geology and Minerals in the Barents Region» / Eds. D.Roberts, Ø.Nordgulen. V. 7, pp. 349–354.

Glaznev, V. N, M. V. Mints and O. M. Muravina (2016). Density modeling of the central part of the East European platform. Vestnik KRAUNTS, Series of "Earth Sciences" T. 29, No. 1, pp. 53–63. (in Russian).

Glaznev, V. N., O. M. Muravina, T. A. Voronova and V. M. Choline (2014). Evaluation of the thickness gravy-active layer of the crust of the Voronezh crystalline massif. Bulletin of VSU, Series "Geology", No. 4, pp. 78–84. (in Russian).

Glaznev, V.N., Mints, M.V., Muravina, O.M., Raevsky, A.B., Osipenko, L.G (2015a). Complex geological–geophysical 3D model of the crust in the southeastern Fennoscandian Shield: Nature of density layering of the crust and crust–mantle boundary. Geodynamics & Tectonophysics. V. 6, № 2, pp. 133–170.

Glaznev, V. N., O. M. Muravina, T. A. Voronova and E. B. Kislova (2015b). Thickness of the gravy-active layer of the upper crust of the Voronezh crystalline massif from the stochastic analysis of the gravity anomalies. The materials of the 42 session of the international seminar named D. G. Uspensky "The theory and practice of geological interpretation of geophysical fields". Perm: MI UB RAS, pp. 46–48. (in Russian).

Glaznev, V. N., T. A. Voronova, I. Yu. Antonova and O. M. Muravina (2015c). Methods and results of the 3D density modelling in studding construction of the upper crust of the Voronezh crystalline massif. The materials of the 42 session of the international seminar named D. G. Uspensky "The theory and practice of geological interpretation of geophysical fields". Perm: MI UB RAS, pp. 49–52. (in Russian).

Glaznev, V. N., V. I. Zhavoronkin, M. V. Mints, O. M. Muravina and N. E. Hovansky (2013) Petrodensity model and the gravitational effect of sedimentary cover of the Voronezh crystalline massif and its borders. The materials of the 40st session of the international seminar named D. G. Uspensky "The theory and practice of geological interpretation of geophysical fields" Moscow: IPE RAS, pp. 107–112. (in Russian).

Glaznev, V. N., V. I. Zhavoronkin and O. M. Muravina (2012) Accounting of the gravitational effect of sedimentary cover of the Voronezh crystalline massif. Materials II school-seminar «Gordinskie chteniy». Moscow: IPE RAS, pp. 45–46. (in Russian).

Kozlov, N. E., N. O. Sorokhtin, V. N. Glaznev, N. E. Kozlova, A. A. Ivanov, N. M. Kudryashov, E. V. Martynov, V. A. Tyuremnov, A. V. Matyushkin and L. G. Osipenko (2006). Geology of Archaean of the Baltic Shield. Saint-Petersburg, Nauka, 329 p. (in Russian).

Muravina, O. M. (2016). Density model of the earth crust of the Voronezh crystalline massif. Bulletin of VSU, Series Geology, No. 1, pp. 108–114. (in Russian).

Muravina, O. M. and V. N. Glaznev (2013) Some of the results of the statistical analysis of Petrophysical parameters of the sedimentary cover rocks of the Voronezh anticline. Materials 14-th International Conference «Physical-chemical and Petrophysical studies in Earth Sciences». Moscow: IPE RAS, pp. 190–193. (in Russian).

Muravina, O. M. and V. N. Glaznev (2014). Structural-parametric models of Petrophysical parameters of sedimentary cover of the Voronezh anticline. News of the SB RANS. V. 44, No 1, pp. 81–87. (in Russian).

Muravina, O. M. and V. N. Glaznev (2015) Methodology for integrated models of the lithosphere platform areas in conditions of incomplete information. Materials III school-seminar «Gordinskie chteniy». Moscow: IPE RAS, pp. 22–26. (in Russian).

Muravina, O. M., V. N. Glaznev and V. I. Zhavoronkin (2013). Petrophysical characteristics of sedimentary cover of Voronezh anticlines. Bulletin of VSU, Series Geology, No. 1, pp. 189–196. (in Russian).

Muravina, O. M. and V. I. Zhavoronkin (2015). Statistical analysis of the digital petrodensity map of the basement rocks for Voronezh crystalline massif. Bulletin of VSU, Series Geology, No. 2, pp. 94–99. (in Russian).

Muravina, O. M., V. I. Zhavoronkin and V. N. Glaznev (2014a). The petrodensity model of the crystalline basement of the Voronezh crystalline massif. The materials of the 41st session of the international seminar named D. G. Uspensky "The theory and practice of geological interpretation of geophysical fields." Yekaterinburg: IGF UB RAS, pp. 171–174. (in Russian).

Muravina, O. M, V. I. Zhavoronkin and V. N Glaznev (2014b). Spatial analysis of the density distribution of Precambrian formations of the Voronezh crystalline massif. Collection of materials of the XV International Conference "Physical-Chemical and Petrophysical Studies in Earth Sciences". Moscow, p. 171–173. (in Russian).

Muravina, O. M, V. I. Zhavoronkin and V. N Glaznev (2015). Correlation analysis of the digital base map of the isodense of the Voronezh crystalline massif and the gravity field. The materials of the XV International Conference "Physical-Chemical and Petrophysical Studies in Earth Sciences". Moscow, pp. 201–203. (in Russian).

Muravina, O. M., V. I. Zhavoronkin and V. N. Glaznev (2016). The Petrodensity Map of the Voronezh crystalline massif. The materials of the 43-th session of the international seminar named D. G. Uspensky "The theory and practice of geological interpretation of geophysical fields". Moscow: IFE RAS, pp. 133–136. (in Russian).

Voronova, T. A. and V. N. Glaznev (2014). Three-dimensional density model of the granite massive from Hoper Megablock (Voronezh crystalline massif). The materials of the 41st session of the international seminar named D. G. Uspensky "The theory and practice of geological interpretation of geophysical fields". Yekaterinburg: IGF UB RAS, pp. 82–84. (in Russian).

Voronova, T. A. and O. M. Muravina (2014). Detailed density modelling of the upper crust for the Voronezh crystalline massif. Bulletin of VSU, Series "Geology", No. 2, pp. 150–154. (in Russian).

Voronova, T. A. and O. M. Muravina (2017). Optimization solving the inverse problem of gravimetry in the construction of detailed density model. The materials of the 44th session of the international seminar named D. G. Uspensky "The theory and practice of geological interpretation of geophysical fields". Moscow: IPE RAS, pp. 114–116. (in Russian).

Chapter 7
Continuous Method for Solution of Gravity Prospecting Problems

I. V. Boikov, A. I. Boikova and O. A. Baulina

Abstract The continuous operator method for solving operator equations is presented. An applications of the continuous operator method for solving linear and nonlinear inverse problems of logarithmic and Newtonian potentials are given.

Keywords Inverse tasks · Gravity prospecting · Continuous method · Nonlinear models

Introduction

The tasks of gravity prospecting can be divided into two large classes: direct and inverse problems.

The inverse problem is modeled by the operator equation

$$K(x) = f, \qquad (7.1)$$

where K is an operator that maps a metric space M to a metric space D. Here M is the metric space of the required model parameters, D is the metric data space.

In further, it is more convenient to consider Eq. (7.1) in Banach spaces, assuming that K is an operator mapping a Banach space X into a Banach space Y.

In gravity prospecting problems under the spaces X and Y, it is natural to understand the spaces of square summable functions.

The inverse problem consists in solving the operator Eq. (7.1). There are two classes of inverse problems: correctly posed and incorrectly posed.

Recall, following (Zdanov 2002), the definitions of correctly and incorrectly posed problems.

I. V. Boikov (✉) · A. I. Boikova · O. A. Baulina
Faculty of Computer Engineering, Penza State University, Penza, Russia
e-mail: i.v.boykov@gmail.com

Definition 1 Problem (7.1) is well posed if the following conditions are satisfied: (7.1) the solution x^* of Eq. (7.1) exists; (7.2) the solution x^* of Eq. (7.1) is unique; (7.3) the solution x^* depends continuously on f.

Definition 2 Problem (7.1) is incorrectly posed if at least one of the conditions listed in Definition 1 is not satisfied.

In the case of linear problems, the correctness of the problem requires the existence of a linear inverse operator K^{-1}.

It is known (Tikhonov 1977; Mudretsova 1990) that most inverse problems of geophysics are incorrect.

This means that, in the case of linear problems, the operator K^{-1} either does not exist or is not bounded.

In the case of non-linear problems of the form (7.1), the connection between the linear invertibility of the Frechet derivative $K'(x_0)$ on the elements $x_0 \in X$ and the ill-posedness of the problem (7.1) is more complicated.

To solve ill-posed problems, a large number of different methods have been proposed, both general and narrowly specific, designed to solve specific problems. Among the methods intended for solving linear ill-posed problems, the Tikhonov regularization method (Tikhonov 1977), the iterative methods—the minimal residual method, the generalized minimal discrepancy method and their various modifications (Zdanov 2002), should be noted first. To solve ill-posed nonlinear problems, we use the method of early descent, the method of directional search, Newton's method, the conjugate gradient method and their modifications (Zdanov 2002).

To implement the methods of speedy descent and the method of searching in the direction it is required that the Frechet derivative of the operator $K(x)$ is not degenerate in some neighborhood of the solution. This requirement imposes certain restrictions on the scope of the method of early descent and the method of searching in the direction.

In order to realize the solution of Eq. (7.1) by the Newton–Kantorovich method

$$x_{n+1} = x_n + [K'(x_n)]^{-1} K x_n, n = 0, 1, \ldots,$$

the continuous invertibility of the operator $K'(x_n)$ is necessary at each step of the iteration process.

Using the modified Newton–Kantorovich method

$$x_{n+1} = x_n + [K'(x_0)]^{-1} K x_n, n = 0, 1, \ldots,$$

the existence of a linear inverse operator on the initial approximation is required.

The existence of a linear inverse operator is also required for the realization of a continuous analogue of the Newton–Kantorovich method

$$\frac{dx(t)}{dt} = [K'(x(t))]^{-1} K(x(t)),$$

ascending to Gavurin (1958) and found wide application in modeling problems of nuclear physics (Zhidkov 1973; Puzynina 2003).

Thus, the application of the Newton–Kantorovich method for solving inverse problems of geophysics imposes fairly rigid restrictions on them.

These restrictions no in the continuous operator method (Boikov 2012).

Let us recall the main points of this method.

Continuous Operator Method

Consider the equation

$$A(x) - f = 0, \qquad (7.2)$$

where $A(x)$ is a nonlinear operator acting from a Banach space X into X.

We denote by x^* the solution of Eq. (7.2).

The paper (Boikov 2012) establishes a connection between the stability of solutions of operator differential equations in Banach spaces and the solvability of operator equations of the form (7.2).

We give the necessary notation:

$$R(a,r) = \{z \in B : \|z-a\| \le r\}, \quad S(a,r) = \{z \in B : \|z-a\| = r\}, \quad ReK = K_R = (K+K^*)/2,$$
$$\Lambda(K) = \lim_{h \to 0}(\|I+hK\| - 1)/h.$$

Here B is a Banach space, $a, z \in B$, K is a linear operator acting from B to B, $\Lambda(K)$ is a logarithmic norm (Daletskii 1970) of the operator K, I is the identity operator.

For the most commonly used spaces, the logarithmic norms are known.

Let $A = \{a_{ij}\}$, $i,j = 1,2,\ldots,n$, be a real matrix in the n-dimensional space R_n of the vectors $x = (x_1, \cdots, x_n)$ with the norm $\|x\|_1 = \sum_{k=1}^{n} |x_k|$, $\|x\|_2 = \left[\sum_{k=1}^{n} \|x_k\|^2\right]^{1/2}$, $\|x\|_3 = \max_{1 \le k \le n} |x_k|$.

The logarithmic norm of the matrix A is (Dekker 1988):

$$\Lambda_1(A) = \max_j \left(a_{jj} + \sum_{i=1, i \ne j}^{n} |a_{ij}|\right), \quad \Lambda_2(A) = \lambda_{max}\left(\frac{A+A^T}{2}\right),$$

$$\Lambda_3(A) = \max_i \left(a_{ii} + \sum_{j=1, j \ne i}^{n} |a_{ij}|\right).$$

Here $\lambda_{max}((A+A^T)/2)$ is the largest eigenvalue of the matrix $(A+A^T)/2$. Equation (7.2) corresponds to the Cauchy problem

$$\frac{dx(t)}{dt} = A(x(t)) - f, \tag{7.3}$$

$$x(0) = x_0. \tag{7.4}$$

Theorem 1 (Boikov 2012). Suppose that Eq. (7.2) has a solution x^* and on any differential curve $g(t)$, located in a Banach space B, the following inequality holds

$$\lim_{t\to\infty} \frac{1}{t} \int_0^t \Lambda(A'(g(\tau)))d\tau \leq -\alpha, \alpha > 0. \tag{7.5}$$

Then the solution of the Cauchy problem (7.3), (7.4) converges to the solution x^* of Eq. (7.2) for any initial approximation.

Theorem 2 (Boikov 2012). Suppose that Eq. (7.2) has a solution x^* and on any differentiable curve $g(t)$, in the ball $R(x^*, r)$, the following conditions are fulfilled:

(1) for any $t(t > 0)$ the following inequality holds

$$\int_0^t \Lambda(A'(g(\tau)))d\tau \leq 0; \tag{7.6}$$

(2) The inequality (7.5) is valid.

Then the solution of the Cauchy problem (7.3), (7.4) converges to the solution of Eq. (7.2).

Remark 1 It follows from (7.5, 7.6) that the logarithmic norm of $\Lambda(A'(g(\tau)))$ can be nonnegative for some values of τ, that is, the Frechet derivative $A'(g(\tau))$ can degenerate into an operator that is identically equal to zero.

Remark 2 The solution of the model example (approximate solution of the hypersingular integral equation) (Boykov 2018) demonstrated the convergence of the iterative process, based on the continuous operator method, with the Frechet derivative vanishing in the initial approximation.

Approximate Solution of Inverse Problems of Gravity Prospecting

We introduce a Cartesian rectangular coordinate system by directing the Oz axis downward.

Let the ore body lies at a depth of H, and its lower surface coincides with the plane $z = H$, and the upper surface is described by the function $z(x, y) =$

7 Continuous Method for Solution of Gravity ...

$H - \phi(x,y)$ where $0 \leq \phi(x,y) \leq H$. Then the gravitational field on the surface the Earth is described by the equation

$$G \int_{-\infty}^{\infty} \int_{-\infty}^{\infty} \int_{H-\phi(\zeta,\eta)}^{H} \frac{\sigma(\zeta,\eta,\xi)\xi d\zeta d\eta d\xi}{((x-\zeta)^2 + (y-\eta)^2 + \xi^2)^{3/2}} = f(x,y,0) \quad (7.7)$$

where G is the gravitational constant; $\sigma(\zeta,\eta,\xi)$ is the density of the body.

It is assumed that, first, the density $\sigma(\zeta,\eta,\xi) \equiv 0$ outside the body; second, the function of density is differentiable with respect to ξ.

To simplify further calculations, we assume that the density does not depend on ξ. Then we arrive at the equation

$$G \int_{-\infty}^{\infty} \int_{-\infty}^{\infty} \sigma(\zeta,\eta) \left[\frac{1}{((x-\zeta)^2 + (y-\eta)^2 + (H-\phi(\zeta,\eta))^2)^{1/2}} \right.$$

$$\left. - \frac{1}{((x-\zeta)^2 + (y-\eta)^2 + H^2)^{1/2}} \right] d\zeta d\eta = f(x,y,0). \quad (7.8)$$

The linearization of Eq. (7.8) leads to the equation

$$G \int_{-\infty}^{\infty} \int_{-\infty}^{\infty} \sigma(\zeta,\eta) \left[\frac{H\phi(\zeta,\eta)}{((x-\zeta)^2 + (y-\eta)^2 + H^2)^{3/2}} \right] d\zeta d\eta = f(x,y,0). \quad (7.9)$$

Below we shall assume that the density is constant and for the convenience of describing the algorithms proposed in this paper we put $G\sigma(\zeta,\eta) = 1/2\pi$.

We represent Eq. (7.9) in the form

$$\frac{1}{2\pi} \int_{-\infty}^{\infty} \int_{-\infty}^{\infty} \left[\frac{H\phi(\zeta,\eta)}{((x-\zeta)^2 + (y-\eta)^2 + H^2)^{3/2}} \right] d\zeta d\eta = f(x,y,0). \quad (7.10)$$

The logarithmic potential problem with the corresponding simplification leads to nonlinear integral equations

$$G \int_a^b \sigma(s) \ln \frac{(x-s)^2 + H^2}{(x-s)^2 + (H-z(s))^2} ds = f(x), \quad (7.11)$$

where $z = z(s)$ is the equation of the profile of an infinitely extended body; H is the depth of occurrence.

Methods for solving the inverse problems of the logarithmic potential are given in (Starostenko 1978).

The linearization of Eq. (7.11) leads (Strakhov 1970) to linear integral equation

$$2G\sigma H \int_a^b \frac{z(\zeta)d\zeta}{(x-\zeta)^2 + H^2} = f(x). \qquad (7.12)$$

Regularizing algorithms for solving Eq. (7.11) were investigated in (Glasko 1970; Tikhonov 1965); iterative methods for solving equations of the form (7.10–7.12) are proposed in the papers (Boikov 1999; 2009, 2013).

A detailed review of the literature is contained in the books (Zdanov 2002; Boikov 2013).

At first we consider the inverse problems of potential theory in a linear formulation.

Consider the equation

$$\frac{1}{\sqrt{2\pi}} \int_{-\infty}^{\infty} \frac{2Hz(s)ds}{(x-s)^2 + H^2} = f(x), \quad -\infty < x < \infty. \qquad (7.13)$$

Let $A-$ be a sufficiently large positive number. We introduce the nodes $x_k = -A + \frac{Ak}{N}$, $k = 0, 1, \ldots, 2N$.

The computational scheme of the collocation method has the form

$$\frac{1}{\sqrt{2\pi}} \frac{A}{N} \sum_{l=0}^{2N-1} \frac{2Hz(x_l)}{(x_k - x_l)^2 + H^2} = f(x_k), k = 0, 1, \ldots, 2N - 1,$$

or, after substituting the values x_k and x_l:

$$\frac{AN}{\sqrt{2\pi}} \sum_{l=0}^{2N-1} \frac{2Hz_l}{(A(k-l))^2 + N^2 H^2} = f(x_k), k = 0, 1, \ldots, 2N - 1. \qquad (7.14)$$

Here $z_l = z(x_l)$, $l = 0, 1, \ldots, 2N$.

We associate the system of Eq. (7.14) with the system of linear differential equations

$$\frac{dz_k(t)}{dt} = -\frac{1}{\sqrt{2\pi}} AN \sum_{l=0}^{2N-1} \frac{2Hz_l(t)}{(A(k-l))^2 + N^2 H^2} + f(x_k), k = 0, 1, \ldots, 2N - 1.$$

$$(7.15)$$

As follows from the results of Section "Continuous Operator Method", if the logarithmic norm of the matrix of the system (7.14) is negative, then the solution $\{z_k(t)\}$, $k = 0, 1, \ldots, 2N - 1$, as $t \to \infty$ tends to the solution of the system of Eq. (7.14).

7 Continuous Method for Solution of Gravity ...

We find a combination of parameters in the system of Eq. (7.14), in which the methods of stability theory guarantee the convergence of solution of the system of differential Eq. (7.15) to the solution of the system of Eq. (7.14).

To do this, we estimate the logarithmic norm of the matrix $B = \{b_{kl}\}$, $k, l = 1, 2, \ldots, 2N$, where

$$b_{kl} = -\frac{2NH}{A} \frac{1}{(k-l)^2 + c^2}, \quad c = \frac{NH}{A}.$$

For definiteness, the system (7.14) will be considered in the space of $2N$-dimensional vectors with the norm $\|x\|_3 = \max_{1 \leq i \leq 2N} |x_i|$.

The diagonal elements of the matrix B are equal to $b_{kk} = -\frac{2A}{NH}$.

We estimate the sum of the modules of the off-diagonal elements of the matrix B. Obviously,

$$\sum_{l=0}^{2N-1}{}' |b_{kl}| = \sum_{l=0}^{k-1} |b_{kl}| + \sum_{l=k+1}^{2N-1} |b_{kl}| = \sum_{l=0}^{k-1} \frac{2NH}{A} \frac{1}{(k-l)^2 + c^2} + \sum_{l=k+1}^{2N-1} \frac{2NH}{A} \frac{1}{(k-l)^2 + c^2}$$

$$= \sum_{j=1}^{k} \frac{2NH}{A} \frac{1}{j^2 + c^2} + \sum_{j=1}^{2N-1-k} \frac{2NH}{A} \frac{1}{j^2 + c^2}$$

$$\leq \int_0^k \frac{2NH}{A} \frac{1}{x^2 + c^2} dx + \int_0^{2N-1-k} \frac{2NH}{A} \frac{1}{x^2 + c^2} dx$$

$$= 2\left(\arctg\frac{k}{c} + \arctg\frac{2N-k-1}{c}\right) \leq 2 \arctg\frac{2A}{H},$$

where $\sum_{l=0}^{2N-1}{}'$ means summation over $l \neq k$.

Thus, if the inequality

$$\frac{2A}{NH} > 2 \arctg \frac{2A}{H} \tag{7.16}$$

is satisfied then the solution of the system of differential equations converges to the solution of the system of Eq. (7.14).

Inequality (7.16) can be replaced by the following simpler inequality

$$\frac{2A}{NH} > \pi, \tag{7.17}$$

Remark 3 Since $h = A/N$ is the step of the computational scheme (7.14), it is always possible to select values of A and N such that conditions (7.16) and (7.17) are satisfied.

Remark 4 The conditions for convergence depend on the concrete spaces in which the system (7.14) is investigated.

Remark 5 A regularization of system (7.14) can be carried out by introducing a nonnegative function $\alpha(t), \lim_{t\to\infty} \alpha(t) = 0$. As a result, we obtain the system equations

$$\frac{dz_k(t)}{dt} = -\alpha(t)z_k(t) - \frac{1}{\sqrt{2\pi}} AN \sum_{l=0}^{2N-1} \frac{2Hz_l(t)}{(A(k-l)^2 + N^2H^2)} + f(x_k), \ k$$
$$= 0, 1, \ldots, 2N-1.$$

As an example, we consider the equation

$$\frac{1}{\sqrt{2\pi}} \int_{-\infty}^{\infty} \frac{2Hz(s)ds}{(x-s)^2 + H^2} = \sqrt{\frac{\pi}{2}} \frac{1+H}{H(x^2 + (H+1)^2)}, \tag{7.18}$$

the exact solution of which is

$$z(s) = \frac{1}{2H} \frac{1}{s^2+1}.$$

With the number of collocation nodes equal to 40, the error in solving the system of Eq. (7.18) is 10^{-3}.

Now apply the continuous operator method to the nonlinear Eq. (7.11).
We approximate the Eq. (7.11) by the simpler nonlinear equation

$$G \int_a^b \sigma(s) \frac{2Hz(s) - z^2(s)}{(x-s)^2 + H^2} ds = f(x). \tag{7.19}$$

This equation is obtained from (7.11) if we restrict ourselves to the second power in the expansion of the function $\ln(1+u)$, $u = \frac{2Hz(s)-z^2(s)}{(x-s)^2 + (H-z(s))^2}$ in a Taylor series and approximating the function $\frac{2Hz(s)-z^2(s)}{(x-s)^2 + (H-z(s))^2}$ by function $\frac{2Hz(s)-z^2(s)}{(x-s)^2 + H^2}$.

The Eq. (7.19) is approximated by the system of nonlinear algebraic equations

$$G \frac{2A}{N} \sum_{l=0}^{N-1} \sigma(x_l) \frac{2Hz(x_l) - z^2(x_l)}{(x_k - x_l)^2 + H^2} = f(x_k), \tag{7.20}$$

where $x_k = -A + 2Ak/N$, $k = 0, 1, \ldots, N$.

The system of Eq. (7.20) is associated with the system of nonlinear differential equations

$$\frac{dz_k(t)}{dt} = -\left(\frac{2AG}{N}\sum_{l=0}^{N-1}\sigma(x_l)\frac{2Hz_l(t) - z_l^2(t)}{(x_k - x_l)^2 + H^2} - f(x_k)\right), \qquad (7.21)$$

$$k = 0, 1, \ldots$$

The Frechet derivative of the matrix on the right-hand side of the system of Eq. (7.21) on the element z_l^*, $l = 0, 1, \ldots, N-1$, is the matrix $B = \{b_{ij}(t)\}$, $i, j = 1, 2, \ldots, N$, where

$$b_{ij} = -\frac{2AG}{N}\sigma(x_{j-1})\frac{2H - 2z_{j-1}^*(t)}{(x_{i-1} - x_{j-1})^2 + H^2}.$$

The convergence of the solution of the system of Eq. (7.21) to the solution of the system of Eq. (7.20) is determined by the conditions imposed on the logarithmic norm of the matrix B.

Example. Consider equation

$$\int_{-\infty}^{\infty} \sigma(s)\frac{2Hz(s) - z^2(s)ds}{(x-s)^2 + H^2} = f(x), \quad -\infty < x < \infty,$$

in which $\sigma(s) = (s^2 + 4)/(s^4 + 2s^2 + 1)$, $f(s) = \frac{\pi}{10}\frac{113x^2 + 5628}{x^4 + 85x^2 + 1764}$.

This equation is approximated by the system of nonlinear algebraic equations

$$\frac{2A}{N}\sum_{l=0}^{N-1}\frac{x_l^2 + 4}{x_l^4 + 2x_l^2 + 1}\frac{1}{(x_k - x_l)^2 + H^2}(2Hz(x_l) - z^2(x_l)) = f(x_k), \qquad (7.22)$$

$$k = 0, 1, \ldots, N-1, A = 10.$$

If the conditions, given in Section "Continuous Operator Method" are met, the solution of the system of ordinary differential equations

$$\frac{dz_k(t)}{dt} = -\left(\frac{2A}{N}\sum_{l=0}^{N-1}\frac{x_l + 4}{x_l^4 + 2x_l^2 + 1}\frac{1}{(x_k - x_l)^2 + H^2}(2Hz_l(t) - z_l^2(t)) - f(x_k)\right),$$

$$(7.23)$$

$k = 0, 1, \ldots, N-1$, converges to the solution of the system of Eq. (7.22).
The exact solution of the system of Eq. (7.22) is $z(s) = (s^2 + 1)/(s^2 + 4)$.
The system of Eq. (7.23) was solved by the Euler method.
With the number of collocation nodes equal to 100, the error in solving the system of Eq. (7.22) is 0.5×10^{-2}.

Newtonian potential.

We construct a computational scheme for the approximate solution of Eq. (7.9). We introduce the system of knots $\{x_{kl} = (x_k, x_l)\}$, $k, l = 0, 1, \ldots, N$, where $x_k = -A + \frac{2kA}{N}$, $k = 0, 1, \ldots, N$, $A-$ a sufficiently large positive number.

An approximate solution of Eq. (7.9) will be sought in the form of a function

$$\varphi_{i,j}(x,y) = \begin{cases} 1, (x,y) \in \Delta_{ij}, \\ 0, (x,y) \in \Omega \setminus \Delta_{ij}, \end{cases}$$

where

$$\Delta_{ij} = ([x_i, x_{i+1}) \times [x_j, x_{j+1})) \cap \Omega, \, i, j = 0, 1, \ldots, N-2,$$
$$\Delta_{ij} = ([x_i, x_{i+1}) \times [x_j, x_{j+1}]) \cap \Omega, \, i = 0, 1, \ldots, N-2, j = N-1;$$
$$\Delta_{ij} = ([x_i, x_{i+1}] \times [x_j, x_{j+1})) \cap \Omega, \, i = N-1, j = 0, 1, \ldots, N-2;$$
$$\Delta_{ij} = ([x_i, x_{i+1}] \times [x_j, x_{j+1}]) \cap \Omega, \, i, j = N-1,$$

The computational scheme of the collocation method has the form

$$\frac{1}{2\pi} \frac{(2A)^2}{N^2} \sum_{l_1=0}^{N-1} \sum_{l_2=0}^{N-1} \frac{H \phi_{l_1 l_2}}{((x_{k_1} - x_{l_1})^2 + (x_{k_2} - x_{l_2})^2 + H^2)^{3/2}} = f(x_{k_1}, x_{k_2}), \, k_1, k_2 = 0, 1, \ldots, N-1, \quad (7.24)$$

where $\phi_{l_1, l_2} = \phi(x_{l_1}, x_{l_2})$, $l_1, l_2 = 0, 1, \ldots, 2N - 1$.

The system of Eq. (7.24) is associated with the following system of ordinary differential equations

$$\frac{dz_{k_1 k_2}(t)}{dt} = -\frac{1}{2\pi} \frac{(2A)^2 H}{N^2} \sum_{l_1=0}^{N-1} \sum_{l_2=0}^{N-1} \frac{z_{l_1 l_2}(t)}{((x_{k_1} - x_{l_1})^2 + (x_{k_2} - x_{l_2})^2 + H^2)^{3/2}} \quad (7.25)$$
$$+ f(x_{k_1}, x_{k_2}), \quad k_1, k_2 = 0, 1, \ldots, N-1.$$

Using the results of Section "Continuous Operator Method", we can show that if the set of parameters A, H, N is such that the logarithmic norm of the matrix of system (7.24) is negative, then the solution of the system of differential Eq. (7.25) as $t \to \infty$ tends to the solution of the system of Eq. (7.24).

To justify the convergence of the system of differential equations for $t \to \infty$, we need to bring the multidimensional matrix (Sokolov 1972) on the left-hand side of the system of Eq. (7.24) to the standard form. Then we need to calculate the logarithmic norm of the constructed matrix multiplied by (-1) and use the stability conditions for the solutions of the differential equations (Boikov 1990, 2008).

Substituting the values x_{k_i}, x_{l_i}, $k, l = 0, 1, \ldots, N-1, i = 1, 2$, in (7.19) we arrive at the system of equations

$$\frac{1}{2\pi}d\sum_{l_1=0}^{N-1}\sum_{l_2=0}^{N-1}\frac{\varphi_{l_1 l_2}}{((k_1-l_1)^2+(k_2-l_2)^2+d^2)^{3/2}}=f(-1+k_1 h,-1+k_2 h), k_1,k_2$$
$$=0,1,\ldots,N-1,$$

(7.26)

where $h=2A/N$, $d=NH/(2A)=H/h$.

Put $M=N^2$. We introduce the designations

$$z_{Nl+k+1}=\varphi_{kl}, \quad l=0,1,\ldots,N-1, \quad k=0,1,\ldots,N-1.$$

We denote by B the matrix $B=\{-b_{ij}\}$, $i,j=1,2,\ldots,N^2$ with elements: $b_{ij}=0$ if $(|k_1-l_1|=1,|k_2-l_2|=1)$, $(|k_1-l_1|=1,k_2=l_2)$, $(k_1=l_1,|k_2-l_2|=1)$; and $b_{ij}=-\frac{d}{2\pi}\frac{1}{((k_1-l_1)^2)+(k_2-l_2)^2+d^2)^{3/2}}$ with the rest (i,j), $i,j=1,2,\ldots,N^2$. Here $i=Nk_1+k_2+1$, $j=Nl_1+l_2+1$, $k_1,k_2,l_1,l_2=0,1,\ldots,N-1$.

The diagonal elements of the matrix B are equal to $b_{ii}=-\frac{1}{2\pi d^2}$.

It is easy to see that the maximum of the sum of the modules of the off-diagonal elements is

$$\Delta=\frac{d}{2\pi}\sum_{l_1=0}^{N}\sum_{l_2=0}^{N}\frac{1}{(l_1^2+l_2^2+d^2)^{3/2}}$$
$$+\frac{d}{2\pi}\sum_{l_1=1}^{N}\sum_{l_2=0}^{N}\frac{1}{(l_1^2+l_2^2+d^2)^{3/2}}$$
$$+\frac{d}{2\pi}\sum_{l_1=0}^{N}\sum_{l_2=1}^{N}\frac{1}{(l_1^2+l_2^2+d^2)^{3/2}}$$
$$+\frac{d}{2\pi}\sum_{l_1=1}^{N}\sum_{l_2=1}^{N}\frac{1}{(l_1^2+l_2^2+d^2)^{3/2}}-\frac{1}{2\pi d^2}.$$

Estimating these sums is a difficult task. In addition, the diagonal elements of the matrix B are sufficiently small in absolute value in the most important case when $h<H$. Therefore, the field of applications of the system (7.20) for solving 3D problems is rather limited.

For a model example, an equation of the form (7.24) was considered. As exact solution of this equation was chosen the function

$$\phi(x,y)=\begin{cases}1, (x,y)\in[-10,10]^2,\\ 0, (x,y)\in(-\infty,\infty)\setminus[-10,10]^2.\end{cases}$$

With the number of collocation nodes equal to 40 for each variable, the error of the solution is 10^{-4}.

Consider a modification of the method, given above.

We fix a natural number q, whose choice will be described below.

As above, we introduce the nodes $x_k = -A + 2Ak/N$, $k = -N, N+1, \ldots, 2N$.

To each node (x_i, x_j), $i,j = 0, 1, \cdots, N$, we put in line piecewise-constant functions:

$$\phi_{i,j}^q(x,y) = \begin{cases} 1, & (x,y) \in \Delta_{ij}^q, \\ 0, & (x,y) \in \Omega \setminus \Delta_{ij}^q, \end{cases}$$

where

$$\Omega = [-A, A]^2;$$

$$\Delta_{ij}^q = ([x_{i-q}, x_{i+q}) \times [x_{j-q}, x_{j+q})) \cap \Omega, \; i,j = 0, 1, \ldots, N-2,$$
$$\Delta_{ij}^q = ([x_{i-q}, x_{i+q}) \times [x_{j-q}, x_{j+q}]) \cap \Omega, \; i = 0, 1, \ldots, N-2, j = N-1;$$
$$\Delta_{ij}^q = ([x_{i-q}, x_{i+q}] \times [x_{j-q}, x_{j+q})) \cap \Omega, \; i = N-1, j = 0, 1, \ldots, N-2;$$
$$\Delta_{ij}^q = ([x_{i-q}, x_{i+q}] \times [x_{j-q}, x_{j+q}]) \cap \Omega, \; i,j = N-1,$$

$$\varphi_{i,j}(x,y) = \begin{cases} 1, & (x,y) \in \Delta_{ij}, \\ 0, & (x,y) \in \Omega \setminus \Delta_{ij}, \end{cases}$$

where

$$\Delta_{ij} = ([x_i, x_{i+1}) \times [x_j, x_{j+1})) \cap \Omega, \; i,j = 0, 1, \ldots, N-2,$$
$$\Delta_{ij} = ([x_i, x_{i+1}) \times [x_j, x_{j+1}]) \cap \Omega, \; i = 0, 1, \ldots, N-2, j = N-1;$$
$$\Delta_{ij} = ([x_i, x_{i+1}] \times [x_j, x_{j+1})) \cap \Omega, \; i = N-1, j = 0, 1, \ldots, N-2;$$
$$\Delta_{ij} = ([x_i, x_{i+1}] \times [x_j, x_{j+1}]) \cap \Omega, \; i,j = N-1,$$

The collocation method for Eq. (7.9) has the form

$$\frac{1}{2\pi} \operatorname{mes} \Delta_{kl}^q \frac{1}{H^2} \phi_{kl}^q(x_{kl}) + \frac{1}{2\pi} \frac{4A^2}{N^2} \sum_{i=0}^{N-1} \sum_{j=0}^{N-1}{}' \frac{H\phi_{ij}(x_{ij})}{((x_k - x_i)^2 + (x_l - x_j)^2 + H^2)^{3/2}} = f(x_{k,l}), \quad (7.27)$$

$$k, l = 0, 1, \ldots, N-1.$$

Here $\sum_i \sum_j{}'$ means summation over squares Δ_{ij} whose intersection measure with the domain Δ_{kl}^q is zero.

We denote by C the matrix representing the left-hand side of the system of Eq. (7.27).

Obviously, one can choose a q such that the logarithmic norm of the matrix $C = \{c_{ij}\}$, $i,j = 1, 2, \ldots, N$, will be negative.

Consider the system of differential equations

$$\frac{d\phi_{kl}(t)}{dt} = -\left(\frac{1}{2\pi H^2}\operatorname{mes}\Delta^q_{kl}\phi_{kl}(t) + \frac{(2A)^2 H}{2\pi N^2}\sum_{i=0}^{N-1}\sum_{j=0}^{N-1}{}' \frac{\phi_{ij}(t)}{((x_k - x_i)^2 + (x_l - x_j)^2 + H^2)^{3/2}} - f(x_{k,l})\right),$$
(7.28)

$$k, l = 0, 1, \ldots, N - 1.$$

From the results of Sect.``Continuous Operator Method' it follows that for q such that the logarithmic norm of the matrix C is negative, the solution of the system of differential Eq. (7.28) converges to the solution of system (7.27).

The work was supported by the Russian Foundation for Basic Research. Grant 16-01-00594.

References

Boikov I.V., Moiko N.V. (1999) An iterative method for solving the inverse gravity problem for a contact surface// Izvestia, Physics of the Solid Earth. –1999. –№ 2. – P. 52–56.
Boikov I.V., Boikova A.I. (2009) A parallel method of solving nonlinear inverse problems in gravimetry and magnetometry// Izvestiya, Physics of the Solid Earth,2009. Volume 45, Issue 3, pp. 248-257.
Boikov I.V., Boikova A.I. (2013) Priblizennie metodi resenia pryamich i obratnich zadach gravimetrii (Approximate methods for solution direct and inverse task of gravity prospecting) Penza: Penzensk. Gos. Univ., 2013. 510 p.
Boikov I.V. (1990) Stability of Solutions of Differential and Difference Equations in Critical Cases// Dokl. Akad. Nauk SSSR, 1990, vol. 314, no. 6, pp. 1298–1300.
Boikov I.V. (2008) Ustoichivost' reshenii differentsial'nykh uravnenii (Stability of Solutions of Differential Equations), Penza: Penzensk. Gos. Univ., 2008. 244 p.
Boikov I.V. (2012) On a continuous method for solving nonlinear operator equations// Differential equations. 2012, V. 48, No 9. P. 1308 – 1314.
Boykov I.V., Roudnev V.A., Boykova A.I., Baulina O.A. (2018) New iterative method for solving linear and nonlinear hypersingular integral equations//Applied Numerical Mathematics. Volume 127, May 2018, Pages 280–305.
Daletskii Yu.L. and Krein M.G. (1970) Ustoichivost' reshenii differtsial'nykh uravnenii v banakhovom prostranstve (Stability of Solutions of Differential Equations in Banach Space), Moscow: Nauka, 1970.
Dekker K. and Verwer J.G. (1988) Stability of Runge-Kutta Methods for Stiff Nonlinear Differential Equations, Amsterdam: North Holland Publishing Co., 1984. Translated under the title Ustoichivost' metodov Runge-Kutty dlya zhestkikh nelineinykh differentsial'nykh uravnenii, Moscow: Mir, 1988.
Gavurin M.K. (1958) Nonlinear Functional Equations and Continuous Analogues of Iteration Methods, Izv. Vyssh. Uchebn. Zaved. Mat., 1958, no. 5, pp. 18–31.
Glasko V.B., Ostromogil'skii A.Kh., Filatov V.G. (1970) Reestablishment of the depth and shape of a contact surface by regularization//USSR Computational Mathematics and Mathematical Physics.1970. V.10, No 5. P. 284–291.
Mudretsova E.A., Veselov K.E. (1990) Gravirazvedka (Gravity prospecting) Moscow: Nedra, 1990. – 607 p.

Puzynina T.P. (2003) Modified Newton Schemes for the Numerical Investigation of Quantum-Field Models, Doctoral (Phys.-Math.) Dissertation, Tver, 2003.

Sokolov N.P. (1972) Vvedenie v teoriu mnogomernich matrith (Introduction to multidimensional matrixes) Kiev: Naukova Dumka. 1972. 177 p.

Starostenko V.I. (1978) Ustoichivie chislennie metodi v zadachach gravimetrii (Stable numerical methods in gravity problems), Kiev: Naukova Dumka. 1978. 226 p.

Strakhov V.N. (1970) Some aspects of the plane gravitational problem.// Izvestia of Academy of Science of USSR. Physics of the Solid Earth. – 1970. – No. 12. – P. 32–44.

Tikhonov A.N., Glasko V.B. (1965) Use of the regularization method in non-linear problems// USSR Computational Mathematics and Mathematical Physics.1965. V. 5, No 3. P. 93–107.

Tikhonov A.N. and Arsenin V.J. (1990) Solutions of Ill-Posed Problems, Wiley, New-York, 1977.

Zhdanov M.S. (2002) Geophysical Inverse Theory and Regularization Problems. – N. Y. : Elsevier 2002. – 610 p.

Zhidkov E.N., Makarenko G.I., and Puzynin M.V. (1973) A Continuous Analogue of Newton's Method in Nonlinear Problems of Physics, Fiz. Elem. Chastits Atomn. Yadra, 1973, vol. 4, no. 1, pp. 127–166.

Chapter 8
The Density Model of the Crystalline Crust the Southwestern Part of the Lipetsk Region

T. A. Voronova, V. N. Glaznev, O. M. Muravina and I. Y. Antonova

Abstract The problems of constructing a detailed three-dimensional density model of the upper crust of the Voronezh crystalline massif using gravimetric and petrophysical data are considered. The obtained model shows the main features of the volumetric structure of the research area.

Keywords Gravity field · Detailed density modeling · Geological structure · Petrophysical database

The density structure of the upper part of the crust is studied on the basis of solving the inverse problem of gravimetry. Technology for construction of detailed density models of the environment based on the interpretation of local anomalies of the gravity field is constantly improving with the advent of new a priori data and new methods of working with information. The results of solving the inverse problem is created some model of the spatial location of the anomalous density objects in the investigated area, which with a guaranteed accuracy satisfies the observed gravity field. The accuracy of the detailed three-dimensional density models of the media can be improved if we perform a simulation taking into account all available a priori information.

The construction of the density model of the upper part of the crystalline crust of the study area was solved on the basis of inversion of local anomalies of the gravity field. The statement of the problem of detailed density modeling within a limited area is schematically shown in Fig. 8.1.

As initial data for density modelling were used: the regional density model of the lithosphere, built for the entire territory of the Voronezh crystalline massif (Glaznev et al. 2016); the regional gravity field corresponding to this model (Muravina 2016); map of region density on a geological basis for scale 1:500,000 (Muravina et al. 2014a, b, 2016a, b; Muravina and Zhavoronkin 2015); the values of the thickness

---

T. A. Voronova (✉) · V. N. Glaznev · O. M. Muravina · I. Y. Antonova
Voronezh State University, Voronezh, Russia
e-mail: voronova28@yandex.ru

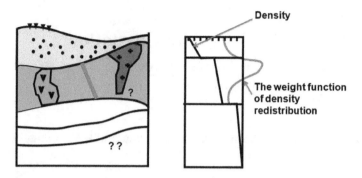

Fig. 8.1 To the solution of the problem detailed density modeling

for "gravity active" layer, obtained from the results of statistical analysis of the anomalous field (Glaznev et al. 2014, 2015a, b).

Three-dimensional density modeling was carried out at a area located in the south-western part of the Lipetsk region for the depth up to 12 km, which corresponds to the position of the "gravity active" layer for the given territory (Glaznev et al. 2014; Glaznev et al. 2015a, b). The position of the upper boundary of the model was established in accordance with the depth of the surface of the crystalline basement. The original model consisted of 10 layers of different thickness, which increased with a depth from 0.5 to 2 km. The layers of the modeling area also differed in the degree of detail of the density representation: the number of small objects decreased with increasing depth of the layer.

Absolute density values were set at the boundaries of the model layers in accordance with the petrophysical data of the region (Muravina et al. 2014a, b). The transition to excess density values was carried out by subtracting the density values of the regional model, which were interpolated to the points of the modeling domain. The three-dimensional regional density model of the central part of the East European Platform defines the distribution of density on the roofs and soles of the upper, middle and lower crust layers, in the transition layer and upper mantle to a depth of 80 km. In calculations, the density of the "gravity" layer, which is the object of detailed investigation, at this stage was assumed equal to the average value for the territory and amounted to 2.72 g/cm^3, which is typical for Archean crystalline rocks (Galitchanina et al. 1995; Kozlov et al. 2006). The solution of the direct problem of gravimetry for the developed model allows to estimate with the necessary accuracy the regional component of the gravity field for any territory within the region. Also at each point in the detailed three-dimensional model, the minimum and maximum density constraints were set.

Inversion of the gravity field in density was carried out on the basis of the quasinormal solution of the inverse problem in a three-dimensional case in Cartesian coordinates (Voronova and Glaznev 2014; Voronova and Muravina, 2014, 2017;

Glaznev et al. 2015a, b). From the observed gravity field of the study area, the influence of the sedimentary cover was excluded. The solution of the inverse problem was carried out using a program based on the modified method of local corrections (Muravina and Glaznev 2015). This method allows you to effectively work with input data represented by a large number of numbers. The stability of the solution is achieved through the organization of the iterative process in such a way that at each calculation point the correction of the original model is consistently performed.

The starting model plays an important role in ensuring the geological content of the solution. This model is built on the basis of a priori information, and generalizes the petrophysical and geological data related to the study area. In solving the inverse problem, the starting model is described by a significant number of parameters and is characterized by a high degree of complexity. In fact, during the process of field inversion, the initial model is refined in the specified limits of the parameter variation in order to obtain a mass distribution equivalent to the observed field.

Using the values of the reduced gravitational field at the grid nodes on the surface and the starting model of the density distribution in the lower half-space, a direct problem for the upper layer was solved. Next, we calculated the discrepancy of the field, which in the next step allowed us to calculate the corresponding equivalent density in a bounded flat layer. At the next stage, the residual of the model was redistributed into the lower half-space in accordance with the weight function. For the model obtained, the gravity field was calculated, which was compared with the initial observed field.

When solving a direct problem, a recursive algorithm was used that ensures high computational speed at a guaranteed level of error (Glaznev and Loshakov 2012; Muravina and Loshakov 2015). The weight function of density redistribution was determined by the initial approximation of the model taken into account the geological considerations and ranges of density variation of the corresponding geological complexes (Muravina et al. 2014b, 2016a, b; Muravina and Zhavoronkin 2015).

The algorithm for solving the inverse problem is shown in Fig. 8.2.

The obtained results generally demonstrate the correspondence of the density model to the initial geological data of the upper part structure of the earth's crust for study area. The value of the RMS error of iterations in the process of solving the inverse problem decreased from 5.90 mGal for the starting model to 0.21 mGal for the final density model (Fig. 8.3).

At the same time, there are a number of contradictions connected with a certain difference in anomalous gravity and anomalous magnetic fields.

The study area is completely covered by magnetic survey, both at the ground and in the aero-variant, which made it possible to clarify the geological features of the obtained density model taking into account the magnetic data.

This work was supported by RFBR, grant No 16-05-00975.

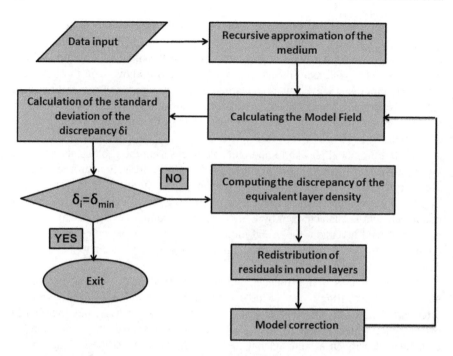

Fig. 8.2 Block diagram of the algorithm for solving the inverse problem

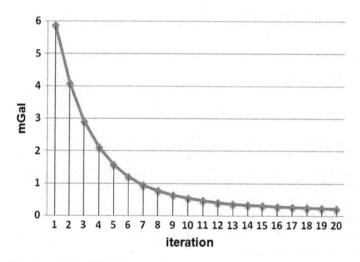

Fig. 8.3 The change in the RMS discrepancy

References

Galitchanina, L. D., V. N. Glaznev, F. P. Mitrofanov, O. Olesen and H. Henkel (1995) Surface density characteristics of the Baltic Shield and adjacent territories. Norges Geologiske Undersøkelse. Special Publication « Geology of the Eastern Finnmark - Western Kola Peninsula Region » . Proceedings of the 1st. International Barents Symposium « Geology and Minerals in the Barents Region »/ Eds. D.Roberts, Ø.Nordgulen. PP. 349–354.

Glaznev, V. N. and G. G. Loshakov (2012). One method of simulation of ore objects using adaptive approximation. Bulletin of VSU, Series "Geology", No. 1, pp. 243–246. (in Russian).

Glaznev, V. N, M. V. Mints and O. M. Muravina (2016). Density modeling of the central part of the East European platform. Vestnik KRAUNTS, Series of "Earth Sciences" T. 29, p. 53–63. (in Russian).

Glaznev, V. N., O. M. Muravina, T. A. Voronova and V. M. Choline (2014). Evaluation of the thickness gravy-active layer of the crust of the Voronezh crystalline massif. Bulletin of VSU, Series "Geology", No. 4, pp. 78–84. (in Russian).

Glaznev, V. N., O. M. Muravina, T. A. Voronova and E. B. Kislova (2015a). Thickness of the gravy-active layer of the upper crust of the Voronezh crystalline massif from the stochastic analysis of the gravity anomalies. The materials of the 42 session of the international seminar named D. G. Uspensky "The theory and practice of geological interpretation of geophysical fields". Perm: MI UB RAS, pp. 46–48. (in Russian).

Glaznev, V. N., T. A. Voronova, I. Yu. Antonova and O. M. Muravina (2015b). Methods and results of the 3D density modelling in studding construction of the upper crust of the Voronezh crystalline massif. The materials of the 42 session of the international seminar named D. G. Uspensky "The theory and practice of geological interpretation of geophysical fields". Perm: MI UB RAS, pp. 49–52. (in Russian).

Kozlov, N. E., N. O. Sorokhtin, V. N. Glaznev, N. E. Kozlova, A. A. Ivanov, N. M. Kudryashov, E. V. Martynov, V. A. Tyuremnov, A. V. Matyushkin and L. G. Osipenko (2006). Geology of Archaean of the Baltic Shield. Saint-Petersburg, Nauka, 329 p. (in Russian).

Muravina, O.M. (2016). Density model of the earth's crust for Voronezh crystalline massif. Bulletin of VSU, Series "Geology" no. 1, pp. 108–114. (in Russian).

Muravina, O. M. and V. N. Glaznev (2015). The method of local correction in the density modeling of the lithosphere. Materials of the VIII scientific readings named after Y. P. Bulashevich "Deep structure, geodynamics, thermal field of the Earth, interpretation of geophysical fields". Ekaterinburg, pp. 249–252. (in Russian).

Muravina, O. M. and G. G. Loshakov (2015). The principles of solving the direct problem of potential in modeling of the lithosphere. Bulletin of VSU, Series "Geology", No. 3, pp. 97–100. (in Russian).

Muravina, O. M. and V. I. Zhavoronkin (2015). Statistical analysis digital fundamentals petrodensity map of the Voronezh crystalline massif. Bulletin of VSU, Series Geology, No. 2. – P. 94–99. (in Russian).

Muravina, O. M., V. I. Zhavoronkin and V. N. Glaznev (2014a). The petrodensity model of the crystalline basement of the Voronezh crystalline massif. The materials of the 41st session of the international seminar named D. G. Uspensky "The theory and practice of geological interpretation of geophysical fields." Yekaterinburg: IGF UB RAS, pp. 171–174. (in Russian).

Muravina, O. M, V. I. Zhavoronkin and V. N Glaznev (2014b). Spatial analysis of the density distribution of Precambrian formations of the Voronezh crystalline massif. Collection of materials of the XV International Conference "Physical-Chemical and Petrophysical Studies in Earth Sciences". Moscow, p. 171–173. (in Russian).

Muravina, O. M., V. I. Zhavoronkin and V. N. Glaznev (2016). The Petrodensity Map of the Voronezh crystalline massif. The materials of the 43-th session of the international seminar named D. G. Uspensky "The theory and practice of geological interpretation of geophysical fields". Moscow: IFE RAS, pp. 133–136. (in Russian).

Voronova, T. A. and V. N. Glaznev (2014). Three-dimensional density model of the granite massive from Hoper Megablock (Voronezh crystalline massif). The materials of the 41st session of the international seminar named D. G. Uspensky "The theory and practice of geological interpretation of geophysical fields." Yekaterinburg: IGF UB RAS, pp. 82–84. (in Russian).

Voronova, T. A. and O. M. Muravina (2014). Detailed density modelling of the upper crust for the Voronezh crystalline massif. Bulletin of VSU, Series "Geology", No. 2, pp. 150–154. (in Russian).

Voronova, T. A. and O. M. Muravina (2017). Optimization solving the inverse problem of gravimetry in the construction of detailed density model. The materials of the 44th session of the international seminar named D. G. Uspensky "The theory and practice of geological interpretation of geophysical fields." Moscow: IPE RAS, pp. 114–116. (in Russian).

Part II
Modern Algorithms and Computer Technologies

Chapter 9
Neural Network Algorithm for Solving 3d Inverse Problem of Geoelectrics

M. I. Shimelevich, E. A. Obornev, I. E. Obornev, E. A. Rodionov and S. A. Dolenko

Abstract The approximating neural network algorithm for solving the inverse problems of geoelectrics in the class of grid (block) models of the medium is presented. The algorithm is based on constructing an approximate inverse operator using neural networks and makes it possible to formally obtain the solutions of the geoelectrics inverse problem with a total number of the sought parameters of the medium $\sim n \times 10^3$. The questions concerning the correctness of the problem of constructing the inverse neural network operators are considered. The a posteriori estimates of the degree of ambiguity in the inverse problem solutions are calculated. The work of the algorithm is illustrated by the examples of 2D and 3D inversions of the synthesized data and the real magnetotelluric sounding data.

Keywords Geoelectrics · Inverse problem · Approximation · A priori and a posteriori estimates · Neural networks

Introduction

Many practical inverse problems of geophysics including geoelectrics are reduced to the numerical solution of a finite-dimensional (in the general case, nonlinear) operator equation on a compact set S_N (Glasko et al. 1976; Dmitriev 2012; Shimelevich and Obornev 2009; Shimelevich et al. 2017):

M. I. Shimelevich · E. A. Obornev · E. A. Rodionov
Russian State Geological Prospecting University MGRI-RSGPU, Moscow, Russia
e-mail: shimelevich-m@yandex.ru

I. E. Obornev (✉) · S. A. Dolenko
Lomonosov Moscow State University Skobeltsyn Institute of Nuclear Physics (MSU SINP), Moscow, Russia
e-mail: o_ivano@mail.ru

$$A_N s = e, \quad s \in S_N \subset R^N, \quad e \in R^M, \quad M \geq N;$$
$$S_N : [s_{\min} \leq s^n \leq s_{\min} + D_s], \quad n = 1, \ldots, N, \qquad (9.1a)$$

where $s = (s^1, \ldots, s^N)$ is the sought vector of the parameters of the medium each of which can vary within a given interval of the values $[s_{\min}, s_{\min} + D_s]$; $e = (e^1, \ldots, e^M)$ is the input data vector; A_N is the operator of numerical solution of the forward problem in a given finite–parameter class of the media. In terms of the projections, system (9.1) has the following form:

$$A^m(s^1, \ldots, s^N) = e^m, \quad m = 1, \ldots, M, \quad M \geq N,$$
$$s_{\min} \leq s^n \leq s_{\min} + D_s, \quad n = 1, \ldots, N, \qquad (9.1b)$$

where $A^m(s^1, \ldots, s^N)$, $m = 1, \ldots, M$ are the coordinate functions of operator A_N. In this work we consider the grid (block) parameterization of the medium (Shimelevich et al. 2017) which is most universal if the a priori infornation is scarce. In this case, the projections s^n of the sought parameter vector of the medium are the values of the sought characteristic of the medium in the cells of a given parameterization grid θ_N. The latter is constructed by merging (by a certain rule) the cells of the initial finite difference grid θ_{N_0} of size N_0 on which the forward boundary problem is solved. In the cases when the additional a priori information about the structure of the studied medium is available, also other types of the parameterization and narrower sets S_N of the admissible solutions are used, for instance, $s_{\min}^n \leq s^n \leq s_{\min}^n + D_s^n$, $n = 1, \ldots, N$. In the general case, the problem of solving Eq. (9.1) at large N is Ill-posed and obtained solutions are practical ambiguity. In (Shimelevich et al. 2017) it was shown that for a given error level δ_0 in the input data it is possible to build the regularized parameterization grid in such a way that the degree of ambiguity of any solutions of the inverse problem (9.1) will not exceed the given value ε_0. Irrespective of the numerical method used for solving Eq. (9.1), its *approximate solution* is a certain *vector function* $\tilde{\psi}(e^1, \ldots, e^M)$ of M variables, which are projections of the input data vector $e = (e^1, \ldots, e^M)$. In the multidimensional inverse problem of geophysics, in the case of a general type of nonlinearity, the problem of calculating the function $\tilde{\psi}(e)$ cannot always be formalized (e.g., because of the presence of local extrema, etc.) and every time it has to be solved *individually for the different input data*. In this work, we consider the approximating approach to constructing the universal neural network (NN) approximator with the use of which the inverse problem is solved for any arbitrary input data vector $e \in R^M$. The general concept of applying the NN technologies in the problems of processing and interpretation of the geophysical data is described in the review (Raiche 1991). The examples of solving the *inverse problem of geoelectrics* with the use of neural networks are presented in (Hidalgo et al. 1994; Poulton et al. 1992; Spichak and Popova 1998; Shimelevitch and Obornev 1998. In these works it was shown that with the aid of NN approximators it is possible to obtain the approximate solutions of the inverse problems of geoelectrics with $N \sim 10 \div 15$ sought parameters. Tuning the general methods of constructing the NN approximators of inversion for the problem to be solved and developing the advanced modifications of the method with the use of Monte-Carlo algorithms allowed us to generalize the NN technique for the case of 2D and 3D media whose geoelectrical properties are determined by $n \times 10^2 \div n \times 10^3$ parameters (Shimelevich and Obornev 2009; Shimelevich et al. 2017).

Approximating Approach to Solving Finite-Dimensional Nonlinear Operator Equations

In the approximating approach, the solution of Eq. (9.1) is searched for in the form of a certain (analytically specified) vector function $\boldsymbol{\Psi} = (\Psi^1, \ldots, \Psi^N)$ of M variables $e^1, \ldots, e^M$, the coordinate functions $\Psi^n(a^{n1}, \ldots a^{nJ}, e^1, \ldots, e^M)$ of which depend on the free coefficients $a^{n1}, \ldots a^{nJ}$, $n = 1, \ldots, N$, composing the matrix $\hat{a} = \{a^{nj}\}$ of size $N_\Psi = N \cdot J$. Vector function $\boldsymbol{\Psi}(\hat{a}, e)$ is referred to as the *approximator of the inversion* for Eq. (9.1). For determining the matrix of the coefficients $\hat{a}$, the problem of *training the approximator* $\boldsymbol{\Psi}(\hat{a}, e)$ is solved. The trained approximator $\boldsymbol{\Psi}(\hat{a}, e)$ is an approximate numerical inverse operator for Eq. (9.1) which is represented in the analytical form. With the use of this approximator, the approximate solution of the inverse problem is solved rapidly and universally in a given class of the media for any arbitrary right-hand side $e \in R^M$, which is the main advantage of the method.

Approximating neural network approach
The approximating neural network (ANN) method pertains to the group of the approximating methods for solving the inverse problems. As the coordinate functions of the approximator of the inversion, the ANN method uses the approximating constructions referred to as the neural networks. The simplest and, at the same time, most commonly used in the practice is three-layer neural network (perceptron) (Haykin 1999). With this NN, the *coordinate functions* Ψ^n of the NN approximator $\boldsymbol{\Psi}(V, W, e)$ are represented in the following form:

$$\Psi^n(V, W, e^1, \ldots, e^M) = \sum_{l=1}^{L} v^{nl} \lambda \left(\sum_{m=1}^{M} w^{lm} e^m \right), \quad n = 1, \ldots, N, \qquad (9.2)$$

where $\lambda(x)$ is a given (in the general case, nonlinear) bounded monotonically increasing differentiable function (activation function), for instance, $\lambda(x) = 1/(1 + e^{-x})$. Coefficients v^{nl}, w^{lm} of matrices V, W determine the free coefficients of a given type of NN approximator $\boldsymbol{\Psi}(V, W, e)$ whereas parameter L specifies the complexity of the letter and the size of the matrix of coefficients V, W. A total number N_Ψ of the free coefficients of the NN approximator $\boldsymbol{\Psi}(V, W, e)$ is $N_\Psi = L(M + N)$.

The extensive practical use of neural network of type (9.2) is, on one hand, due to the simplicity of neural network construction and, on the other hand, it is caused by the fact that the arbitrary continuous function of M variables can be approximated by the three-layer neural network of the form (9.2) with any prescribed accuracy with sufficiently large L (Cybenko 1989).

The Scheme of ANN Algorithm for Solving the Inverse Problem of Geoelectrics

The algorithm of the ANN method includes the following main blocks:

I. Numerical solution of the forward problem of geoelectrics on a given numerical grid.
II. Constructing the model grid (block) class of the media based on designing a regularized parameterization grid $\theta_{N_{max}}$ of optimal dimension N_{max} at which the degree of practical ambiguity of the inverse problem solution does not exceed a given value ε_0 with a given error level δ_0 in the input data (Shimelevich et al. 2013, 2017).
III. Solving the problem of training the NN approximator of the inversion in a given class of the media. With the use of the forward operator A_N, the set $\mathbb{Q}_{bs}$ of the basic solutions of the (forward and inverse) problems for Eq. (9.1) is formed. This set is referred to as a *training set*. Training the NN approximator on set $\mathbb{Q}_{bs}$ is reduced to solving the nonlinear multi-extremum problem of conditional optimization (Haykin 1999) on the set $\mathbb{Q}_{bs}$. Rigorous substantiation of the solution techniques for these problems is in most cases difficult. Therefore, for fitting the standard optimization methods for solving the training problem, these methods are combined with the informal, heuristic approaches. Among the latter, the method of stochastic gradient with *back propagation error* (BPE) algorithm (Werbos 1974; Haykin 1999) is most common in the practice.

For reducing the dimensionality of the training problem, data are preliminarily compacted and the optimal dimensions of the input and output NN vectors are determined with the account for the specificity of the problem to be solved (Dolenko et al. 2009). In the case of the complex nonlinear probelms, three-layer networks are not the optimal approximating constructions; hence, multilayer networks are used in the in the practice Haykin (1999). The best results achieved in our works concerning the considered problems of geoelectrics based on the numerical experimenrts were obtained with five-layer networks with $L_k = 32, 16, 8$; $k = 1, 2, 3$, neurons in the hidden k th layers. The calculations also show that the practical reasonable interpolation properties of the NN approximators for the probelms of geoelectrics can be achieved with the training sets as large as $I_{bs} \sim 10000$ for 2D problems and $I_{bs} \sim 20000 \div 25000$ for 3D problems.

IV. Conducting the inversion of the observed data and calculating the residual δ of the solution of the inverse problem.

V. When necessary, constructing the adjusting NN approximators of the inversion and conducting additional iterations of the solution of inverse problem (Shimelevich et al. 2017).

VI. Calculating the a posteriori estimates for the degree of ambiguity (error) $\beta_1^i(s_{\delta 1}, \delta)$ of solution $s_{\delta 1}$ of the inverse problem which is determined with the actual residual δ (Shimelevich et al. 2017). The estimates determine the maximal deviations of δ-equivalent solutions of the inverse problem from the obtained solution $s_{\delta 1}$ for each i th level of the parameterization grid.

Program Codes

The computations for the forward 3D problems of geoelectrics are conducted with the use of MTD3FWD program developed by Mackie et al. (1994) with the input and output procedures modified for the purposes of mass-parallel computations. Supercomputing clusters are used for constructing the sets of basic solutions of the problems.

The NN approximators of the inversions were trained using the Fortran-77 program codes of (Lönnblad et al. 1992) adapted for the particular features of the problem to be solved. The solutions to the technical problems associated with selecting the optimal parameters of the NN construction were obtained with the use of the program complex of (Dolenko et al. 2009) which is based on CUDA technology for graphical processors GPU.

Numerical Examples

In this section we present the examples of solving the 2D and 3D inverse problems of geoelectrics for the synthesized data and field measurements. For illustrating the work of the method, we constructed 2D and 3D inversion approximators for solving of geoelectrics problems with a depth of investigation down to 5 km. As the a priori information for solving the inverse problems we only specified the range of variations in log resistivity ρ in the real media: $\lg \rho = 0 \div 4$; the first approximation was not specified.

Examples of NN Inversion of the Synthesized 2D and 3D Data

Figure 9.1 shows the results of the NN inversion of the synthesized data for the 2D model of the medium specified on the regularized 2D parameterization grid including five levels. The number of the sought parameters in the model is $N_2 = 315$. The mean errors $\bar{\varepsilon}_i$ of the solution from the parameters for each ith grid level were estimated by the formula $\bar{\varepsilon}_i = \frac{1}{N_i \cdot D_s} \sum_{n=1}^{N_i} |\Delta s_n|$ where Δs_n is the difference between the true and obtained parameters with index n; N_i is the number of the parameters in the ith level; $\bar{\varepsilon}$ is the mean error over all the levels. The estimate of the degree of ambiguity β_1^i of the solution for each level and the mean estimate over all

Fig. 9.1 The results of the inversion for 2D model: **a** initial model; **b** the result of the inversion on the first iteration; **c** the result of the inversion after the fifth iteration; $\overline{\delta}$ is actual residual; $\overline{\varepsilon}$ is mean error of the solution over all the grid levels

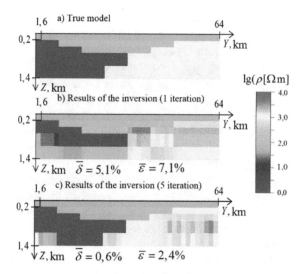

Table 9.1 The estimation results for the inversion for 2D model shown in Fig. 9.1

No. of grid level	First iteration		Fifth iteration	
	$\overline{\varepsilon}_i$ (%)	$\overline{\beta}_1^i$ (%)	$\overline{\varepsilon}_i$ (%)	$\overline{\beta}_1^i$ (%)
1	0.29	3.90	0.03	0.89
2	1.46	7.86	0.06	1.78
3	7.24	12.9	1.52	2.95
4	7.36	15.7	2.40	5.71
5	19.1	23.9	7.71	9.85
Average	$\overline{\varepsilon} = 7.1$	$\overline{\beta}_1 = 12.9$	$\overline{\varepsilon} = 2.4$	$\overline{\beta}_1 = 4.2$
Residual	$\overline{\delta} = 5.1$		$\overline{\delta} = 0.6$	

the levels $\overline{\beta}_1$ are presented in the Table 9.1.

Figures 9.2–9.3 show the results of NN inversion of the synthesized data for the 3D models specified on the regularized 3D parameter grid which includes five levels. The number of the sought parameters in each model is $N = 532$. The results of the NN inversion for 3D models are presented in the averaged (smoothed) form on the initial finite-difference grid. The results of estimating the inversion are illustrated in Table 9.2.

From the presented results it can be seen that highest errors of the inversion are associated with the fifth level. The highly conductive target inclusions (rectangular blocks shown in blue) are reconstructed reasonably accurately which allows us to reconstruct the position and contours of the studied objects. The a posteriori estimates of the ambiguity $\overline{\beta}_1$, adequately reflect the results of the inversion; however, their values are overestimated since they determine the maximal deviations of the equivalent solutions from the obtained one.

9 Neural Network Algorithm for Solving ...

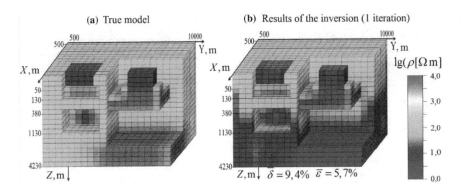

Fig. 9.2 Initial (true) model M1 **a** with the cut out frontal fragments and **b** the result of the inversion

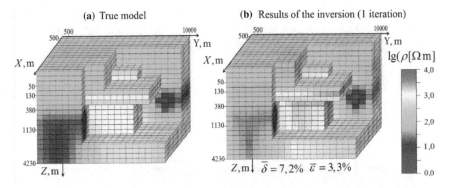

Fig. 9.3 Initial (true) model M2 **a** with the cut out frontal fragments and **b** the result of the inversion

Table 9.2 The estimation results for the inversion for 3D model shown in Figs. 9.2–9.3

No. of grid level	Model M1 (Fig. 9.2)		Model M2 (Fig. 9.3)	
	$\bar{\varepsilon}_i$ (%)	$\bar{\beta}_1^i$ (%)	$\bar{\varepsilon}_i$ (%)	$\bar{\beta}_1^i$ (%)
1	1.4	9.4	2.1	3.5
2	4.5	23.7	3.3	13.7
3	4.9	19.9	3.6	23.7
4	6.1	36.5	3.6	28.5
5	11.6	38.5	3.9	30.0
Average	$\bar{\varepsilon} = 5.7$	$\bar{\beta}_1 = 25.6$	$\bar{\varepsilon} = 3.3$	$\bar{\beta}_1 = 19.88$
Residual	$\bar{\delta} = 9.36$		$\bar{\delta} = 7.23$	

Results of 2D NN inversion of the field MT data.

As the example illustrating the NN inversion of the practical observations, we present a segment of **profile 2DV** from 490 to 1100 km (Feldman et al. 2008). The result of the NN inversion of the field observations (preprocessed for eliminating the shift effects and random outliers by the technique (Feldman et al. 2008)) is shown in Fig. 9.4a. Figure 9.4b shows the results of the independent inversion carried out by EMGEO (Feldman et al. 2008).

During solving the inverse problem, N = 1580 parameters were determined within the considered 2D area shown in Fig. 9.4a. The first approximation was not prescribed. The solution residuals calculated from separate segments of the profile are shown in Fig. 9.3a. The estimates of the degree of ambiguity in the inverse problem solution on these segments vary from 6 to 22%. The overall residual over the entire segment of the profile is $\bar{\delta} = 11\%$, and the degree of ambiguity of the inverse problem solution (mean over all the grid levels) is $\overline{\beta_1} = 21.6\%$. For comparison, the obtained geoelectrical section (Fig. 9.4a) is superimposed with the independent seismic data (black contours). It can be seen that the boundaries of the trough-like geolelectrical structures which are identified by the NN inversion are reasonably well consistent with the seismic boundaries.

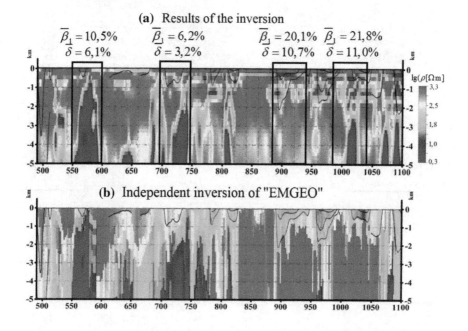

Fig. 9.4 Geoelectric section along the regional profile 2DV, segment 1 (490–1100 km): **a** NN inversion, 1580 parameters determined; **b** independent EMGEO inversion. Black boxes mark the separate segments for the detailed analysis

Conclusions

(1) The approximating neural network method and its modifications allow the stable approximate solutions of the inverse 2D and 3D coefficient problems of geoelectrics to be found in a formalized way in the class of grid models of the media on the *regularized* parameterization grid with a reasonable practical accuracy *without specifying the first approximation*. The number of the parameters of the medium that are determined by the inversion is $\sim n \times 10^3$.

(2) For the obtained approximate solutions of the inverse problem, the estimates of the degree of ambiguity (errors) can be calculated, which do not depend on the particular method of the inversion.

(3) The approaches and methods developed in this work largely rely on the modern computing capabilities such as supercomputing cluusters and technologies for mass-parallel computations.

Acknowledgements The research was carried out using supercomputers at Joint Supercomputer Center of the Russian Academy of Sciences (JSCC RAS). This study was supported by the Russian Science Foundation (project no. 14-11-00579).

References

Cybenko, G. (1989) Approximation by superpositions of a Sigmoidal Function // Mathematics of Control, Signals, and Systems, V.2, pp. 303-314, 1989.

Dmitriev V.I. (2012) Obratnye zadachi geofiziki [Inverse Problems o f Geophysics]. Moscow, MAKS Press Publ., 340 p., 2012.

Dolenko S., Guzhva A., Persiantsev I., Obornev E. and Shimelevich M. (2009) Comparison of adaptive algorithms for significant feature selection in neural network based solution of the inverse problem of electrical prospecting// In: C. Alippi et al (Eds.): ICANN 2009, Part II. Lecture Notes in Computer Science, 2009 - V. 5709 - pp. 397405.- Springer-VerlagBerlinHeidelberg.

Feldman I.S., Okulesky B.A., Suleimanov A.K. (2008) Elektrorazvedka metodom MTZ v komplekse regionalnyh neftegazopoiskovyh rabot v evropejskoj chasti Rossii [Electrical exploration by MTS method in a complex of regional oil and gas prospecting works in the European part of Russia // Journal of Mining Institute, St. Petersburg, 176: 125-131, 2008.

Glasko V.B., Gushhin G.V. and Starostenko V.I. (1976) O primenenii metoda reguljarizacii A.N. Tihonova k resheniju nelinejnyh sistem uravnenij[On the application of the method of regularization of A.N. Tikhonov to the solution of nonlinear systems of equations] // ZhVM and MF, 16 (2): 283-292, 1976.

Haykin S. (1999) Neural networks: A Comprehensive Foundation. 2nd ed. Pearson Education, 823 p., 1999.

Hidalgo H., Gómez-Treviño E. and Swiniarski R. (1994) Neural Network Approximation of a Inverse Functional // IEEE International Conference on Neural Networks, V. 5, pp. 3387-3392, 1994.

Lönnblad L., Peterson C., and Rögnvalsson T. (1992) Pattern recognition in high energy physics with artificial neural networks—JETNET 2.0 // Computer Physics Communications, Vol. 70, 1, pp. 167-182, 1992.

Mackie R.L., Smith J.T. and Madden T.R. (1994) Three-dimensional electromagnetic modeling using finite difference equations: the magnetotelluric example // Radio Science, 29, pp. 923–935, 1994.

Poulton M., Sternberg B., and Glass C. (1992) Neural network pattern recognition of subsurface EM images // Journal of Applied Geophysics, V.29, Is.1, pp. 21–36, 1992.

Raiche A. (1991) A pattern recognition approach to geophysical inversion using neural nets // Geophysics J. Int., 105, pp. 629–648, 1991.

Shimelevitch, M. and Obornev, E., The Method of Neuron Network in Inverse Problems MTZ, Abstr. of the 14th Workshop on Electromagnetic Induction in the Earth, Sinaia, Romania, 1998.

Shimelevich M.I. and Obornev E.A. (2009) An approximation method for solving the inverse mts problem with the use of neural networks. Izvestiya - Physics of the Solid Earth 45(12), 1055–1071, 2009.

Shimelevich M.I., Obornev E.A., Obornev I.E., and Rodionov E.A. (2017) The neural network approximation method for solving multidimensional nonlinear inverse problems of geophysics. Izvestiya - Physics of the Solid Earth 53(4): 588–597, 2017.

Shimelevich M.I., Obornev E.A., Obornev I.E., and Rodionov E.A. (2013) Numerical methods for estimating the degree of practical stability of inverse problems in geoelectrics. Izvestiya - Physics of the Solid Earth 49(3): 356–362, 2013.

Spichak, V.V., and Popova, I.V. (1998) Application of the neural network approach to the reconstruction of a three-dimensional geoelectric structure. Izvestiya - Physics of the Solid Earth 34(1): 33–39, 1998.

Werbos P.J. (1974) Beyond regression: New tools for prediction and analysis in the behavioral sciences. Ph.D. thesis, Harvard University, Cambridge, MA, 1974.

Chapter 10
Optimization of Computations for Modeling and Inversion in NMR T2 Relaxometry

L. Muravyev, S. Zhakov and D. Byzov

Abstract Great success is achieved currently in the using of NMR relaxometry for detecting and distinguishing of reservoir fluids, for example, free and bound water, oil. NMR data enable petrophysicists, specialists in the development of deposits and geologists to study the types of fluids and their distribution in a reservoir that has been opened by a well. NMR allows identifying the intervals in which hydrocarbons are present and predict their recoverability. The investigations carried out in this work are aimed at the optimizing of calculating time for the integrals arising in the NMR forward and inversion problems, while preserving the predetermined error. The method of the Legendre polynomial expansion application for the solution of the problem of modeling relaxation curves in the NMR method is described. This tool makes it possible to reduce significantly the computational complexity of the relaxation curve calculation, and hence the calculation time in comparison with numerical integration methods. In addition, numerical methods do not allow to pre-select the parameters for partitioning a segment to achieve a given error. Since the method described in this paper uses an analytic expression for the integral, the calculation accuracy depends only on the integration error. The given approximation error is achieved due to the choice of the maximum degree of the polynomial at the stage of calculating the coefficients of the series of the Legendre polynomials.

Keywords NMR · Petrophysics · Relaxation time · CPMG

L. Muravyev (✉)
Ural Federal University, Ekaterinburg, Russia
e-mail: mlev@igeoph.net

S. Zhakov
Institute of Metal Physics of Ural Branch of Russian Academy of Sciences, Ekaterinburg, Russia

L. Muravyev · D. Byzov
Institute of Geophysics Ural Branch of RAS, Ekaterinburg, Russia

Introduction

Impulse nuclear magnetic resonance relaxometery (NMR) is one of the most perspective methods for studying of petrophysical properties of reservoir rocks (Farrer, Becker 1971). Measurement result received by the spin echo method (Carr-Purcell-Meiboom-Gill sequence) is a relaxation curve, a superposition of decaying exponential signals (Coates el al. 2000). The transverse relaxation time distribution T2 makes it possible to determine the type of formation fluid. Since the fluids enclosed in small pores are close to the rock matrix surface, they are characterized by a short relaxation time T2, and free fluid in large pores has long T2 time. Therefore, analysis of the T2 distribution makes it possible to distinguish the type of fluid in the rock: movable and irreducible components. The equipment for these experiments are NMR logging instruments for measurements in wells and laboratory NMR-relaxometers for the study of core samples and drill mud (Gang et al. 2006).

A NMR-relaxometer with a constant polarizing field was manufactured in Russia. The value of the magnetic field is comparable with the field of nuclear magnetic logging instruments, which allows to use relaxometer data for preliminary tuning and correct interpretation of the results obtained during NMR well logging (Muravyev and Zhakov 2016). The developed optimal magnetic system of the relaxometer provides a small instrument size and weight, make it transportable, applicable for working on a well during drilling (Mirotchnik 2004).

Theory of the NMR Method

The precession signal is observed by the spin-echo method (Farrer Becker 1971), by means of a sequence of radio-frequency pulses: the Carr-Purcell-Meiboom-Gill sequence. The rate of decay of the resulting signal's amplitude is the relaxation time T_2. The relaxation time obtained in the experiment depends on the interaction of the fluid molecules with each other and at the interface with the sample matrix. Surface relaxation plays the main role in the pore environment, so the relaxation time depends on the ratio of the square of pore surface to the volume of the pores. The real sample contains a set of pores of various sizes, thus the relaxation curve observed in the NMR experiment is the sum of several decaying components. By decomposing this curve into the spectrum—a set of exponentials, we can investigate the physical processes in the medium. Experimentally confirmed that the shape of the spectrum of relaxation times coincides with the distribution of pore size in the sample (Coates el al. 2000) (Fig. 10.1).

The porosity measured by the NMR method and the pore size distribution obtained by the inversion are used to estimate the permeability of the medium. The reliability of this decomposition directly affects the result of determining the amount of free fluid. This parameter characterizes the recoverability of hydrocarbon raw materials.

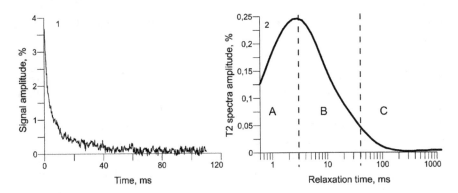

Fig. 10.1 Impulse NMR experiment results: **a** relaxation curve, **b** its inversion: distribution by relaxation times (continuous spectrum); fluid components: A—clay-bound water, B—capillary bound water, C—movable (free fluid index)

In the general case, for a fluid in a pore medium, there is a continuous distribution of the relaxation times $f(T_2)$. It determines the form of the relaxation curve observed in the NMR experiment (Coates el al. 2000):

$$R(t) = \int f(T_2) e^{(-\frac{t}{T_2})} dT_2 \qquad (10.1)$$

The unknown function $f(T_2)$ contains the information about the pore medium: pore size distribution in the sample, as well as the content of the free and bound fluid. Thus, from the experimentally observed relaxation curve $R(t)$ we must obtain a continuous function $f(T_2)$ or a discrete set of amplitudes A_i and relaxation times T_{2i}. There are several approaches to relaxation curves interpretation. For example, the direct solution of the integral equation by the Tikhonov regularization method, the inverse Laplace transform, the Prony filtering method. Also, it is possible to find spectra as a set of test functions representing various types of fluid in a pore medium (clay-bound, capillary-bonded, free fluid).

The algorithms based on the expansion of the curve as a set of exponents defined on a regular grid of times T_2 are used more often in practice. The optimal values of amplitude coefficients can be determined by one of the minimization methods (Himmelblau 1972, Lawson and Hanson 1974, Salaezar-Tio and Sun 2010). Various regularization methods can be used to suppress the influence of noise in the original signal and to obtain a physically justified solution. Some algorithms are implemented as software products: the Stanley Provencher (1982) method used by Bruker, the UPEN method by Borgia et al. (1998), developed at the University of Bologna.

Forward Modelling of Relaxation Curves Using Legendre Polynomials

Since the integral (10.1) needs to be calculated many times in solving an inverse problem, it is suggested to approximate $A(T)$ by an optimal polynomial $A_m(T)$ of degree m in order to save processor time and preserve the required accuracy, and to obtain the expression for the integral (10.1) via analytic functions. We use the well-known property of the partial sum of the expansion of a Lipschitz function in a series in the Legendre polynomials. This sum defines the polynomial of the chosen degree, optimally approximating decomposable function (in the sense of the Euclidean norm in L2).

In practice, the relaxation curve is selected by a set of trial finite functions f(T) with a support in the interval [a, b] ($\alpha_i > 0$, $h_i > 0$):

$$A(T) = \sum_{i=0}^{N} \alpha_i f\left(\frac{T-\mu_i}{h_i}\right) \quad (10.2)$$

We approximate $f(T)$ by partial sum of Legendre polynomials:

$$f_m(T) = \begin{cases} \sum_{n=0}^{m} c_n P_n\left(\frac{T-\mu}{h}\right), & T \in [a,b] \\ 0, & T \notin [a,b] \end{cases}, \quad (10.3)$$

where $\mu = \frac{a+b}{2}$, $h = \frac{b-a}{2}$, $b > 0$. $P_n(x)$—The Legendre polynomial of degree n, whose formula is:

$$P_n(x) = \frac{1}{2^n} \sum_{k=0}^{[n/2]} (-1)^k C_n^k C_{2n-2k}^n x^{n-2k} \quad (10.4)$$

The coefficients of the expansion are calculated by the formula:

$$c_n = \frac{2n+1}{2} \int_{-1}^{1} f(hT + \mu) P_n(T) dT \quad (10.5)$$

It should be noted that the values of c_n remain constant for affine transformations of the interval $[a, b]$. Thus, they can be calculated once for the trial function $f(T)$ in the sum (10.2).

The square of the discrepancy norm for the approximation used is calculated by the formula:

$$(\Delta f_m)^2 = \|f - f_m\|_{L_2[a,b]}^2 = \int_a^b (f(T))^2 dT - 2h \sum_{n=0}^{m} \frac{c_n^2}{2n+1} \quad (10.6)$$

The necessary approximation error is achieved by choosing the maximum exponent of the degree of Legendre polynomials.

The optimal polynomial $A_m(T)$ for approximating $A(T)$:

$$A_m(T) = \sum_{i=0}^{N} \alpha_i f_m\left(\frac{T - \mu_i}{h_i}\right) \quad (10.7)$$

Since $A_m(T)$ epends linearly on $f_m(T)$, the error of approximation of the Lipschitz function $A(T)$ by the polynomial $A_m(T)$ can be limited as follows:

$$\|A - A_m\|_{L_2} \leq K \Delta f_m,$$

where K—some constant coefficient.

Substituting (10.3) in (10.7) and then in (10.1), we obtain the approximation $s_m(t)$ for the integral:

$$s_m(t) = \int_0^\infty A_m(T) e^{-\frac{t}{T}} dT = \sum_{i=0}^{N} \alpha_i \int_{\max\{0, ah_i + \mu_i\}}^{bh_i + \mu_i} e^{-\frac{t}{T}} \sum_{n=0}^{m} c_n P_n\left(\frac{T - \mu h_i - \mu_i}{hh_i}\right) dT$$

$$(10.8)$$

Reducing silimar terms before the same powers in the integrand of formula (10.8), we obtain:

$$s_m(t) = \sum_{i=0}^{N} \alpha_i \sum_{n=0}^{m} \frac{\beta_{m,n}}{h^n h_i^n} \int_{\max\{0, ah_i + \mu_i\}}^{bh_i + \mu_i} e^{-\frac{t}{T}} (T - \mu h_i - \mu_i)^n dT, \quad (10.9)$$

where coeficients $\beta_{m,n}$ for even and odd n are expressed:

$$\beta_{m,2n} = \sum_{k=n}^{[m/2]} \frac{(-1)^{k-n}}{2^{2k}} c_{2k} C_{2k}^{k-n} C_{2n+2k}^{2n}$$

$$\beta_{m,2n+1} = \sum_{k=n}^{[(m-1)/2]} \frac{(-1)^{k-n}}{2^{2k+1}} c_{2k+1} C_{2k+1}^{k-n} C_{2n+2k+2}^{2n+1}$$

The integral of (10.9) can be expressed in terms of analytic functions by means of formula:

$$\int_0^x x^n e^{-\frac{t}{x}} dx = xe^{-\frac{t}{x}} \sum_{k=0}^{n} \frac{(-1)^k (n-k)!}{(n+1)!} t^k x^{n-k} + \frac{(-1)^{n+1}}{(n+1)!} t^{n+1} E_1\left(\frac{t}{x}\right)$$

where $E_1(x)$—first-order integral exponential function.

A rough upper bound for the error in approximating the integral (10.1) by the function $s_m(t)$ is given by:

$$\|s - s_m\|_{L_2} \leq K \Delta f_m \cdot \sqrt{\frac{1}{N} \sum_{i=0}^{N} \mu_i \sum_{i=0}^{N} h_i}$$

Inversion of Relaxation Curves

For the inversion of relaxation curves we propose the Fletcher-Reeves minimization (conjugate gradients) with different types of the regularizing terms for a more stable solution and for achievement of the required spectrum smoothness (Himmelblau 1972).

For practical NMR spectra approximation with sufficient accuracy we used the function $f(T_2)$, belonging to classes: discrete, piecewise-constant and piecewise-linear.

The discrete version of the inverse problem is described above by the formula (10.2), with the set of relaxation times T_{2i} being defined as the partition of the interval between the prescribed boundaries T_{2min} and T_{2max}. The fixed values of T_{2i} can be distributed both uniformly and logarithmically uniformly. With this approach, only the set A_i, the amplitudes corresponding to the times T_{2i}, is to be selected.

In piecewise-linear and piecewise-constant variants of interpretation, a uniform or log-uniformly uniform grid of times T_2 is set similarly. In the case of piecewise linear selection, the spectrum is sought as a piecewise linear function: on each interval $[T_{2i}, T_{2i+1}]$, the spectrum graph is a segment whose ordinates of ends are A_i and A_{i+1}, respectively. For a piecewise constant interpretation, the spectrum is given as a piecewise constant function on each interval $[T_{2i}, T_{2i+1}]$ having the value A_i. In both cases, the integral (10.1) can be written in terms of analytic functions: a set of integral exponents.

For testing of the developed algorithm, we used two test dataset types. Model relaxation curves computed by the simulation program, based on pre-defined different spectra variants. Practical data—archive NMR signals of core samples of different strata and deposits of Western Siberia, obtained on a portable NMR-relaxometer.

Conclusion

We present a method for optimizing the calculating time for numeric solving of NMR forward and inversion problems using Legendre polynomial expansion and Fletcher-Reeves minimization (conjugate gradients). This method is implemented in software for processing and geological-geophysical interpretation of relaxation curves, which is taking into account medium models and preliminary geological information. The developed method is realized for a high-performance computer complex including an array of NVIDIA video adapters.

Reliable NMR data during studies of the petrophysical properties of reservoir rocks is allowing to refine the empirical relationships between the T2 distribution and the pore characteristics of the geological media. Core and mud studies using modern NMR equipment allow more accurate interpretation of NMR logging results and other well survey complexes. It may increase the reliability of the analysis of reservoir properties of core samples, and provide the validity of assessments of reserves and oil recovery ratio.

Acknowledgements The authors are grateful to Corresponding Member of RAS, prof P.S. Martyshko for general guidance and V.A.Vavilin and V.M.Mursakayev for a valuable discussion.

References

Borgia, G. C., Brown, R. J. S., Fantazzini P. (1998). Uniform-Penalty Inversion of Multiexponential Decay Data. Journal of magnetic resonance 132, p. 65–77.
Coates, G.R., Xiao Lizhi, Prammer, M.G. (2000). NMR Logging. Principles & applications. Hulliburton Energy Services Publishing, Houston (USA). 234 p.
Farrer, T. C., Becker, E. D. (1971). Pulse and Fourier Transform NMR: Introduction to Theory and Methods. New York and London, Academic Press. 118 p.
Gang Yu, Zhizhan Wang, K. Mirotchnik, Lifa Li (2006). Application of Magnetic Resonance Mud Logging for Rapid Reservoir Evaluation. Poster presentation at AAPG Annual Convention, Houston, Texas, April 9–12.
Himmelblau, D. (1972). Applied nonlinear programming. McGraw-Hill. 498 p.
Mirotchnik, K., Kryuchkov, S., Strack, K. (2004). A Novel Method to Determine NMR Petrophysical Parameters From Drill Cuttings. SPWLA 45th Annual Logging Symposium. Pare MM.
Muravyev, L.A., Zhakov, S.V. (2016). Methodical issues of investigations with laboratory NMR relaxometer. Geoinformatics 2016: XVth International Conference on Geoinformatics - Theoretical and Applied Aspects. Ukraine. https://doi.org/10.3997/2214-4609.201600527
Lawson, C.L., Hanson, R.J. (1974). Solving Least Squares Problems, Prentice-Hall.
Provencher, S.W. (1982). CONTIN: A general purpose constrained regularization program for inverting noisy linear algebraic and integral equations. Comput. Phys. Commun. 27, 229.
Rafael Salaezar-Tio, Bogin Sun (2010). Monte Carlo optimazation-inversion methods for NMR. Petrophysics, vol.51, no.3,. pp. 208–218.

Chapter 11
Field of Attraction of Polyhedron and Polygonal Plate with Linear Density Distribution

K. M. Kuznetsov, A. A. Bulychev and I. V. Lygin

Abstract New presentation of field of attraction elements (potential and its first derivatives) is demonstrated for such important approximating models as polyhedron and polygonal plate with density that changes in accordance with linear laws. It is shown that these elements are defined by elements of models' fields with known analytical representations (polyhedron, polygonal plate and material segment with constant density) and additional integrals for which explicit analytical expressions exist.

Keywords Gravimetry · Direct gravity problem · Linear density distribution

Attraction Potential of Polyhedron with Linear Density Distribution

Attraction potential of polyhedron having Q faces and density $\delta(M) = a_x \xi + a_y \eta + a_z \zeta + a_0$ in a point $M_0(x, y, z)$ can be written as

$$V(M_0) = \frac{1}{2} \sum_{q=1}^{Q} \left[(\zeta^q - z^q) V_q(M_0) - a_q V_q^R(M_0) \right] \quad (11.1)$$

In this formula, it is assumed that the new coordinate system associated with the qth face is introduced; its Oz^q axis coincides with direction of outward normal $\mathbf{n}_q$ to this face; ζ^q and z^q are coordinates of the plate position and the point M_0 in this coordinate system; $V_q(M_0)$ is the potential of the qth plate having surface density $\delta_q(M^q) = a_x^q \xi^q + a_y^q \eta^q + a_0^q$:

K. M. Kuznetsov (✉) · A. A. Bulychev · I. V. Lygin
Department of Geophysical Methods of Earth Crust Study, Faculty of Geology,
Lomonosov Moscow State University, Moscow, Russian Federation
e-mail: kirillkuz90@gmail.com

© Springer Nature Switzerland AG 2019
D. Nurgaliev and N. Khairullina (eds.), *Practical and Theoretical Aspects of Geological Interpretation of Gravitational, Magnetic and Electric Fields*, Springer Proceedings in Earth and Environmental Sciences, https://doi.org/10.1007/978-3-319-97670-9_11

$$V_q(M_0) = V_q(M_0^q) = G \int_{S_q} \delta_q(M^q) \frac{1}{r_{MM_0}} dS. \quad (11.2)$$

M^q and M_0^q are positions of points M and M_0 in the coordinate system associated with this face. Value a_q is determined as scalar product of vectors $(a_x \mathbf{i}_x + a_y \mathbf{i}_x + a_z \mathbf{i}_x)$ and $\mathbf{n}_q$. V_q^R is an integral looking like

$$V_q^R(M_0) = G \int_{S_q} r_{MM_0} dS. \quad (11.3)$$

Field of Attraction Created by Polyhedron with Linear Density Distribution

Field of attraction of polyhedron with linear density distribution can be written as

$$\mathbf{g}(M_0) = -\sum_{q=1}^{Q} \mathbf{n}_q V_q(M_0) + \mathbf{a} V^0(M_0). \quad (11.4)$$

As in the previous section, here $\mathbf{n}_q$ is a unit normal vector to the qth face; $V_q(M_0)$ is potential of the plate coinciding with the qth face and having linear variation of surface density $\delta_q(M)$; $V^0(M_0)$ is potential of polyhedron having unit density, $\mathbf{a} = a_x \mathbf{i}_x + a_y \mathbf{i}_y + a_z \mathbf{i}_z$.

Thus, field of a polyhedron having linearly varying density is determined by potentials of its faces with linear surface density distributed over them and potential of the polyhedron with constant unit density.

Expression for potential of homogeneous polyhedron has been obtained by Strakhov V.N (Strakhov and Lapina 1983, 1986b); therefore we need to determine the expressions for potential of polygonal plate with linear surface density.

Attraction Potential of N-Cornered Plate with Linear Surface Density Distribution

Let us assume that the plate lies in the plane Oxy, and the axis Oz coincides with the normal direction to this plate. The normal direction is determined by the order of traversal of the polygon sides. Surface density of such a model is described with the following expression:

11 Field of Attraction of Polyhedron and Polygonal ...

$$\delta_S(M) = \delta_S(\xi, \eta, \zeta) = a_x\xi + a_y\eta + a_0. \tag{11.5}$$

Plate potential can be written in the following form:

$$V_S(M_0) = \sum_{\nu=1}^{N} (\xi^\nu - x^\nu) V_\nu(M_0) - \sum_{\nu=1}^{N} a_\nu V_\nu^R(M_0) - (\zeta - z) g_z^s(M_0). \tag{11.6}$$

Here again, as in Section Attraction Potential of Polyhedron with Linear DensityDistribution, it is assumed that the new coordinate system is introduced associated with νth side of the plate, where axis Oy^ν coincides with νth side of polygon and is oriented along the direction of its traversal; axis Ox^ν coincides with the direction of vector $\mathbf{n}_\nu$, i.e., with the vector of outward normal to the νth side of the plate that lies in its plane; ξ^ν and x^ν are coordinates of νth side position and of point M_0 in this coordinate system; $V_\nu(M_0)$ is potential of the segment (rod) coinciding with νth side of the plate having the density distribution $\delta_\nu(M^\nu) = a_y^\nu \eta^\nu + a_0^\nu = \delta_\nu(\eta^\nu)$:

$$V_\nu(M_0) = G \int_{\Gamma_\nu} \frac{\delta_\nu(M)}{r_{MM_0}} dl. \tag{7}$$

$a_\nu = (a_x \mathbf{i}_x + a_y \mathbf{i}_y) \cdot \mathbf{n}_\nu$ is a parameter, while value $V_\nu^R(M_0)$ is determined by the integral

$$V_\nu^R(M_0) = V_\nu^R(M_0^\nu) = G \int_{\eta_1^\nu}^{\eta_2^\nu} r_{MM_0} d\eta^\nu, \tag{11.8}$$

that has the explicit analytical solution:

$$V_\nu^R(M_0) = G\frac{1}{2}\left[(\eta_2^\nu - y^\nu)r_2^\nu - (\eta_1^\nu - y^\nu)r_1^\nu + \left((\xi^\nu - x^\nu)^2 + (\zeta - z)^2\right)V_\nu^0(M_0)\right] \tag{11.9}$$

Value of ζ corresponds to coordinate of the plate position along the Oz axis; z is coordinate of point M_0; $g_z^s(M_0)$ is value of vertical component of attraction of plate having linear surface density distribution.

Field of Attraction of N-Cornered Plate Having Linear Surface Density Distribution

4.1. Horizontal components of plate's field of attraction in point M_0 can be written in the following form:

$$g_x^s \mathbf{i}_x + g_y^s \mathbf{i}_y = -\sum_{v=1}^{N} \mathbf{n}_v V_v(M_0) + \mathbf{a} V_S^0(M_0). \tag{11.10}$$

Here $V_S^0(M_0)$ is potential of plate having constant unit surface density; vector $\mathbf{a} = a_x \mathbf{i}_x + a_y \mathbf{i}_y$; $\mathbf{n}_v$ is the vector of outer normal to vth side, which lies in the plane of the plate; $V_v(M_0)$ — is potential of material segment (rod) coinciding with vth side of the plate and having density $\delta_v(M)$ that varies in accordance with a linear law.

4.2. Vertical component of the plaste attraction can be written in the form

$$g_z^s(M_0) = \frac{1}{(\zeta - z)} \sum_{v=1}^{N} (\xi^v - x^v) \left(I_{11}^v - \left(a_y^v I_{121}^v + \left(a_y^v y^v + a_0^v \right) I_{122}^v \right) \right)$$

$$- \frac{|\zeta - z|}{(\zeta - z)} \sum_{v=1}^{N} a_v I_2^v, \tag{11.11}$$

where

(a) I_{11}^v coincides with potential of material segment $V_v(M_0)$: $I_{11}^v = V_v(M_0)$;

(b)
$$I_{121}^v = G\left((r_2^v - r_1^v) - |\varsigma - z| \ln \frac{|\zeta - z| + r_2^v}{|\zeta - z| + r_1^v} \right). \tag{11.12}$$

Here, r_1^v and r_2^v are distances from the beginning and the end of the vth segment to the point M_0;

(c) Expression for I_{122}^v has been determined by Strakhov V.N. (Strakhov and Lapina 1986a):

$$I_{122}^v = G\left(\ln \frac{(\eta_2^v - y^v) + r_2^v}{(\eta_1^v - y^v) + r_1^v} - 2|\varsigma - z| \frac{1}{(\xi^v - x^v)} \operatorname{arctg} \frac{(\xi^v - x^v)(w_2^v - w_1^v)}{w_2^v w_1^v + (\xi^v - x^v)^2} \right)$$

$$\tag{11.13}$$

Note that the first component in the obtained expression corresponds to attraction potential of material segment $V_v^0(M_0)$; parameters w_1^v, w_2^v are determined by the expressions

$$w_1^v = (\eta_1^v - y^v) + r_1^v + |\zeta - z|, \quad w_2^v = (\eta_2^v - y^v) + r_2^v + |\zeta - z|.$$

(d) I_2^v can be written in the form

$$I_2^v = G\left[|\xi^v - x^v|\text{arctg}\left(\frac{\eta^v - y^v}{|\xi^v - x^v|}\right) - |\xi^v - x^v|\text{arctg}\left(\frac{|\zeta - z|(\eta^v - y^v)}{|\xi^v - x^v|r_{MM_0}}\right)\right.$$
$$\left. + |\zeta - z|\ln((\eta^v - y^v) + r_{MM_0}) + (\eta^v - y^v)\ln(|\zeta - z| + r_{MM_0}) - \eta^v\right]_{\eta_1^v}^{\eta_2^v}.$$
(11.14)

Attraction Field of Material Segment Having Linear Density Distribution

5.1. Attraction field of material segment having linear density distribution is discussed in details in the work of Strakhov V. N.[Strakhov 1985]. Let us give an expression for potential of a material segment located on the Oy axis and having density varying in accordance with the linear law $\delta(\eta) = a_y + a_0$

$$V_v(M_0) = a_y\sqrt{x^2 + (\eta - y)^2 + z^2}\Big|_{\eta_1}^{\eta_2} + (a_y y + a_0)\ln((\eta - y) + r_{MM_0})\Big|_{\eta_1}^{\eta_2}. \quad (11.15)$$

In general, potential of a material segment (rod) $V_v(M_0)$ located along the Oy axis can be written as

$$V_v(M_0) = a_y(r_2 - r_1) + (a_y y + a_0)V_v^0(M_0). \quad (11.16)$$

where r_1 and r_2 are distances from the rod's starting and ending points to the point of calculation M_0; $V_v^0(M_0)$ is potential of the rod having a constant unit density. Note that the same potential can be written in the following form (Strakhov 1985):

$$V_v^0(M_0) = \ln\frac{r_1 + r_2 + L}{r_1 + r_2 - L}, \quad (11.17)$$

where L is the rod length.

Let us discuss the integral $V^R(M_0)$ (11.8). In the coordinate system associated with the polygonal plate sides and using notation $(\xi^v - x^v)^2 + (\zeta - z)^2 = b^2$, this integral can be written as follows:

$$V_v^R(M_0) = V_v^R(M_0^v) = \int_{\eta_1^v}^{\eta_2^v} r_{MM_0} d\eta^v = \int_{\eta_1^v}^{\eta_2^v} \sqrt{b^2 + (\eta^v - y^v)^2} d\eta^v$$

$$= \frac{1}{2}\left((\eta^v - y^v)\sqrt{b^2 + (\eta^v - y^v)^2} + b^2 \ln\left[(\eta^v - y^v) + \sqrt{b^2 + (\eta^v - y^v)^2}\right]\right)_{\eta_1^v}^{\eta_2^v}$$

$$= \frac{1}{2}\left[(\eta_2^v - y^v)r_2^v - (\eta_1^v - y^v)r_1^v + b^2 V_v^0(M_0)\right]$$

(11.18)

where r_1^v and r_2^v are distances from the rod ends to the point M_0.

The integral (11.3) is necessary to calculate values of attraction potential. It is possible to obtain an analytical solution for this integral, but it can also be determined using numerical integration.

5.2. Let us consider vertical component of polygonal plate lying in the Oxy plane and having linear surface density distribution $\delta_S(M)$ (11.11):

$$g_z^s(M_0) = \frac{|\zeta - z|}{(\zeta - z)} \oint_{\partial S} \left(\delta_S(M) \frac{\partial \ln(|\zeta - z| + r_{MM_0})}{\partial \mathbf{n}}\right.$$

$$\left. - \ln(|\zeta - z| + r_{MM_0}) \frac{\partial \delta_S(M)}{\partial \mathbf{n}}\right) dl.$$

(11.19)

We can write this expression in the form

$$g_z^s(M_0) = \frac{|\zeta - z|}{(\zeta - z)} \sum_{v=1}^{N} \int_{\Gamma_v} \delta_S(M) \frac{\partial \ln(|\zeta - z| + r_{MM_0})}{\partial \mathbf{n}_v} dl$$

$$- \frac{|\zeta - z|}{(\zeta - z)} \sum_{v=1}^{N} \int_{\Gamma_v} \ln(|\zeta - z| + r_{MM_0}) \frac{\partial \delta_S(M)}{\partial \mathbf{n}_v} dl = I_1 - I_2.$$

(11.20)

Further transformations again will be based on introducing the auxiliary coordinate systems associated with vth sides of polygonal plate. As before, axis Ox^v will coincide with the direction of outward normal $\mathbf{n}_v$ to the vth side; the axis Oy^v will coincide with the vth side; and in this coordinate system we can write

$$\delta_S(\xi, \eta) = \delta_v(M^v) = a_y^v \eta^v + a_0^v = \delta_v(\eta^v).$$

(11.21)

5.3. Component I_1 in (11.20) will be described by the equation

$$I_1 = \frac{|\zeta - z|}{(\zeta - z)} \sum_{v=1}^{N} \int_{\eta_1^v}^{\eta_2^v} \delta_v(\eta_v) \frac{\partial \ln(|\xi - z| + r_{MM_0})}{\partial \xi^v} d\eta^v \qquad (11.22)$$

$$= \frac{1}{(\zeta - z)} \sum_{v=1}^{N} (\xi^v - x^v) \left(I_{11}^v - I_{12}^v \right).$$

The integral I_{11}^v coincides with potential of vth material segment having density $\delta_v(\eta^v)$:

$$I_{11}^v = \int_{\eta_1^v}^{\eta_2^v} \delta_v(\eta^v) \frac{1}{r_{MM_0}} d\eta^v = V_v(M_0); \qquad (11.23)$$

For integral I_{12}^v, let us carry out the following transformations:

$$I_{12}^v = \int_{\eta_1^v}^{\eta_2^v} \delta v(\eta^v) \frac{1}{|\zeta - z| + r_{MM_0}} d\eta^v$$

$$= \int_{\eta_1^v}^{\eta_2^v} \left(a_y^v(\eta^v - y^v) + a_y^v y^v + a_0^v \right) \frac{1}{|\zeta - z| + r_{MM_0}} d\eta^v \qquad (11.24)$$

$$= a_y^v \int_{\eta_1^v}^{\eta_2^v} \frac{(\eta^v - y^v)}{|\zeta - z| + r_{MM_0}} d\eta^v + \left(a_y^v y^v + a v_0 \right) \int_{\eta_1^v}^{\eta_2^v} \frac{1}{|\zeta - z| + r_{MM_0}} d^v$$

$$= a_y^v I_{121}^v + \left(a_y^v y^v + a_0^v \right) I_{122}^v;$$

$$I_{121}^v = \int_{\eta_1^v}^{\eta_2^v} \frac{(\eta^v - y^v)}{|\zeta - z| + r_{MM_0}} d\eta^v = r_{MM_0} \big|_{\eta_1^v}^{\eta_2^v} - |\zeta - z| \ln(|\zeta - z| + r_{MM_0}) \big|_{\eta_1^v}^{\eta_2^v} \qquad (11.25)$$

$$= (r_2^v - r_1^v) - |\zeta - z| \ln \frac{|\zeta - z| + r_2^v}{|\zeta - z| + r_1^v}.$$

Solution of the integral I_{122}^v was obtained by Strakhov V.N. [Strakhov and Lapina 1986a, Bulychev, 2010]:

$$I_{122}^v = \int_{\eta_1^v}^{\eta_2^v} \frac{1}{|\xi - z| + r_{MM_0}} d\eta^v \qquad (11.26)$$

$$= \ln \frac{(\eta_2^v - y^v) + r_2^v}{(\eta_1^v - y^v) + r_1^v} - 2|\xi - z| \frac{1}{(\xi^v - x^v)} \operatorname{arctg} \frac{(\xi^v - x^v)(w_2^v - w_1^v)}{w_2^v w_1^v + (\xi^v - x^v)^2}$$

Note that the first component in the obtained expression (11.23) corresponds to attraction potential of material segment $V_v^0(M_0)$; the parameters w_1^v, w_2^v are determined by the expressions

$$w_1^v = (\eta_1^v - y^v) + r_1^v + |\xi - z|, \quad w_2^v = (\eta_2^v - y^v) + r_2^v + |\xi - z|. \qquad (11.27)$$

Thus, the first part of the term $g_z^s(M_0)$ can be written in the following form:

$$\begin{aligned}
I_1 &= \frac{1}{(\zeta - z)} \sum_{v=1}^{N} (\xi^v - x^v)(I_{11}^v - I_{12}^v) \\
&= \frac{1}{(\zeta - z)} \sum_{v=1}^{N} (\xi^v - x^v)\left(I_{11}^v - \left[a_y^v I_{121}^v + \left(a_y^v y^v + a_0^v\right) I_{122}^v\right]\right),
\end{aligned} \qquad (11.28)$$

where integrals I_{11}^v, I_{121}^v, and I_{122}^v are determined by formulas (11.23, 11.25, 11.26), respectively.

5.4. Let us consider the second component I_2 in (20):

$$I_2 = \frac{|\zeta - z|}{(\zeta - z)} \sum_{v=1}^{N} \int_{\Gamma_v} \ln(|\zeta - z| + r_{MM_0}) \frac{\partial \delta_S(M)}{\partial \mathbf{n}_v} dl. \qquad (11.29)$$

Given that $\frac{\partial \delta_S(M)}{\partial \mathbf{n}_v} = \operatorname{grad}_{\xi,\eta} \delta_S(M) \cdot \mathbf{n}_v = (a_x \mathbf{i}_x + a_y \mathbf{i}_y) \cdot \mathbf{n}_v = a_v$, expression for I_2 can be presented in the following form:

$$\begin{aligned}
I_2 &= \frac{|\zeta - z|}{(\zeta - z)} \sum_{v=1}^{N} \int_{\Gamma_v} \ln(|\zeta - z| + r_{MM_0}) \frac{\partial \delta_S(M)}{\partial \mathbf{n}_v} dl \\
&= \frac{|\zeta - z|}{(\zeta - z)} \sum_{v=1}^{N} a_v \int_{\eta_1^v}^{\eta_2^v} \ln(|\zeta - z| + r_{MM_0}) d\eta^v = \frac{|\zeta - z|}{(\zeta - z)} \sum_{v=1}^{N} a_v I_2^v.
\end{aligned} \qquad (11.30)$$

Denoting $b^2 = (\xi^v - x^v)^2 + (\xi - z)^2$, we can write solution for this integral (the solution was obtained using the integration software provided at the website www.webmath.ru):

$$I_2^v = \int_{\eta_1^v}^{\eta_2^v} \ln(|\zeta - z| + r_{MM_0}) d\eta^v = \left[\sqrt{b^2 - |\zeta - z|^2} \arctg\left(\frac{\eta^v - y^v}{\sqrt{b^2 - |\zeta - z|^2}} \right) \right.$$

$$- \sqrt{b^2 - |\zeta - z|^2} \arctg\left(\frac{|\zeta - z|(\eta^v - y^v)}{\sqrt{b^2 - |\zeta - z|^2}\sqrt{b^2 + (\eta^v - y^v)^2}} \right)$$

$$+ |\zeta - z| \ln\left[2\left((\eta^v - y^v) + \sqrt{b^2 + (\eta^v - y^v)^2} \right) \right]$$

$$\left. + (\eta^v - y^v) \ln\left(|\zeta - z| + \sqrt{b^2 + (\eta^v - y^v)^2} \right) - \eta^v \right]_{\eta_1^v}^{\eta_2^v}$$

(11.31)

After some simplifications we obtain

$$I_2^v = \left[|\xi^v - x^v| \arctg\left(\frac{\eta^v - y^v}{|\xi^v - x^v|} \right) - |\xi^v - x^v| \arctg\left(\frac{|\xi - z|(\eta^v - y^v)}{|\xi^v - x^v| r_{MM_0}} \right) \right.$$

$$\left. + |\zeta - z| \ln((\eta^v - y^v) + r_{MM_0}) + (\eta^v - y^v) \ln(|\zeta - z| + r_{MM_0}) - \eta^v \right]_{\eta_1^v}^{\eta_2^v}$$

(11.32)

Finally, for vertical component of horizontal polygonal plate having linear surface density distribution and taking into account (11.20–11.25), the following expressions can be written:

$$g_z^s(M_0) = \frac{1}{(\zeta - z)} \sum_{v=1}^{N} (\xi^v - x^v) \left(I_{11}^v - \left(a_y^v I_{121}^v + \left(a_y^v y^v + a_0^v \right) I_{122}^v \right) \right)$$

$$- \frac{|\zeta - z|}{(\zeta - z)} \sum_{v=1}^{N} a_v I_2^v v$$

(11.33)

Conclusion

The main result of this work is demonstration of how elements of attraction fields (potential and its first derivatives) for such important approximating models as polyhedron and polygonal plate with density varying in accordance with linear law can be represented by model field elements having known analytical representation (polyhedron, polygonal plate, and material segment with constant density) and additional integrals, for which an explicit analytical expressions exist. The obtained representations can be the basis for computer programs in the case of media with gradient variation of density.

References

Bulychev A.A., Lygin I.V., Melikhov V.R. (2010) Numerical methods for forward solution of gravimetry and magnetometry problems (Compendium of Lectures). Moscow, Faculty of Geology, Lomonosov Moscow State University, 2010, 164 pp. (geophys.geol.msu.ru/STUDY/facultet/forward08_03_2011.pdf) (in Russian).

Strakhov V.N., Lapina M.I. (1983) Forward and inverse problems of gravimetry and magnetometry for arbitrary homogeneous polyhedrons. Theory and practice of gravity and magnetic fields interpretation in the USSR (Proceedings of III-rd All-Union Workshop and School)/ edited by Starostenko V.I., Kyiv, Naukova Dumka, 1983, pp. 3–86. (in Russian).

Strakhov V.N. (1985) On the problem of forward modelling in gravimetry and magnetometry for material rod having polynomial density.// Geophysical Journal, 1985, Vol. 7, № 1, pp. 3–9. (in Russian).

Strakhov V.N., Lapina M.I. (1986) Gravity forward modelling for horizontal homogeneous polygonal plate.// Geophysical Journal, 1986, Vol. 8, № 4, pp. 20–31. (in Russian).

Strakhov V.N., Lapina M.I. (1986) Forward problems of gravimetry and magnetometry for homogeneous polyhedrons.// Geophysical Journal, 1986, Vol. 8, № 6, pp. 20–31. (in Russian).

Chapter 12
Allowance for the Earth's Surface Topography in Processing the Magnetic Field Measurements

A. S. Dolgal

Abstract The questions concerning the allowance for the Earth's surface topography in the measurements of the magnetic field in mountainous regions are considered. The new method is suggested for estimating the influence of the sharply contrasting relief composed of the strongly magnetized rocks on the results of the field observations. The method uses the decomposition of the magnetic field and elevation data in the empirical modes with subsequent application of the method of group allowance for arguments. The results of reducing the airborne magnetic survey carried out n the northwestern part of the Siberian platform above the copper-nickel ore deposit are presented.

Keywords Magnetic prospecting · Terrain topography · Reduction

In magnetic prospecting, the questions associated with the allowance correction for surface topography of the Earth composed of magnetized rocks have been explored much less than the analogous questions in gravity prospecting. However, the influence of the Earth's surface topography on the results of the magnetic measurements can be very significant. For instance, in one area of the 1:50,000 airborne magnetic survey within the Putorana plateau where elevations H of surface topography range from 40 to 1600 m, the variations ΔT in the amplitude of the anomalous magnetic field reach ∼2000 nT and a clearly pronounced spatial correlation is observed between the magnetic anomalies and geomorphological features (Fig. 12.1).

The necessity of taking into account the distortions of magnetic measurements by the terrain topography can arise in very different physical and geological conditions and in different types of magnetic surveys (Rempel 1980; Nusipov and Akhmetov 1991; Dolgal and Khristenko 1997; Yurovskikh 1997). These distortions are contributed by two factors: (1) the anomalous effect of the magnetic masses

---

A. S. Dolgal (✉)
Perm Federal Research Center, Ural Branch, Russian Academy of Sciences,
Perm, Russia
e-mail: dolgal@mi-perm.ru

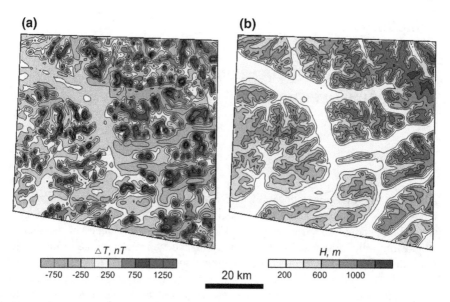

Fig. 12.1 The influence of the surface topography of the region composed of the rocks of basalt formation on the magnetic field: **a**—map of isodynamic lines of the anomalous magnetic field $(\Delta T)_a$; b—map of the horizontal terrain contours

located between the Earth's surface and the conditional surface bounding these masses at the depth; and (2) the different distances between the points of magnetic field measurements and the studied perturbing objects due to the different heights (z-coordinates) of the observation surface—the height-difference effect (Dolgal 2002). These distortions can be suppressed by two methods: (a) by calculating the corrections by solving the direct forward problem from the digital elevation model of the terrain and (b) by recalculating (reducing) the field onto the horizontal surface (or onto a smoothed synthesized surface) with the use of equivalent sources (Aronov 1976; Pilkington and Urquhar 1990; Strakhov 1992).

Let us explain the influence of the two cited factors—the effect of topographic masses and the anomalous vertical gradient – by the model example. Figure 12.2a shows the graph of the theoretical field ΔT of the horizontal magnetic plate at a height of 250 m calculated by solving the direct problem. The "observed" field ΔT for the presented cross section (Fig. 12.2b) corresponds to the conditions of the ground magnetic survey conducted on the curved Earth's surface. The anomalous effect of the plate was reconstructed by subtracting the component caused by the magnetic topography (Fig. 12.2d) from the observed field and subsequent recalculation of the difference field (Fig. 12.2c) to a depth of 250 m by the approximation technique. The obtained anomaly (Fig. 12.2e) is close to the theoretical one (Fig. 12.2a); in contrast to the difference field, its configuration shape excludes of the possibility of the oblique position of the anomalous object.

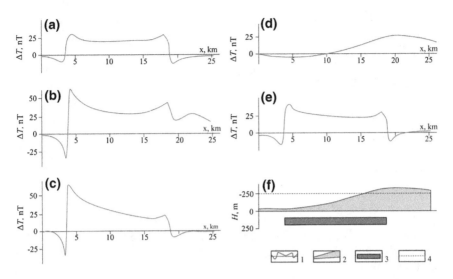

Fig. 12.2 Model example: reconstruction of the magnetic anomaly of the plate from the values of the field ΔT specified on the surface of the magnetic relief: **a**—anomaly ΔT from the plate on the horizontal profile $H = -250$ m; **b**—total anomaly ΔT from the plate and magnetic rocks forming the topography; **c**—anomaly ΔT from the plate on the Earth's surface; **d**—anomaly ΔT from the magnetic rocks forming the topography; **e**—reconstructed anomaly ΔT from the plate on the profile $z = 250$ m; **f**—cross section: 1, graphs of the magnetic field ΔT; 2, magnetic rocks composing the relief ($J = 0.5$ A/m); 3, plate ($J = 1$ A/m); 4, level $H = -250$ m

The physical meaning of calculating the topographic corrections in gravity and magnetic measurements is identical: it consists in quantitative calculation of the anomalous effects produced by the rocks composing the relief with the given spatial distribution of their physical properties (density or magnetization). The significant difficulties associated with taking into account the influence of the topography on the observed magnetic field in the real physical–geological conditions are caused by the wide range of the variations in the petromagnetic characteristics of the rock (Dolgal 2001). The problem of determining the magnitude of laterally varying effective magnetization $J_{eff} = J_{eff}(x, y)$ in the conditions of the visually noticeable influence of terrain topography on the magnetic field is typically solved by constructing the correlation dependences between the amplitude of magnetic anomalies ΔT and height marks H of the Earth's surface topography. Compared to the standard statistical processing procedures, certain advantages are offered by the algorithm of step-by-step calculation of the linear regression dependence between δT_P (at J_{eff} = const) and ΔT developed by the author with the subsequent rejection of the parameter values having the largest deviations from the selected regression equation (Table 12.1; Fig. 12.3).

For suppressing the magnetic anomalies of "topographic" origin, the author developed the RELMAG program package which was approbated and recommended for adoption by the Scientific methodical council on geological and geophysical technologies in exploration and prospecting for solid minerals of the

Table 12.1 Characteristics of the results of determining the effective magnetization J_{ef} for the synthetic example

Initial data	Statistical parameters			
	Minimum	Maximum	Mean	RMS
"Observed" magnetic field ΔT, nT	−147	540	10.0	±86.2
Random noise ε, nT	−25	25	0.3	±14.6
Topographic correction δT_p, nT	−53	124	0.4	±27.6
Initial magnetization, A/m	0.204	3.78	1.40	±0.79
Calculated magnetization, A/m	0.550	4.15	1.79	±0.85
Fisher criterion F	1.2	55.0	4.85	±7.84

Ministry of Natural Resources and the Environment of the Russian Federation (St. Petersburg, September 29, 2000). In 2006–2007, on the order of OOO Sibir'geofizika, the RELMAG program package was updated and expanded (Dolgal and Chervonyi 2008).

A new method for correcting the anomalous magnetic field data for the influence of the terrain topography was suggested in 2017. This method relies on the hypothesis that the hidden peculiarities of the correlation between the geophysical fields and anomaly-forming objects can probably be revealed within certain intervals of spatial frequencies where these peculiarities manifest themselves fairly distinctly (Dolgal et al. 2017). Let us consider the example of applying this method for approximate calculating the effect of the Earth's surface topography on the results of the large-scale airborne magnetic survey conducted in the northwestern part of the Siberian platform. The plateau-basalts that are widespread in the survey area and have a maximal thickness more than 3000 m are the main impediment in geophysical prospecting for ultrabasic intrusions hosting copper-nickel ore bodies.

The influence of the rugged topography with the height contrast up to 600–800 m which is formed by the strongly magnetized rocks (J_{eff} = 3–5 A/m) manifests itself by the spatial correlation between the increases of the amplitude of the magnetic field ΔT and the increase of the elevation marks H. However, due to the complicated physical and geological situation, the dependence of ΔT on H is nonlinear. In the presented example the values of the anomalous geomagnetic field ΔT are specified on the latitudinal profile with a length of 60 km intersecting the copper-nickel ore deposit. The amplitude of the field varies from −627 to 512 nT; the spacing between the points is 100 m (Fig. 12.4).

As a result of empirical mode decomposition (ERMD), different frequency components were obtained for the magnetic field above the tuff lava stratum (Intrinsic Mode Function IMF $\Delta T_i, i = \overline{1,7}$) and for the topographic heights (IMF $H_j, j = \overline{1,4}$) as well as the residuals $r_{\Delta T}$ and r_H, respectively. For the low-frequency part of the ΔT signal, the mode mixing effect was revealed. This effect manifests itself as the intersection of the frequency intervals for functions $\Delta T_i, i = \overline{4,7}$. For suppressing this effect, the total IMF $\Delta T_4^* = \sum_{i=4}^{7} \Delta T_i$ was formed. The relationship between the modal components of the field and the topography is illustrated by

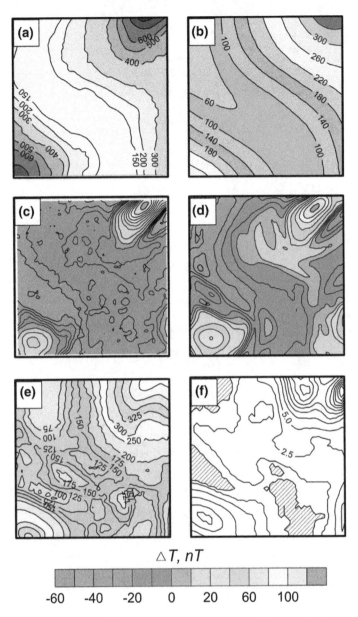

Fig. 12.3 Synthetic example of determining the magnetization of the rocks in the upper part of the geological section with the use of linear regression. Initial data: **a**—terrain topography, m; **b**—magnetization of the rocks J, $\times 10^{-2}$ A/m; **c**—"observed" magnetic field ΔT, nT; **d**—topographic correction δT_p, nT; results of the calculations: **e**—calculated magnetization of the rocks J, $\times 10^{-2}$ A/m; **f**—Fisher criterion. Note: within the hachured areas the statistical correlation between ΔT and δT_p is not revealed

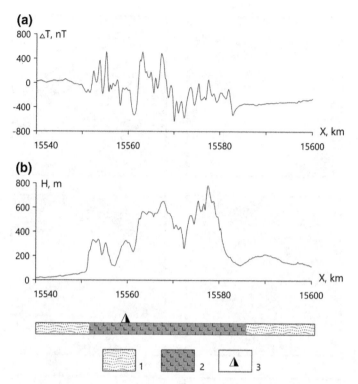

Fig. 11.4 Anomalous magnetic field (**a**) and Earth's surface topography (**b**): 1—sedimentary rocks; 2—tuff-lava stratum; 3—copper-nickel ore deposit

Table 12.2 Pair coefficients of linear correlation K between the magnetic field, topographic heights, and results of EMD

Parameter	ΔT	ΔT_1	ΔT_2	ΔT_3	ΔT_4^*	$r_{\Delta T}$
H	0.295	0.078	0.092	0.171	0.381	−0.076
H_1	0.224	0.644	0.141	0.0193	−0.070	−0.043
H_2	0.338	0.082	0.614	0.235	−0.064	0.075
H_3	0.317	−0.029	0.110	0.303	0.177	0.184
H_4	0.546	−0.105	−0.076	0.168	0.881	0.167
r_H	−0.092	0.052	−0.053	−0.026	0.068	−0.268

the table of pair coefficients of linear correlation (Table 12.2). More than 85% of the coefficients of correlation K are not significant. The significance was estimated by the Student's t-test for the confidence probability of 99.9% ($\nu = 339$). Generally, the correlation between the heights H and magnetic field ΔT is low ($K = 0.295$); however, the correlation between the IMFs of these parameters is higher (K reaches 0.881).

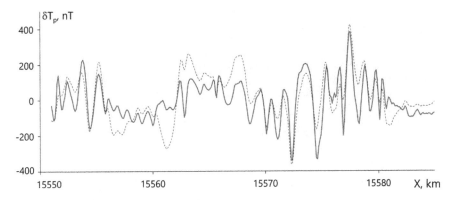

Fig. 12.5 Magnetic field above the tuff-lava stratum δT_P caused by the influence of the terrain topography estimated by the EMD and MGAA methods (red line) and by solving the direct problem of magnetic prospecting

At the second line of the selection, with the consideration of all the obtained IMFs, the linear regression model characterizing the topography-related component of the magnetic field δT_P, was constructed by the method of group allowance for arguments (MGAA): $\delta T_P = a_1 + a_2 Z_1 + a_3 Z_2$, where $Z_1 = b_1 + b_2 H_3 + b_3 H_4 + b_4 H_3 H_4$, $Z_2 = c_1 + c_3 H_1 + c_4 H_1 H_4$, and a,b,c are the coefficients.

The obtained values fairly well agree with the independent estimates calculations of the topographic corrections calculated by solving the direct problem of magnetometry (Fig. 12.5). These calculations used the digital elevation model for the Earth with laterally varying magnetization $J_{eff} = J_{eff}(x,y)$; it was assumed that the topography effect is limited to a zone with a radius of 20 km.

It should be noted that the both methods for calculating the δT_P corrections are approximate; however, the EMD- and MGAA-based technique discussed above is less burdensome and does not require the a priori information about the rock magnetic properties of the studied objects.

References

Aronov V.I., (1976) Obrabotka na EVM znachenii anomalii sily tyazhesti pri proizvol'nom rel'efe poverkhnosti nablyudenii (Computer Processing of Gravity Anomalies in Case of Arbitrary Topography of the Observation Surface), Moscow: Nedra, 1976.

Dolgal A.S. and Khristenko L.A. (1997), Allowance for the influence of topography in processing the data of magnetic measurements, Geofizika, 1997, no. 1, pp. 51–57.

Dolgal A.S. (2001), Finding the corrections for terrain topography in magnetic survey, in Geologiya i mineral'nye resursy Tsentral'noi Sibiri (Geology and Mineral Resources of Central Siberia), Krasnoyarsk, 2001, pp. 183–189.

Dolgal A.S. (2002), Komp'yuternye tekhnologii obrabotki i interpretatsii dannykh gravimetricheskoi i magnitnoi s"emok v gornoi mestnosti (Computerized Technologies for Processing and Interpreting the data of Gravity and Magnetic Surveys in Mountainous Regions), Abakan: MART, 2002.

Dolgal A.S. and Chervonyi N.P. (2008), Allowance for the effect of Earth's surface topography in airborne magnetic measurements, Geoinformatika, 2008, no. 8, pp. 58–66.

Dolgal A.S., Muravina O.M., and Hristenko L.A. (2017) The reduction of the magnetic field within development areas of the plateaubasalts, Geoinformatics 2017, 15–17 May 2017, Kyiv, Ukraine, 11143_ENG. Electronic publication (CD).

Nusipov E.N. and Akhmetov, E.M. (1991) Topographic correction in magnetic prospecting, in Razvitie metodov obrabotki i interpretatsii geofizicheskoi informatsii (Development of the Methods for Processing and Interpreting the Geophysical Indormation), Alma-Ata: Kazakh. politekhn. Inst., 1991, pp. 60–70.

Rempel, G.G. (1980) Topical questions of the technique of introducing the corrections for terrain topography into the data of gravity and magnetic prospecting, Izv. Akad. Nauk SSSR, Ser. Fiz. Zemli, 1980, no. 12, pp. 60–70.

Strakhov V.N. (1992), Algorithms for reducing and transforming the gravity anomalies specified on the physical surface of the Earth, in Interpretatsiya gravitatsionnykh i magnitnykh anomalii (Interpretation of the Gravity and Magnetic Anomalies), Kyiv: Nauk. Dumka, 1992, pp. 4–81.

Yurovskikh V.N. (1997), Calculating the vector of effective magnetization of the trap relief from airborne magnetic data, in Geofizicheskie issledovaniya v Srednei Sibiri (Geophysical Studies in Central Siberia), Krasnoyarsk, 1997, pp. 277–281.

Pilkington M. and Urquhart W.E.S. (1990) Reduction of potential field data to a horizontal plane, Geophysics, 1990, vol. 55, no. 5, pp. 549–555.

Chapter 13
Interpretation Algorithms for Hydrocarbon Deposits

Yuri V. Glasko

Abstract This article has summarized two models of hydrocarbon deposits and several algorithms of interpretation for the models. The algorithms of introcontinuation, balayage, concentration are considered for 2D and 3D cases. The algorithms are realized on a mesh and calculated by computer. The algorithm of introcontinuation uses finit-difference continuation of the field to lower half-space and finit-difference variations of V. M. Berezkin method of full normalized gradient. The algorithm of balayage is based on the balayage-method of H. Poincare and numerical realization of the method on the mesh in terms of D. Zidarov approach. The algorithm of concentration is statistical regularization for distance (discrepancy or smoothing functional) with reiterative balayage. Software package includes the algorithms and it is used for model cases and practice cases for oil and gas deposits.

Keywords Introcontinuation · Balayage · Sweeping · Concentration

Introduction

The application of numerical algorithms together with geological prospecting reduces a cost of exploration activity. At the same time, new investigations of structures and genesis of the hydrocarbon deposits (Megeria et al. 2012; Lukin 2018) provide a priori information for the problems of interpretation and the computational algorithms.

In the article we consider 3 interlocking numerical algorithms for geological prospecting of oil and gas.

The concept of the singular points of the complete normalized gradient is realized by effective V. M. Berezkin method for the geophysical fields (Berezkin 1988). The method is developed in the works of E. V. Bulychev, V. G. Filatov,

Y. V. Glasko (✉)
Research Computing Center, M.V. Lomonosov Moscow State University,
Moscow, Russia
e-mail: glasco@yandex.ru

M. L. Ovsepian (Filatov et al. 2011) et al. The method lets define the elements of the epigenetic mineral formation. On the other hand, this experimental method lets solve Poisson problem with several sources for lower half-space. In the article we consider new 2D and 3D iteration finite-difference modifications of the algorithm.

Consider Poisson problem for the bounded domain (cube). In the case, we define anomaly sources by procedure of concentration (Glasko, 2015). The procedure uses iteration loops of the numerical balayage for mass. This approach is realized by the algorithm with statistical regularization for the inverse problem. The corresponding direct Dirihlet problem and boundary parabolic problems are considered in the fundamental works of H. Poincare (Bogolubov et al. 1974), V. N. Strakhov (Strakhov 1977, 1978), D. Zidarov (Zidarov 1968). The numerical realization of the algorithm is effective. The application of the concentration for oil and gas accumulations is economically.

Physical-Geological Models of Hydrocarbon Deposits and Effect of Epigenetic Mineral Formation

The algorithms of the article used for two models of hydrocarbon deposit.

The first model hydrocarbon deposit of anticlinal type (Lobanov et al. 2009) involves deposit with cover, subvertical side zones, close layer and terrigenous layer. For the physical-geological model we consider geomagnetic model. This model take account secondary mineral formation phenomenon. The basis for the secondary mineral formation is migration hydrocarbon compounds (C_nH_m) and reactive non-hydrocarbon components (H_2S, CO_2, H_2, CO and others) from the oil deposit. Magnetic logging reveals high magnetic susceptibility of the secondary magnetic minerals (patent USA N4729960, Foote RS, date of publication 08.03.1988). The secondary minerals are concentrated in the terrigenous layer, in the cover of deposit and the close layer. In case of epigenetic (not syngenetic) manner of aggregation the secondary minerals it can be asserted that we have real oil deposit (Lobanov et al. 2009).

The second model (Megeria et al. 2011, 2012) is development of the conception of subvertical side zones with improved productivity. The model has multilayer disposal of the oil deposits along of geosoliton tube with small diameter (approximately 50 m). The model accounts for migration of fluids (geosolitons) from Earth's mantle. This migration regenerates oil's resources (Megeria et al. 2012; Lukin 2018).

Introcontinuation

Introcontinuation is called the process of continuation of the gravity field or the magnetic field from day surface to lower half-space through the gravity and magnetic bodies.

Singular point is called the point which to arrise in the process of the introcontinuation. The continuing field soars and disintegrates as ones to the singular point is approached. The procedure of numerical continuation of the field is stable for depth 0.7–0.8 to the upper singular point. In this connection, V. M. Berezkin proposed metod of complete normalized gradient (CNG). The point is CNG has maximal mean in the singular point (Berezkin 1988).

We consider new 2D and 3D varients of the method CNG ($G_H(x,z)$, $G_H(x,y,z)$) (Glasko 2016). Our main interest is finite-difference realization of the method on the mesh (FCNG). Correspondingly, the continuation of the field is consided for finite-difference approximation of Laplas equation. It has been suggested 2 varients 2D algorithm for finite-difference gradient with introcontinuation (FGRIN). First varient has not the boundary conditions on the vertical boundaries. In this case, FGRIN is realized on the triangular grid. For each step we lose left and right means of continued field.

3D algorithm was proposed on base of profile 2D algorithm. The algorithm is oriented for area processing of the field by using set of intersection perpendicular profiles. The algorithm includes procedure with finite-difference approximation Poisson integral. This procedure increasing of effect of the anomaly is used for calculating of the field on step of the grid upwards. The field calculating on one step upwards of the day surface presets initial values for iteration loop of introcontinuation to lower half-space (or lower quadrant of the space) and FCNG.

The technique does not require big volume of a priori information for geological reduction. Thus this technique belongs to set of techniques on base of conception of singular points of potential fields.

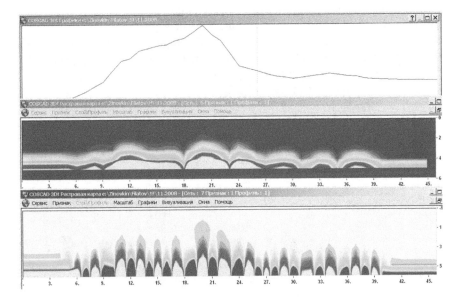

Fig. 13.1 Azevo-Soloushinsky oil Mound of Tatarstan. Results of processing by FCNG (Filatov et al. 2011)

The algorithms for 2D and 3D cases have been applied to set of models (2 foundation steps, gravity anomaly from sphere, 3 sources) and practical materials for oil and gas deposits in West Siberia, Volga Region, Tatarstan (Fig. 13.1), Uzbekistan, Far East (Filatov et al. 2011).

Balayage Method and the Algorithm

Introcontinuation of the field is conducted on base of the Dirichlet problem for domain S with boundary condition is assigns on $\Gamma \equiv \partial S$. The boundary condition is assigns the potential of gravity (or magnetic) field on the boundary Γ. The domain S includes anomaly source $\Omega = \bigcup_{i=1}^{N} \Omega_i$. The domain S is lower half-space.

$$\Delta u(X, z) = 0, (X, z) \in S, \quad (13.1)$$

$$u(X, z)|_{\Gamma} = u_0(X), \Gamma = \{X_1 \leq X \leq X_2, z = 0\} \quad (13.2)$$

$u(X, z)$ is potential of the field. For other model $u(X, z)$ and $u_0(X)$ are mass distribution (charge distribution).

Instead of lower half-space S we can consider bounded domain V with boundary $\Gamma \equiv \partial V$. This problem has solution. The fact of existence of the solution was proved by H. Poincare on base of balayage method (Bogolubov et al., 1974). Principe of the method consists in filling the domain V by spheres and sweeping density from inside domain to boundaries of the spheres. This is iteration process. Condition of end of the process is all density is on the part of boundaries of the spheres approximating the boundary of the domain V.

In the works of Strakhov (1977, 1978) was proposed the model of balayage (sweeping) for mass distribution. The model is based on the parabolic equation with real-time dummy and moving boundary. The model describes balayage principle.

Consider 3D model for the potential $U(X, t)$:

$$\Delta U(X, t) = U_t(X, t), \quad X \in V, \quad t \in (0, T) \quad (13.3)$$

$$U(X, 0) = U_0(X), \quad X \in V \quad (13.4)$$

$$\left.\frac{\partial U(X, t)}{\partial n}\right|_{\Gamma} = U_\Gamma(X, t)|_{\Gamma} \quad (13.5)$$

$$U_0(X) = \begin{cases} U_0^{in}(X), & X \in \Omega \\ U_0^{ex}(X) = U_0^{ex}(X'), & X, \quad X' \in V \setminus \Omega \end{cases} \quad (13.6)$$

$$U_0^{in}(X) > U_0^{ex}(X) \quad (13.7)$$

Here $U_0^{in}(X)$ is inner potential, $U_0^{ex}(X)$ is external potential for the domain Ω. We note that potential of gravitational field stating through the density by integral (sum) (Yagola et al. 2014).

The model for distribution densities of the masses (the charges) ($u(X,t)$) has view:

$$\Delta u(X,t) = u_t(X,t), \quad X \in V, \quad t \in (0,T) \tag{13.8}$$

$$u(X,0) = \begin{cases} \delta(X), & X \in \Omega \\ 0, & X \in V \backslash \Omega \end{cases} \tag{13.9}$$

$$\left.\frac{\partial u(X,t)}{\partial n}\right|_\Gamma = u_\Gamma(X,t)|_\Gamma \tag{13.10}$$

The model can be numerical realized by finite-difference approximation or by numerical realization balayage method on the grid on base of D. Zidarov works (Zidarov 1968). The numerical realization of balayage method is very effective and it uses geology right conditions for soluting. So, condition of piecewize continuity of the density is geology right.

The paper addresses numerical realization the model (13.8)–(13.10) by iteration balayage process for the grid, where the spheres are approximated by the scheme with templete "cross" for 2D and 3D cases. Correspondingly, we use 5 and 7 points schemes (with center node). We note that at the begining of the iterative process all density is concentrated into the domain V_h, at the end of the process all density is concentrated on the boundary (Γ_h) of the domain V_h. The iteration loop for the density $u(X)$ of the cube V_h on each of the iteration (n) consists of 3 nested loops for $= 2, \ldots, I-1, j = 2, \ldots, J-1, k = 2, \ldots, K-1$:

$$u_{i-1,j,k}^n = u_{i-1,j,k} + 1/6 u_{i,j,k}, u_{i+1,j,k}^n = u_{i+1,j,k} + 1/6 u_{i,j,k}, \tag{13.11}$$

$$u_{i,j-1,k}^n = u_{i,j-1,k} + 1/6 u_{i,j,k}, u_{i,j+1,k}^n = u_{i,j+1,k} + 1/6 u_{i,j,k}, \tag{13.12}$$

$$u_{i,j,k-1}^n = u_{i,j,k-1} + 1/6 u_{i,j,k}, u_{i,j,k+1}^n = u_{i,j,k+1} + 1/6 u_{i,j,k}, u_{i,j,k}^n = 0 \tag{13.13}$$

$$u_{i-1,j,k} = u_{i-1,j,k}^n, u_{i+1,j,k} = u_{i+1,j,k}^n, u_{i,j-1,k} = u_{i,j-1,k}^n, u_{i,j+1,k} = u_{i,j+1,k}^n \tag{13.14}$$

$$u_{i,j,k-1} = u_{i,j,k-1}^n, u_{i,j,k+1} = u_{i,j,k+1}^n, u_{i,j,k} = u_{i,j,k}^n \tag{13.15}$$

The truncation condition of the iteration loop is the discrepancy between initial and final (boundary) masses less than the accuracy ε.

The iteration loop for 2D case consist 2 nested loops for $i = 2, \ldots, I-1$, $k = 2, \ldots, K-1$:

$$u_{i-1,k}^n = u_{i-1,k} + 1/4u_{i,k}, \quad u_{i+1,k}^n = u_{i+1,k} + 1/4u_{i,k}, \qquad (13.16)$$

$$u_{i,k-1}^n = u_{i,k-1} + 1/4u_{i,k}, \quad u_{i,k+1}^n = u_{i,k+1} + 1/4u_{i,k}, \quad u_{i,k}^n = 0, \qquad (13.17)$$

$$u_{i-1,k} = u_{i-1,k}^n, \quad u_{i+1,k} = u_{i+1,k}^n, \quad u_{i,k-1} = u_{i,k-1}^n, \quad u_{i,k+1} = u_{i,k+1}^n \qquad (13.18)$$

$$u_{i,k} = u_{i,k}^n \qquad (13.19)$$

Besides, the conformal mapping can be used for 2D balayage in the domain V (Glasko 2012). We have composite 3 conformal mappings: the map of the domain V to the rectangle V_1 with coordinates: $\left(-\frac{a}{2},0\right)$, $\left(\frac{a}{2},0\right)$, $\left(-\frac{a}{2},b\right)$, $\left(\frac{a}{2},b\right)$; the map of V_1 to upper half of the plane $W : z > 0$; the map of W to the circle with radius 1. These conformal mappings express in terms of the fraction-linear functions (maps 1 and 3) and elliptic sine (map 2). The density is calculated by double integral for normal derivative of Green function.

Algorithm Concentration of the Density

Inverse problem of interpretation consists in determination density and geometrical characteristics of anomaly source by observed anomaly potential field (Lavrentiev et al. 2010, 2011). One of approaches to interpretation consists in introcontinuation with using Berezkin VM method CNG. Second approach consists in concentration of the potential with the boundary of the domain inside the domain. The lower half-space S may be approximated by the domain V. Let us consider the cube V. The concentration of anomaly potential (more exactly the local component of the ones) consists in determination of the density and the geometrical characteristics the anomaly source (Glasko 2015).

The concentration consists in multiple balayage of the potential to achieve given distribution on the boundary $\Gamma \equiv \partial V$ of the domain V. The density and the geometrical characteristics of defined domain are changing in the iterative process.

Here, we consider the problem of the concentration with balayage for the mass. This mass induced the given anomaly. Thus the distribution of the density given on the boundary of the domain V. 3D model of the concentration has view:

$$\Delta u(s,t) = -u_t(s,t), \quad s \equiv (x,y,z) \qquad (13.20)$$

$$\int_0^T \frac{\partial u}{\partial n}\bigg|_\Gamma = u(s,0) \qquad (13.21)$$

$$u(s,0) = \begin{cases} 0, & s \in V \setminus \partial V \\ \delta_\Gamma(s^*) \equiv \delta(s,0), & s \in \partial V \end{cases} \qquad (13.22)$$

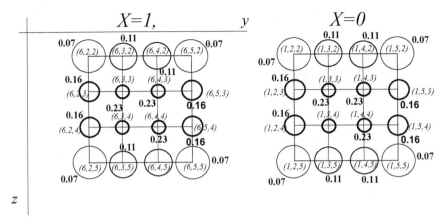

Fig. 13.2 Map of distribution densities on the inside domains the cube faces X = 1, X = 0. The distribution of densities is result of balayage process

Let us define the distribution of the density for $t = T$:

$$u(s,T) = \begin{cases} \delta(s), & \forall s \in \Omega \\ 0, & \forall s \in V \setminus \Omega \end{cases}, \qquad (13.23)$$

where Ω—the source of anomaly.

We note that the potential of the field may be computed for given the distribution of the density. On other hand, the distribution of the density may be computed for given the potential of the field. Consequently, we can formulate the problem of the concentration for the density and for the potential alike.

The algorithm of concentration consists of 2 steps. On first step we determine the geometry of the object Ω by observed the distribution of the density on the

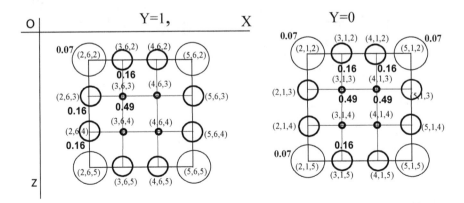

Fig. 13.3 Map of distribution of densities on the inside domains the cube faces Y = 1, Y = 0. The distribution is result of balayage process

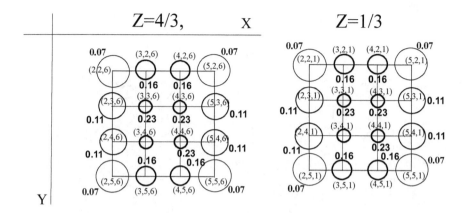

Fig. 13.4 Map of distribution of densities on the inside domains the cube faces $Z = 4/3$, $Z = 1/3$. The distribution is result of balayage process

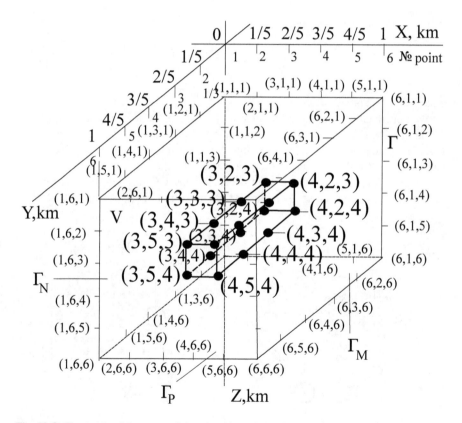

Fig. 13.5 The result of first step of the algorithm of density concentration. X, Y, Z-axes (the distance is measured in kilometers). Γ—the cube (V) faces. We consider the cube in mesh space. The mesh step equal 1/5

13 Interpretation Algorithms for Hydrocarbon Deposits

Table 13.1 The algorithm of concentration of the densities for model 7 (step 2): statistical regularization by Monte-Carlo method and minimization of the discrepancy. δ—error of the input data, ε_{MC}—accuracy of Monte-Carlo method, ε— error of the result

δ	ε_{MC}	$\varepsilon\ (\%)$
[0, 5%]	$10^{-3}/5$	1.7
[6, 10%]	10^{-3}	4.7
[11, 15%]	$10^{-2}/5$	6.4
[16, 20%]	$10^{-2}/3$	8.5

boundary $\Gamma \equiv \partial V$ of the domain V including Ω. On second step we estimate the density of the source Ω. Computing experiment revealed that the maximum values of the density are observed on nearest to the source sides of the boundary. For second step a priori information is segment of means of oil density (gas density). First step uses topological interpretation of maps for balayage density. Second step includes iteration loop for numerical realization of balayage method for density (for 3D—(12.11–12.15), for 2D—(12.16–12.19)). This loop is minimization of the square of the distance (or discrepancy) between calculated value and specified value of the density on the domain V boundary. In this case, the value of the domain Ω density is defined from given segment with using Monte-Carlo method. In the general case, the anomaly source: $\Omega = \bigcup_{i=1}^{N} \Omega_i$.

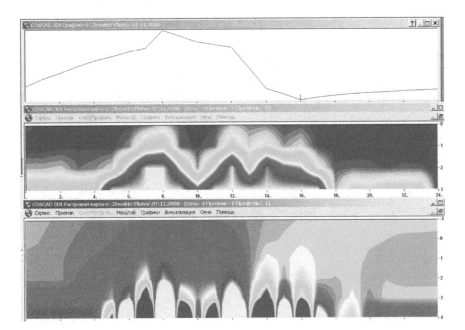

Fig. 13.6 Results of interpretation for Urtabulak -gas deposit of Uzbekistan (Lavrentiev et al. 2010)

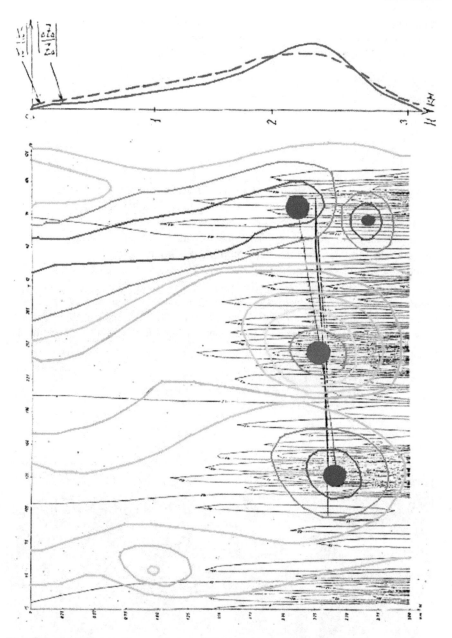

Fig. 13.7 South-Surgut oil accumulation of West Siberia. The deposit configuration is fixed by izolines of FCNG and by concentration of gravitational and magnetic fields (Lobanov et al. 2009)

We consider the computing experiments for set of the objects in $V = [0; 1\ \text{km}] \times [0; 1\ \text{km}] \times [1/3\ \text{km}; 4/3\ \text{km}]$. At first step we determine the objects: the point-source (model 1), 2 point-sources at the different depth (model 2),

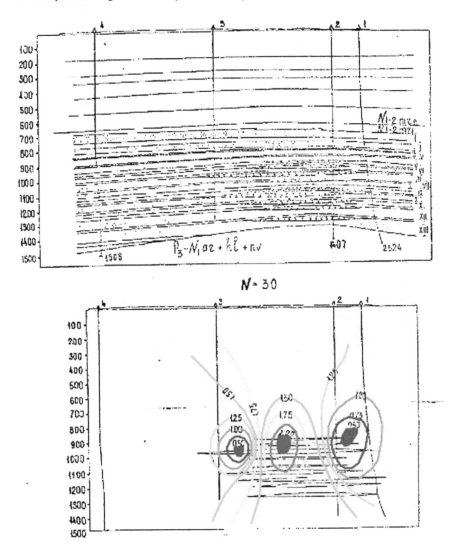

Fig. 13.8 East-Lugovskoie gas deposit of Sakhalin. The geological profile (by A. I. Yurochko) and field of FCNG (Lavrentiev et al. 2010)

two cylinders at the different depth (models 3, 4), the horizontal layer (model 5), the cube (model 6), the parallelepiped extended along of Y-axis (model 7), two cubes at the same depth (model 8). At second step we determine the densities of the objects. If we use lower permissible bound (value of oil's density) times of computing is shorten. For general case $[0.4 \text{ g/cm}^3; 1 \text{ g/cm}^3]$ we may cut of a majority of non-perspective sources.

The results of the algorithm of concentration for model 7 are represented by Figs. 13.2, 13.3, 13.4, 13.5 and by Table 13.1.

The software packages for 2.5D and 3.5D cases include components Sweeping, Concentration, Separation of fields. The packages are used for interpretation of data for oil and gas deposits of Uzbekistan (Fig. 13.6), Siberia (Fig. 13.7), Far East (Fig. 13.8).

Acknowlegements I am grateful to Prof. I. N. Korchagin and to Academician of NAS of Ukraine V. M. Shestopalov for useful recommendations and new geological information about conception of Earth Degassing, to Prof. V. M. Megeria for new information about structures of oil and gas deposits, to Prof. I. E. Stepanova for her interest to algorithms of the work and to Prof. A. G. Yagola for support in discussing the results. I am sincerely grateful to N. Matveeva (secr. Uspensky Seminar 2018) for her organize work.

References

Berezkin VM (1988) Full gradient technique in geophysical survey. Nedra, Moscow (Russia).
Bogolubov NN, Arnold VI, Pogrebynsky IB (eds) (1974) Henri Poincare Selected Works, vol 3. Nauka, Moscow.
Filatov VG, Ovsepian ML, Glasko YV et al (2011). The Application of Procedure FCNG-FGRIN for Models and Hydrocarbon Deposits. In: Nikitin AA, Petrov AV, Megeria VV et al. Optimal Filtering and Introcontinuation of Geophysical Fields for Finding Oil and Natural Gas with Respect to Secondary Magnetic Minerals Formation. NT Press, Moscow, p 66–97.
Glasko YV (2012) One Problem of Equivalent Redistribution of Mass. Izv., Phys. Solid Earth 48 (2): 174–179.
Glasko YV (2015) The Problem of Concentration of Mass. Izv., Phys. Solid Earth 51(2): 191–196.
Glasko YV (2016) 2D and 3D algorithms of introcontinuation. Numerical Methods and Programming 17: 291–298. http://num-meth.srcc.msu.ru.
Lavrentiev MM, Starostenko VI, Filatov VG, Megeria VM, Lobanov AM, Ovsepian ML, Glasko YV et al (2010). Application of Regularization in Gravity Prospecting and Magnetic Prospecting in the Search for Hydrocarbon Deposits. Russian State Geological Prospecting University (MGRI-RGGRU), Moscow.
Lavrentiev MM, Filatov VG, Glasko YV (2011) Method of Regularization for Inverse Problems of Geophysics (2011). In: Nikitin AA, Petrov AV, Megeria VM, Filatov VG, Lobanov AM The Application of Regularization and Optimal Filtration Geophysical Data in the Search for Hydrocarbon Deposits. NT Press, Moscow, p 10–28.
Lobanov AM, Filatov VG, Petrov AV, Ovsepian ML, Glasko YV et al (2009). Introcontinuation and Epigenetic Magnetic Minerals Formation for Finding Oil and Natural Gas. MGRI-RGGRU, Moscow.
Lukin AE (2018) Earth degassing, genesis of hydrocarbons and oil-and-gas potential. In: Shestopalov VM (ed) Essay On Earth Degassing. National Academy of Sciences of Ukraine, Kyiv, p 187–280.
Megeria VM, Starostenko VI, Nikitin AA, Petrov AV et al (2011). Application Geo-soliton Conception of Earth Degassing, Regularization and Optimal Filtering of Geophysical Evidence in the Search for Hydrocarbon Deposits. NT Press. Moscow.
Megeria VM, Filatov VG, Starostenko VI, Korchagin IN, Lobanov AM, Glasko YV et al (2012). Geosolitonic concept and prospects of application of non-seismic methods for prospecting hydrocarbons accumulations. Kyiv: Geophysical journal 34(3): 4–21.
Strakhov VN (1977) About balayage of masses Poincare and its using for solution gravimetry direct and inverse problems. DAN USSR 236(1): 54–57.

Strakhov VN (1978) Theory of plain problem gravimetry and magnetometry – "analytical world" is gerenated by H. Poincare balayage. Izv. AN USSR, Phys. Solid Earth. 2: 47–73.

Yagola AG, Wang Yanfei, Stepanova IE, Titarenko VN (2014) Inverse problems and the methods of solution. Geophysics applications. BINOM, Moscow.

Zidarov D (1968) About some inverse problems of potential fields and their application to questions of geophysics. BAN, Sofia.

Chapter 14
Features of Localization of the Poles of the Gravity Potential Regarding to the Field Sources and the Practical Implementation of the «Polus» Method

G. Prostolupov and M. Tarantin

Abstract The principle of the vector method "Polus" is described. There are different types of points of intersection of the vectors called poles: in particular the first, second and third orders; definite and indefinite; principal and conjugate. Described in details the peculiarities of their localization on the sources of the gravitational field, the efficiency in solving the inverse problem of gravimetry. An example of the practical use of the program Polus2D on the model of the Verkhnekamsk potassium salts deposit is presented. Concluded the greatest efficiency of poles localization of second and third orders for the observed gravitational field interpretation.

Keywords Gravimetry · Geology · Gradient · Inverse problem

Introduction

The method "Polus" is based on the properties of full vectors of potential gradient to be aimed in the direction of the perturbing masses, i.e., for at least a couple of vectors to converge towards a positive source and to disperse in the upper half-space in the case of negative one, the intersection point vectors (pole) spatially coincides with the source and with the center the ball-type (or point) source. Further research led to the use not only the points of intersection of vectors of the potential gradient of the first derivative but also the second and third derivative as a meaningful interpretation unit. In the gravimetric practice it is common to use the first derivative of the gravity V_z due to the fact that the vertical derivative of potential is directly measured by the gravimeter, and horizontal components can be measured or calculated. In the method Polus all of the first derivatives of the potential V are used, as they are the components of the vector pointing exactly toward the source

---

G. Prostolupov (✉) · M. Tarantin
Mining Institute UB RAS, Perm, Russian Federation
e-mail: genagravik@gmail.com

© Springer Nature Switzerland AG 2019
D. Nurgaliev and N. Khairullina (eds.), *Practical and Theoretical Aspects of Geological Interpretation of Gravitational, Magnetic and Electric Fields*, Springer Proceedings in Earth and Environmental Sciences, https://doi.org/10.1007/978-3-319-97670-9_14

(not only V_z, as is customary). Derivatives of the potential, V_z, V_x, V_y, and the higher derivatives V_{zx}, V_{zy}, V_{zz}, V_{zzx}, V_{zzy}, V_{zzz} are calculated using the algorithm of the point source approximations of the field (Dolgal 1999). According to the theory, the calculated gravity field, including derivatives everywhere outside of the anomalous masses will coincide with the real if it would be able to measure (Mikhailov and Diament 2006).

In previous reports (Prostolupov 2016) some specific properties of the first-order poles P_I^{++}, P_I^{--}, the second-order P_{II}^{++}, P_{II}^{--}, and third-order P_{III}^{++}, P_{III}^{--} regarding the sources of the gravitational field were described. According to the classification of poles, there are also uncertain poles of the first-order P_I^+, P_I^-, second-order P_{II}^+, P_{II}^-, third-order P_{III}^+, P_{III}^- and neutral P_I°, P_{II}°, P_{III}°.

Features of Definite Poles

The definite positive poles P^{++} are formed in the lower half-space at the intersection of pairs of down-directed ('positive') vectors. The definitely negative poles P^{--} are formed in the upper half-space at the intersection of pairs of up-directed ('negative') vectors.

The usefulness of definite (apparent) poles of three orders of magnitude P_I, P_{II}, P_{III} is not in doubt. Let's consider the example of their effectiveness for localization of an inclined beam with the dip angle of 45° (Fig. 14.1). Gravity effect Δg and source configuration are taken from the Atlas of theoretical curves for interpretation of magnetic and gravity anomalies (Mikov 1956). The localization of the timber used the principle of complementary effect of explicit poles P_I, P_{II}, P_{III}, described earlier (Prostolupov 2017). Poles of different orders are located with an offset relative to the origin, but better reflect the object of study when aggregated. Some offset of the poles from the center to the dip of beam is noticeable, especially for the poles of the first order P_1. Previously, the efficiency of explicit poles was considered in the solution of inverse problem of localization of objects of ball-type, stem, horizontal plate area and reef.

Features of Uncertain Poles

The uncertain positive poles P^+ are formed in the lower half-space at the intersection of a pair of linear extensions of 'negative' vectors in the opposite direction. Uncertain negative poles P^- are formed in the upper half-space at the intersection of the linear extensions of 'positive' vectors in the opposite direction.

Let's consider the distribution of uncertain poles of the first order in the example of the anomalous field inversion with 4 sources (Fig. 14.2) located at 300 m depth and spaces between them 3, 2 and 1.5 km. Definite poles P_I^{++} were localized in

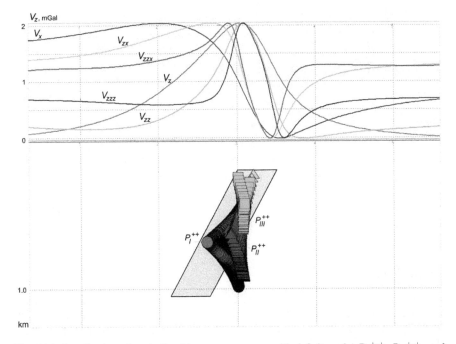

Fig. 14.1 Localization of an inclined beam type source with definite poles $P_I{}^{++}$, $P_{II}{}^{++}$, and $P_{III}{}^{++}$

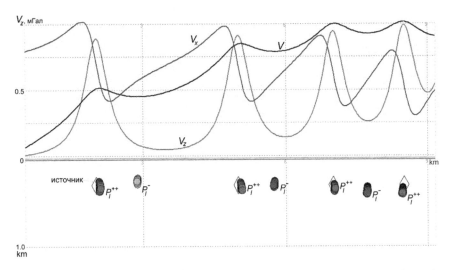

Fig. 14.2 The model of the gravitational effect of 4 sources and distribution of the first order definite positive $P_I{}^{++}$ and uncertain negative $P_I{}^{--}$ poles

the vicinity of sources. Uncertain poles are located approximately at the same depth between the sources.

The horizontal offset of the first-order uncertain poles P_I to the left of the source body can be explained by the nature of the graph of V. Increasing graph causes the left offset of uncertain poles P_I, decreasing causes the right one. This dependence can be seen in Fig. 14.2.

The uncertain poles of the second and third order P_{II} and P_{III} are formed outside of the interpretation region (its depth exceeds 1/3 of the linear size of the survey region) due to the rapid changing nature of graphs of higher derivatives V_{zz}, V_{zzz} and V_{zx}, V_{zzx}—a sharp increase (decrease) alternating smooth components. The nature of the pole location relative to the sources can be described as either random variation in the case of poles of the second type, the set is concentrated between the sources with the opposite excess density. The neutral poles P_o is also consistent with the nature of noise.

Poles Basic and Conjugated

Localization of poles regarding its sources can be divided into basic and conjugated. This applies to all three orders of definite poles P_I, P_{II}, P_{III}. It should be noted that the above-considered uncertain poles are always conjugated.

Let's consider the localization of poles P_{III} in the case of 5 sources (Fig. 14.3). Sources are located at 500 m depth at a horizontal distance from each other 3, 2, 1.5

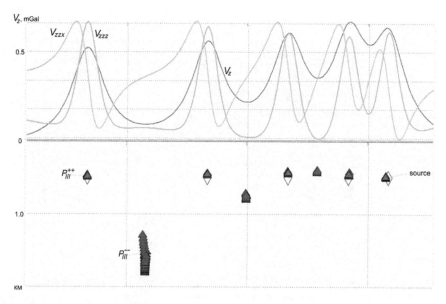

Fig. 14.3 The model of the gravitational effect of 5 sources and the distribution of third order basic positive P_{III}^{++} and conjugated negative P_{III}^{--} poles

and 1 km. Positive poles $P_{III}{}^{++}$ localized in the vicinity of sources are basic ones. Poles of opposite sign $P_{III}{}^{--}$ are localized in the space between basic and are conjugated.

Depth of the conjugated poles depends on the horizontal distance between the sources. The larger is distance, the deeper go conjugated poles. This is understandable, due to the fact that the medium in the horizontal direction between, for example, positive sources, is an extended negative source, and the more this distance is, the more localized negative conjugated poles comes. If there is only one body then conjugated poles are not formed.

Practice

The possibility of practical application of the program Polus2d in geological interpretation of detail gravity field measured on the Verkhnekamsk potassium salt deposit is considered in the example of a schematic density model simulating the structure of the deposit region (Fig. 14.4). In the carnallite zone, coating of rock salt and the lower part of the terrigenous-carbonate strata decompression of 0.1–0.2 g/cm^3 is modeled. The gravitational effect of the entire model can be seen at the graph V_z. The method of point-source approximations of calculated model of V_z gives the values of the derivatives of the potential V_z, V_x, V_y, and the higher derivatives V_{zx}, V_{zy}, V_{zz}, V_{zzx}, V_{zzy}, V_{zzz}. According to it, vectors and their intersections in space are constructed, resulting in an explicit negative poles $P_I{}^{--}$, $P_{II}{}^{--}$, and $P_{III}{}^{--}$. As can be seen in Fig. 14.4 the poles of the second ($P_{II}{}^{--}$) and third ($P_{III}{}^{--}$) orders have the best coincidence with the negative areas of the model density. The poles of the first order are localized outside the considered depressurization.

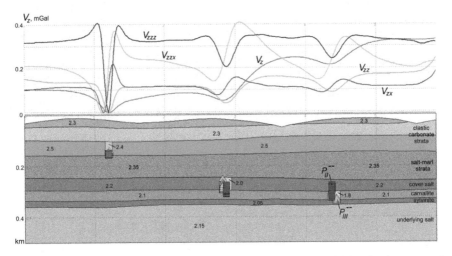

Fig. 14.4 Density model of the Verkhnekamsk potassium salts deposit with low density areas and points of localization of poles $P_{II}{}^{--}$, and $P_{III}{}^{--}$

Conclusion

Based on the studies of the nature of the localization of the poles of the sources definite poles of the second (P_{II}^{++}, P_{II}^{--}) and third (P_{III}^{++}, P_{III}^{--}) orders can be recommended for using in the interpretation process.

Overall, despite the considerable potential for the development of the polar method as one of the branches of the vector gravimetry, we can talk about the completion of the computer program, especially its two-dimensional version "Polus2d", and capability real field data interpretation, which is confirmed by practical examples and certificate of state registration (Tarantin and Prostolupov 2017).

Acknowledgements The work is supported by RFBR grant No. 16-45-590426.

References

Dolgal A. S. (1999). Approximation of the geopotential fields by equivalent sources for solving practical problems. Geophysical journal 4, 71–80. (in Russian).
Mikov D. S. (1956). Atlas of the theoretical curves for interpretation of magnetic and gravity anomalies. Tomsk. Gosgeoltekhizdat. 137 p. (in Russian).
Mikhailov V. O. and M. Diament (2006). Some aspects of interpretation of tensor gradiometry data. Izvestiya. Physics of the Solid Earth 12. V. 42. 971–978.
Prostolupov G. V. (2016). Study of localization properties of poles regarding sources of potential fields. Voronezh. VSU. 187–190. (in Russian).
Prostolupov G. V. (2017). Determination of basic parameters of the sources of the gravitational field by the "Polus" method. Moscow. IPE RAS. 319–324. (in Russian).
Tarantin, M. V. and G. V. Prostolupov (2017). Polar transformation of gravimetric data "Polus2d". Certificate of state registration of computer programs № 2017610475. 11 Jan 2017.

Chapter 15
Two Approaches to the Solution of Inversion Problem in the Bear Experiment

A. A. Zhamaletdinov, M. S. Petrishchev and V. Yu. Semenov

Abstract There is completed the interpretation of the results of the BEAR experiment on synchronous magnetotelluric sounding of the lithosphere of the Fennoscandian Shield on the network of 150 × 150 km. The experiment has been made 20 years ago, in 1998, but till that time no common decision on the deep structure of lithosphere is achieved because of super complicated primary data. Two possible approaches to the solution of the problem are applied in the article. The first approach, purely formal. It applies to the OCCAM inversion technique. The approache is based on the phase values of the impedance with reference to the global magnetovariational sounding. The second is phenomenological approach. It is based on the use of a priori information, which makes it possible to regularize the solution of the inverse one-dimensional problem. Based on the results of phenomenological processing, a quasi-three-dimensional model of the electrical conductivity of the lithosphere is constructed. There are detected two anomalies of reduced resistivity are established in the interval of depths of 30–60 km that spatially coinciding with the regions of the Moho boundary submerging up to 55–60 km.

Keywords Fennoscandian shield · Magnetotelluric sounding · Lithosphere
Occam inversion · Phenomenological interpretation

Presenting author: A. A. Zhamaletdinov.

A. A. Zhamaletdinov (✉) · M. S. Petrishchev
St. Petersburg Branch of Pushkov Institute of Terrestrial Magnetism, Ionosphere and Radio Wave Propagation of the Russian Academy of Sciences,
St. Petersburg, Russia
e-mail: abd.zham@mail.ru

A. A. Zhamaletdinov
Geological Institute of the Kola Science Center of the Russian Academy of Sciences, Apatity, Russia184209

V. Yu. Semenov
Geophysical Institute of the Polish Academy of Sciences, Warsaw, Poland
e-mail: sem@igf.edu.pl

Introduction

The Fennoscandian crystal shield is one of the most representative and extensive reference polygons in area for studying the structure and deep structure of the continental lithosphere. Data on its structure are of interest for obtaining a priori (comparative) data on the deep structure of platform areas, where the resolving power of geophysical methods is strongly limited by the influence of sedimentary covers. At the moment the data of the deep electromagnetic soundings are practically not used in the practice of complex geophysical interpretations. There are schemes of electrical conductivity compiled by generalization of ground and aerial reconnaissance data (Zhamaletdinov and Hjelt 1986; Korja et al. 2002; Zhamaletdinov and Kovtun 1993). But they give an idea of the electrical conductivity of structures emerging on the surface of the day or the results of deep sounding at individual points. In this paper we are attempted to compile the consolidated quasi-three-dimensional model of the deep electrical conductivity of the Fennoscandian shield (from first kilometers to 100–140 km) that is based on the results of a unique BEAR experiment.

Analysis of the Experimental Material

The BEAR experiment (Baltic Electromagnetic Array Research) was performed in June-July 1998. It was carried out by synchronous magnetotelluric sounding (MTS) with the use of 50 stations spread over a network of 150 × 150 km in the territory of the Fennoscandian shield in Sweden, Finland and Russia (Korja 2000). Five stations were damaged due to thunderstorm activity. The arrangement of 45 measuring MTS stations is shown in Fig. 15.1a. The measurements were performed using low-frequency ferro-probe stations of the LEMI series (Korepanov 2002) and 24-bit Riftec loggers. 5 components of the electromagnetic field (E_x, E_y, H_x, H_y, H_z) were measured with a sampling step of 2 s. Effective impedance apparent resistivity and phase curves for the 11 most representative points reflecting the main features of the experimental results are presented in Fig. 15.1b, c.

First of all, in Fig. 15.1b we should note a wide spread of apparent resistivity values, covering almost 5 orders of magnitude—from tenths to hundreds of thousands of ohmmeters. The second feature is that the phase curves of the effective impedance in Fig. 15.1c are fairly compact with a spread of 20°–25° and are in good agreement with the generalized results of phase measurements in the field of controlled sources (CS normal) and magnetovariation data (Global). In the diagram of apparent resistivity curves (Fig. 15.1b), the generalized results of CS normal and Global, together with the imaginary band that connects them, cover effective resistivity curves with the highest resistivity values.

The processing and interpretation of the experimental results were conducted between 1999 and 2002 by international cooperation of scientists from Russia,

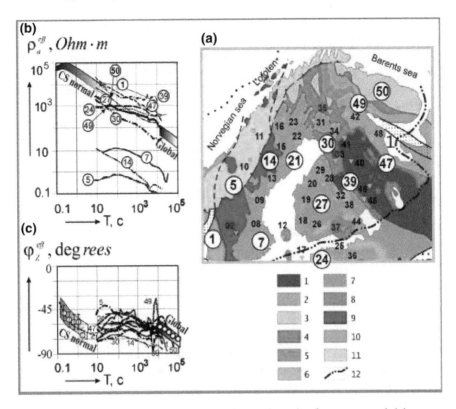

Fig. 15.1 The BEAR experiment. *a*— location of measuring points *b*—apparent resistivity curves (effective). *c*— phase of impedance, Legend: 1–5–Archaean comlexes (1—Karelian terrain, 2—Belomorides, 3—Murmansky terrain, 4—greenstone belts, 5—granulites): 6–10—Proterozoic comlexes (6—argilliters, 7—Svekofenides, 8—rapakivi granites, 9—granites, 10—gotides); 11—kaledonides; 12—the boundary of Russian plateform

Ukraine, Finland, Sweden and Germany with the financial support of the INTAS grant, the BEAR-PMI project (Processing, Modeling, Interpretation). The coordinator of the project is T. Korja (Finland), co-coordinators in the sections—M. Smirnov (processing), I. Varentsov (numerical modeling), L. Pedersen and A. Zhamaletdinov (interpretation). The results of processing and interpretation of data on the project are reflected in a series of domestic and foreign publications (Varentsov et al. 2002; Vardyanants and Kovtun 2009). However, the generally accepted geoelectric model of the structure of the lithosphere has not been received to date. The main reason for this was a strong horizontal heterogeneity of the upper part of the section and a rare network of observations that prevented the use of adequate methods of combating geological noise.

Below we have presented two possible approaches to solving the inverse problem of the BEAR experiment. The first approach, formal, is to apply OCCAM inversion (Constable et al. 1987). Its basis is the support for the phase values of the impedance with reference to the results of magnetovariance sounding. The second approach is phenomenological. It is based on the use of a priori information, which makes it possible to regularize the solution of the inverse one-dimensional problem. Both approaches are designed for the maximum possible use of only magnetic components, since the electric field is subject to strong influence of static distortions.

Processing with Occam Inversion

The basis of Occam inversion is the support for the phase values of the impedance. It is known that the phases of impedance, like the phase of resistivity, reflect only the qualitative features of the geoelectric section. The equation for the phase front has the form

$$z = (\omega t - \varphi_{H_x} + const) \cdot \frac{\lambda}{2\pi}. \tag{15.1}$$

From the analysis of Eq. (15.1), one can see that at each fixed moment of time the phase front is a horizontal plane. Therefore, the wave is called flat. When the time is increasing the depth of z is also increasing. However, the exact depth z can not be established from the phase curve. This means that in the impedance phase, it is impossible to determine the apparent resistivity values and, consequently, the parameters of the geoelectric section can not be investigated quantitatively. For the transition to quantitative interpretation it is necessary at least one of the sections of the transfer function to link it to the values of the apparent resistivity module. In the present study it is performed over the low-frequency part using the results of magnetovariation sounding. In this case, the support goes to the conductivity of the Earth's mantle. In Fig. 15.2a, we give examples of data processing and solving a one-dimensional problem using Occam inversion for four points of BEAR experiment. The position of the points is shown in Fig. 15.1a. The points are chosen with the maximum spread, reaching 5 orders of magnitude—from tenths of an ohmmeter at point 5 to hundreds of thousands of ohms at point 5.

From analyzing of Fig. 15.2, it can be seen that the curves of apparent resistivity have a closely related form. Geoelectrical sections based on the results of inversion have similar features, close in shape and amplitude values to the normal section, obtained from the results of the FENICS experiment. The inhomogeneity of the upper geoelectric section actually disappeared. This led to the loss of some information but it is allowed to obtain more unambiguous information about the deep geoelectric section at the same time. Unfortunately, the phase curves in many cases underwent severe distortions and this limited the possibilities for applying the results of Occam inversion to construct a quasi-three-dimensional model of the electrical conductivity of the lithosphere. The second approach, based on the phenomenological principle, was more acceptable for this task.

15 Two Approaches to the Solution of Inversion Problem ...

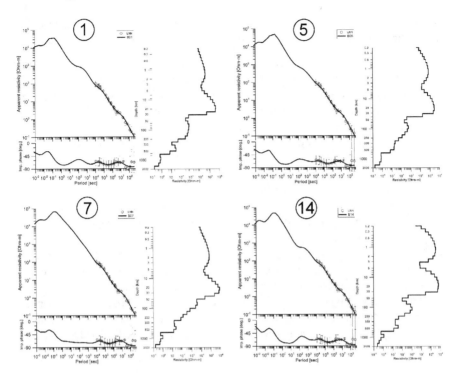

Fig. 15.2 Occam inversion of BEAR data on the example of points 1, 5, 7 and 14. Location of points is shown on the Fig. 15.1a

Phenomenological Interpretation

From Fig. 15.1 it can be seen that the majority of the effective apparent resistivity curves (Fig. 15.1b) are conformal and are located below the dashed line connecting the "normal" band with the global curve of the "Global" magnetovation sounding. The "Global" curve passes through the generally accepted coordinate $T = 10^4$ s and $\rho = 100$ Ωm (Rokityansky and Kulik 1981; Olsen 1998). It can also be noted that the shape of most impedance phase graphs (Fig. 15.2c) remains almost unchanged. They are located with a parallel offset in amplitude within 20-30 degrees. The noted behavioral features of the experimental MTS data on the BEAR array made it possible to assume that the main reason for the displacement of the MTS amplitude curves on the resistivity scale is the influence of static distortions produced by electron-conducting, sulfide-carbon rocks developed within the upper crustal thickness of 10–15 km (Zhamaletdinov 1990). Based on this hypothesis, a phenomenological interpretation of the results of the BEAR experiment is based on the application of five normalizing postulates based on the results of deep soundings using powerful controlled sources (Zhamaletdinov 1990, 2005). The postulates are given below. They can not be taken as absolute truth in the last resort, because they reflect the experimental experience and intuitive considerations of the author.

1. Crustal anomalies of electrical conductivity are concentrated exclusively within the upper stratum of the Earth's crust. Therefore, their influence is static in the frequency range of the BEAR experiment (0.1–0.001 Hz). Consequently, they affect only the electric components of the field and do not affect the phase impedance curves.
2. The upper part of the Earth's crust with a thickness of 10–12 km has an average electrical resistivity of 10^4 Ωm and an average longitudinal conductivity of 1 S. It has a high horizontal heterogeneity due to the widespread development of crustal conduction anomalies in it.
3. The middle part of the Earth's crust (from 10–12 to 20–30 km) is characterized by high resistivity and high horizontal uniformity. The concept of a "normal" geoelectric section is applicable to it.
4. In the depth range from 20–30 to 200–300 km there are possible the horizontal changes in the resistivity due to changes in the geothermal regime of the lithosphere, composition, and rheological properties of rocks.
5. Below 400 km the rock properties do not change in the horizontal direction. All curves of apparent resistivity in the region of long periods (10^4 s and more) converge to the global magnetovariation curve (Rokityansky 1971).

Results

The results of constructing a quasi-three-dimensional model using the phenomenological postulates are given in Fig. 15.3. The model is a section of isolines of the resistivity in the depth range 0–10, 10–30, 30–60 and 60–140 km. A depth section of 1–10 km displays a scheme of crustal anomalies of electrical conductivity that replicates and partially complements the schemes given in (Korja et al. 2002) and (Zhamaletdinov and Kovtun 1993) (see Fig. 15.1).

General decrease in rock resistivity from 10^6 to 10^4 Ωm is observed in the interval 10–30 km to 10^3–10^2 Ωm in the range 60–140 km in the remaining sections. The decrease in resistivity with depth is associated with an increase in temperature with increasing depth. The most poorly conducting part of the lithosphere is located in the extreme north-west, near the Kola Peninsula and Karelia, where the most ancient rocks of the lower Archaean dominate. At the depth section of 30–60 km, an extensive region of low resistivity is distinguished, coinciding with the region of the Moho boundary immersion up to 60 km (Pavlenkova 2006; Sharov and Mitrofanov 2014).

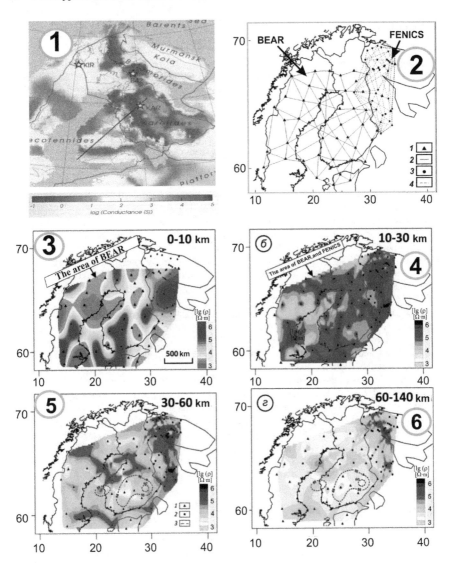

Fig. 15.3 The model of Fennoscandian lithosphere electrical conductivity by results of phenomeno-logical interpretation of the BEAR experiment data 1—The map of crustal anomalies of electrical conductivity compiled from MT-MV data (Korja et al. 2002); 2—The net work of 2D profiles on the BEAR and FENICS arrays taken in consideration for 3-D modeling; 3–6—Sections at different intervals of depth from results of 3D modeling of the BEAR data: 3—The map of crustal anomalies of electrical conductivity compiled from BEAR data; 4—The conductivity layer on the level 10–30 km; 5—same, level 30–60 km; 6—same, level 60–140 km

References

Constable, S. C., Parker, K. L., and C.G. Constable (1987). Occam's inversion a practical algorithm for generating smooth models from EM sounding data. Geophysics, 52, 289–300.

Korepanov, V.Ye. (2002). Electromagnetic sensors for microsatellites. Sensors, Proc. of IEEE, 1718–1722.

Korja, T. (2000). The BEAR Working Group. Lithosphere. Program and Extended Abstracts (Inst. of Seismology, Univ.of Helsinki, Helsinki), Report S-41.

Korja, T., Engels, M., Zhamaletdinov, A.A., Kovtun, A.A., Palshin, N.A., Smirnov, M.Yu., Tokarev, A.D., Asming, V.E., Vanyan, L.L., Vardaniants, I.L. and the BEAR WG. (2002). Crustal conductivity in Fennoscandia – a compilation of a database on crustal conductivity in Fennoscandian shield. Earth, Planets, Space. 54, 535–558.

Olsen N., "The Electrical Conductivity of the Mantle beneath Europe Derived from C-Responses from 3 to 720 km" Geophys. J. Int., No. 133, 298–308 (1998).

Pavlenkova, N.I. (2006). Structure of the lithosphere of the Baltic Shield according to the DSS / Struktura litosfery Baltiiskogo Shchita po dannym GSZ. Structure and Dynamics of the Lithosphere of Eastern Europe. Moscow, Geokart, GEOS. (in Russian).

Rokityansky, I.I. (1971) Deep magnetotelluric sounding in the presence of distortions from horizontal inhomogeneities / Glubinnye magnitotelluricheskie zondirovaniya pri nalichii iskazhenii ot gorizontalnyh neodnorodnostei // Geophysical collection. Kiev, Naukova Dumka, 43, 71–78. (in Russian).

Rokityansky, I.I., Kulik, S.N., and D.A. Rokityanskaya (1981). Ladoga anomaly of electrical conductivity / Ladozhskaya anomaliya elektroprovodnosti. Geophys. Journal. Ukrainian Academy of Sciences. 3 (2), 97–99. (in Russian).

Sharov, N.V., and F.P. Mitrofanov (2014). High-speed heterogeneity of the lithosphere of the Fennoscandian (Baltic) shield / Skorostnye neodnorodnosti litosfery Fennoskandinavskogo Shchita. Reports of the Academy of Sciences. 454 (2), 221–224. (in Russian).

Vardaniants, I.L. and A.A. Kovtun (2009). Investigation of the possibility of the presence of the asthenosphere in the territory of the Fennoscandian shield according to BEAR / Issledovanie vozmozhnosti pristutstviya astenosfery na territorii Fennoskandinavskogo Shita po dannym BEAR. Complex geological and geophysical models of ancient shields. Ed. Yu.L. Voitekhovsky. Apatity, 15–18. (in Russian).

Varentsov, Iv.M., Engels, M., Korja, T., Smirnov, M.Yu. and the BEAR Working Group. (2002). The generalized geoelectric model of Fehnnoscandia: a challenging database for long period 3D modeling studies within Baltic electromagnetic array research (BEAR). Fizika Zemli, 10, 64–105.

Zhamaletdinov, A,A,, and S.E. Hjelt (1986). About the models of electrical conductivity of the Baltic shield / O modelyah elektroprovodnosti Baltiiskogo Shchita // The deep electrical conductivity of the Baltic Shield. Petrozavodsk. Ed. KarFAN USSR. 56–69. (in Russian).

Zhamaletdinov A.A. (1990). Model of the electrical conductivity of the lithosphere from the results of studies with controlled field sources (Baltic Shield, Russian Platform) / Model elektroprovodnosti litosfery po rezultatam issledovanii s kontroliruemymi istochnikami polya (Baltiiskii Shchit, Russkaya platforma). Leningrad, Nauka publishing, 159 p. (in Russian).

Zhamaletdinov, A.A., and A.A. Kovtun (1993). Scheme of electrical conductivity of the north-eastern part of the Baltic Shield. Parameters of the "normal" section / Skhema elektroprovodnosti severo-vostochnoi chasti Baltiiskogo Shchita. Parametry "normalnogo" razreza. The structure of the lithosphere of the Baltic Shield, (ed. N.V. Sharov). Moscow, MGK RAS, 86–88. (in Russian).

Zhamaletdinov, A. A. (2005). Khibiny MHD Experiment: The 30th Anniversary. Izvestiya, Physics of the Solid Earth, 41 (9), 737–742.

Chapter 16
About the Numerical Decision of Problem Dirihle for Equation Laplas in a Rectangle in Researches Under the Decision of a Return Problem in Geophysics

Z. Z. Arsanukaev

Abstract In article it is shown on the basis of results of computing experiments on analytical continuation of preset values of a field located on a full contour, that a major factor influencing a difference in accuracy for areas in the bottom semispace near to a surface of the Earth and near to the top edge of revolting bodies, incorrect statement of problem Dirihle is.

Analytical continuation, discrete equation Laplas

From classical mathematics the exact decision of problem Dirihle for equation Laplas in a rectangle is known which is formulated so is known: on set values of continuous function on a rectangle contour to find the decision of equation Laplas in a rectangle interior. For example, the exact decision of problem Dirihle equation Laplas (16.1) in rectangle G (G = {(x,): $0 \leq x \leq a, 0 \leq y \leq b$}) on method Furje, in the following statement:

$$\frac{\partial^2 u}{\partial x^2} + \frac{\partial^2 u}{\partial y^2} = 0,$$
$$u(0, y) = 0, u(x, b) = \psi(x) (0 \leq x \leq a), \tag{16.1}$$

$$u(x, 0) = \varphi(x), u(x, b) = \psi(x), (0 \leq x \leq a), \tag{16.2}$$

where u (x, y)—required function, (16.2)—regional conditions, is given a kind of infinite series Furje (16.3). (Zhukova et.al 2001)

$$u(x, y) = \sum_{k=1}^{\infty} \left(A_k e^{\frac{\pi k y}{a}} + B_k e^{\frac{-\pi k y}{a}} \right) \sin \frac{\pi k x}{a}, \tag{16.3}$$

Z. Z. Arsanukaev (✉)
D. Mendeleev University of Chemical Technology of Russia, Moscow, Russia
e-mail: zaindy@mail.ru
URL: https://www.muctr.ru/

Factors A_k, B_k are calculated accordingly under formulas (16.4), (16.5):

$$A_k = \frac{2}{a}\left(e^{\frac{\pi k b}{a}}\int_0^a \psi(x)\sin\frac{\pi k x}{a}dx - \int_0^a \varphi(x)\sin\frac{\pi k x}{a}dx\right)/(e^{\frac{2\pi k b}{a}}-1) \qquad (16.4)$$

$$B_k = \frac{2}{a}\left(e^{\frac{2\pi k b}{a}}\int_0^a \varphi(x)\sin\frac{\pi k x}{a}dx - e^{\frac{\pi k b}{a}}\int_0^a \psi(x)\sin\frac{\pi k x}{a}dx\right)/(e^{\frac{2\pi k b}{a}}-1) \qquad (16.5)$$

Thus, keeping in the right part corresponding final number of members of a number, it is possible to receive with any beforehand set accuracy of value of required function u (x, y). (Goloskokov 2004) In some special cases when the functions representing regional conditions (16.2), have a special appearance (as a rule, in the form of periodic functions), the decision (16.3) registers in the form of the sum of several harmonics as in this case all factors A_k, B_k are nulled except several. In present period high-efficiency computing means there was other possibility of the numerical decision of problem Dirihle, and also with any beforehand set accuracy (Boglaev 1990), with use of the following approach. In a rectangle the grid, usually uniform in both directions, and, thus, is entered the continuous two-dimensional geometrical space is replaced with net space. The continuous functions representing regional conditions, are replaced with net functions, and continuous equation Laplas with use discrete relations for replacement of the second derivatives is led to the discrete equation.

Thus, the continuous problem of a finding of values of function u (x, y) in a rectangle on function preset values u (x, y) on all contour of a rectangle with use of a method of grids and discrete equation Laplas is reduced to a problem of drawing up and the decision of systems of the linear algebraic equations. The number of the equations here is equal to number of unknown persons, and thus a matrix of systems square so the system dares any direct method, for example method Gauss. Last scheme of the decision of problem Dirihle for equation Laplas in a rectangle can be used in geophysics at the decision of a return problem taking into account that out of revolting weights the abnormal gravitational potential and its derivatives satisfy to equation Laplas. But attempt directly to use the specified settlement scheme lead to certain difficulties. In—the first, in geophysics gravitational field preset values never are known on all contour of the horizontal layer set in the form of a rectangle in the bottom semispace, and as a rule only for the top part of a contour and are in result shootings of value of a gravitational field on a day surface of the Earth. So statement of problem Dirihle at the decision of equation Laplas in a rectangle here is incorrect. The incorrectness directed by problem Dirihle at application of a net method leads to that systems of the linear algebraic equations arising here will be not predetermined with rectangular matrixes, and their decision is unstable. In—the second, it is necessary to develop indirect methods of the steady decision of systems of the linear algebraic equations of the big usages with rectangular matrixes. The specified difficulties have been successfully overcome in the previous researches of the author (Arsanukaev 2001, 2009a, b, 2010) and other researchers (Strachov and Strachov 1999). As a result of the computing experiments spent for revolting bodies with various density and geometry of a surface on various modelling and practical examples, it has been established, that set on a surface of the Earth of value of a gravitational field with use of discrete equation Laplas analytically proceed in the bottom semispace with high accuracy up to the top edge of revolting bodies. The technique of an estimation of accuracy of values of the field received as a result of analytical continuation has been developed, and by means of this technique has been established, that in some cases at the task of an optimum step of a grid and lengths of a profile the modelling examples relative errors of values of a field received as a result of analytical continuation, in comparison with the exact have in the bottom semispace of value of an order 10^{-5} on depths of equal 1–2 steps of a grid, an order 10^{-3} on a mark equal to half of distance to the top edge and an order 10^{-2} near to top edge; it means, that the values of a field received as a result of analytical continuation, differ from exact values accordingly on 1/1000%, on 1/10% and for some percent. It is necessary to notice, that these values of a field are received at the decision of problem Дирихле still in incorrect statement: in the developed technology of analytical continuation preset values settle down at 2 levels z = 0, z = −h (in гравиметрии axis Oz is directed downwards) i.e. only on one party of a rectangle in which equation Laplas dares. Stability of the decision of system of the linear algebraic equations arising here is reached at the expense of sharing of several discrete approximations of operator Laplasa. But there is obscure a question, nevertheless than the difference in accuracy of values of a field in the bottom semispace, received as a result of analytical continuation, near to a surface of the Earth and near to the top edge of revolting bodies—an incorrectness of statement of a problem, an error of used discrete schemes, an error of a used iterative method for the decision of systems of the linear algebraic equations, is caused by affinity of special points (the top features of revolting bodies). In this connection a number of computing experiments on modelling examples for revolting bodies with various geometry of a surface has been spent at various steps of a grid and at the task of values of a gravitational field on all contour of a rectangle in which internal area with use of discrete equation Лапласа preset values analytically proceeded. Gravitational field preset values in modelling conditions were in a kind of the exact decision of a direct problem for revolting bodies. Results of calculations for one of modelling revolting bodies in the form of a vertical layer (a two-dimensional case) with section 4.8 × 4.0 km with homogeneous density in equal 1 g/sm^3, with step of a grid of 0.2 km, with length of a profile on which preset values equal 32 km and distance from a surface of the Earth to the top edge of section equal 4 km settle down (the size of a matrix of system of linear algebraic equations A = 6042 × 2862 arising here, and a vector of the right part f = 6042 × 1) are resulted in Table 16.1. The analysis of results of computing experiments on modelling examples, at the task of values of a gravitational field for a full contour, shows, that values of a gravitational field in a rectangle can be received by means of discrete equation Лапласа with any beforehand set accuracy up to the top edge of a revolting body (Table 16.1 see).

Table 16.1 Analytical continuation to a mark of 3.8 km at field preset values on 2 horizontal top, 2 vertical lateral, and 1st horizontal bottom profiles

Mode for field preset values	Depth, in km	Relative error
		Profile in length of 32 km
		Template "a direct cross" + "a slanting cross"
Without complication by a hindrance	0.2	1.073628 E−005
	0.4	2.941497 E−005
	0.6	5.376914 E−005
	0.8	8.206198 E−005
	1.0	1.129267 E−004
	1.2	1.452448 E−004
	1.4	1.780613 E−004
	1.6	2.106502 E−004
	1.8	2.425003 E−004
	2.0	2.732097 E−004
	2.2	3.026575 E−004
	2.4	3.309432 E−004
	2.6	3.538232 E−004
	2.8	3.851298 E−004
	3.0	4.112802 E−004
	3.2	4.353962 E−004
	3.4	4.521158 E−004
	3.6	4.369747 E−004

Such accuracy is reached at the expense of reduction of a step of a grid. Results of calculations also confirm high efficiency of an iterative method of the decision of systems of the linear algebraic equations, developed by academician V. N. Strahovym. Thus, distinction in accuracy of values of a field in the bottom semispace received as a result of analytical continuation, near to a surface of the Earth and near to the top edge of revolting bodies, speaks an incorrectness of statement of a problem in resulted above the discrete scheme.

References

Arsanukaev Z.Z. (2009) Analytical continuation of preset values of a gravitational field in discrete statement through sources in a two-dimensional case. Magazine « Bulletin KPAUNC. Sciences about the Earth » 2009 №1. Release 13. p. 47–57.

Arsanukaev Z.Z. (2010) About the decision of a problem of recalculation downwards preset values of a gravitational field with use of software package GrAnM. Materials of 37th session of the International seminar of D.G.Uspensky « questions of the theory and б.практики geological interpretation gravitational, magnetic and electric fields » . On January, 25-29th, 2010 Moscow. Institute of Physics of the Earth of the Russian Academy of Sciences, 2010.- p. 29–34.

Arsanukaev Z. Z. (2009) Allocation of revolting objectsa modern method of recalculation of a gravitational field in the bottom semispace. Magazine « Bulletin KPAUNC. Sciences about the Earth 2009. №1. Release 21. p. 231–241.

Boglaev J.P. (1990) Calculus mathematics and programming. Moscow "Higher school".1990. p. 543.

Goloskokov D.P. (2004) The equations of mathematical physics. Moscow. St.-Petersburg. 2004. p 538.

Strachov V.N, Strachov A.V. (1999) The Basic methods of a finding of the steady approached decisions of systems of the linear algebraic equations arising at the decision of Geophysics problems. II M: Institute of Physics of the Earth of the Russian Academy of Sciences 1999. p. 51.

Strachov V.N, Arsanukaev Z.Z. (2001) Use of a method of discrete potential in Geophysics problems //Questions of the theory and practice of geological interpretation gravitational, magnetic and electric fields: materials of 28th session of the International seminar it. Д. G. Uspenskogo, Kiev, on January, 20th - on February, 2nd, 2001 M: Institute of Physics of the Earth of the Russian Academy of Sciences the Russian Academy of Sciences, 2001. p. 102–104.

Zhukova G.S, Chechetkina E.M., Bogin E.S. (2001) Differentsialnye of the equation in private derivative: Educational grant/ RCTU. TH., 2001..p.197.

Chapter 17
Calculation of Spherical Layer with Variable Density Gravity Field

K. M. Kuznetsov, A. A. Bulychev and I. V. Lygin

Abstract Current article considers efficient algorithm for calculating of gravity effect from a spherical layer with variable density. It is based on using of fast discrete convolution. Features of its software implementation are also considered.

Keywords Gravimetry · Direct gravity problem · Earth's sphericity accounting
Fast discrete convolution · Gravitational potential

Today satellite methods in essence changed the volume and the nature of knowledge of gravity field of Earth. It is bound to development satellite altimetry, which allows to define excesses sea's and ocean's heights over a relevancy ellipsoid, and GRACE, GOCE satellite missions intended for studying of gravity field changings nature at the flight altitude both in space, and in time. New opportunities with uniform methodical positions had appeared to make the comparative analysis of a structure of an upper part of Earth (for example, various earth's plates), to estimate density inhomogeneities of deep spheres of Earth that is important as for planetary geology, tectonics, and for assessment of a stationarity of dynamic characteristics of the rotating planet. However the solution of such problems is impossible without creation of an efficient computing algorithm for direct gravity effect calculation into account sphericity of Earth.

Significant interest in creation of such algorithms and computing software products appeared still in 70—the 80th years of the last century (Hellinger 1983; Johnson and Litehiser 1972). One of approaches for direct calculations of gravity field on a spherical surface was offered in works of V.I. Starostenko and his colleagues (Starostenko et al. 1986) and gained further development in works (Starostenko et al. 2013; Bychkov et al. 2015; Khohlova 2015). It is based on masses approximation by a set of rectangular spherical prisms with constant density

---

K. M. Kuznetsov (✉) · A. A. Bulychev · I. V. Lygin
Department of Geophysical Methods of Earth Crust Study,
Faculty of Geology, Lomonosov Moscow State University,
Moscow, Russian Federation
e-mail: kirillkuz90@gmail.com

(Fig. 17.1). However, expression of gravity field elements for such prism has no terminating analytical decision by means of the elementary functions that leads to complication of a computing algorithm.

At the same time in works (Bulychev et al. 2002) the algorithm of the solution of a direct problem of a gravimetry taking into account sphericity of Earth based on approximation of spherical prism by polyhedron was offered. However, even in this case direct gravity effect calculation for the layer covering all Earth will be rather laborious and demand the considerable time expenditure, especially for a case when it is necessary to calculate effect of a layer on all spherical surface with regular latitude and longitude stride. Considerably it is possible to reduce computing time due to application of an algorithm of fast discrete convolution (Bulychev et al. 1998; Kuznetsov et al. 2017) based on discrete Fourier transform.

Let's consider a computing algorithm in more detail. Main problem is in calculation of potential and gravity effect on the surface of the sphere with radius R_0. Let's assume that a spherical layer with the thickness dR has variable density. Information about this density is set in the form of the grid file with a constant step on latitude $d\varphi$ and longitude $d\lambda$. Latitude range is −90° to 90°, and longitude range is 0°–360°. Then this layer can be presented as set of spherical prisms and every one of them has the constant density. Sizes of these prisms are defined by their geographical location and distance from the center of the sphere. It should be noted, firstly, that within one width belt all these prisms are identical, and, secondly, belt prisms on northern latitude φ coincide with prisms of the southern latitude belt by size at latitude $-\varphi$. These prisms can be approximated by polyhedrons, which all

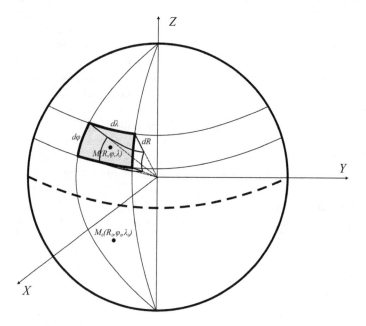

Fig. 17.1 Position of spherical prism

17 Calculation of Spherical Layer with Variable Density Gravity …

sides represent by quadrangular plates. Top and bottom passes sides of these prisms through tops according to an external and internal surface of a spherical prism. Polar prisms which bases are polygons are exceptions (for example if information on density of a layer is set with detail in 1°, then the number of the parties of this polygon N_λ will be equal to 360).

The main flow chart of a calculation algorithm for gravity field elements on the sphere is submitted in the Fig. 17.2. It consists of the following stages:

Step 1. Effect of polar prisms calculation.

1.1. Geographic coordinates of polar prism's vertices are set. So latitude coordinates of northern polar prism's top upper bound are $90° - d\varphi/2$, and longitude coordinates are changes from $-d\lambda/2$ to $d\lambda$. These tops are at distance R from sphere center. Latitude and longitude of bottom bound's tops coincide with the coordinates of the upper, but they will be located at a distance $(R-dR)$, where dR—thickness of the spherical layer.

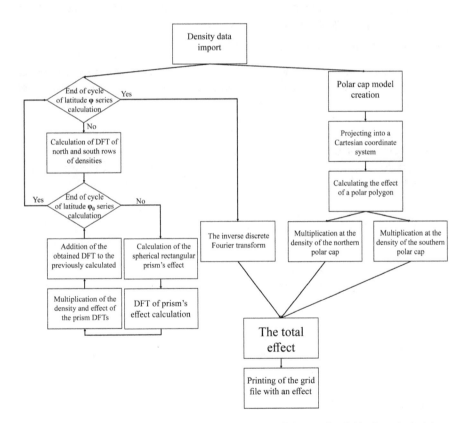

Fig. 17.2 Flow chart of the algorithm for the elements of the gravity field of a spherical layer calculation

1.2. Coordinates of polar prism's tops converts from the geographic coordinate system into Cartesian coordinates with the origin in the center of the sphere; axis Oz, coincides with the polar axis; axis Ox lies in the equatorial plane and passes through the zero meridian; axis Oy is directed to the east.

1.3. Coordinates of M_0 points, for which the effect of the polar prisms will be calculated, are set. In our case, it is convenient to arrange these points along the zeroth meridian with latitude increments $d\varphi$ and at a distance R_0 from sphere center. Coordinates of latitude of calculated points M_0 are from $-90°$ to $90°$. These coordinates are also converted into a Cartesian coordinate system.

1.4. The effect of the polar prism is calculated at the calculated points. For a specific latitude, this effect will be the same for all calculated points of given latitudinal belt. In this case, both values of gravity potential and gravity field component directed towards to center of sphere are calculated. Formulas and an algorithm for calculating these elements are given in the works of Strakhov and Lapina (1986), Blokh (2009). Note that these calculations are performed for only one prism, for example, the north prism. The effect of the southern prism, taking into account its location, will coincide with effect of the northern one, multiplied with corresponding density.

Calculation of the effect of latitude belts.

2.1. Let's define the effect of the latitude belt at the calculated points located on the polar axis. To do this, we define coordinates of the approximation prism with unit density. Vertices of this prism's top surface are located at a distance R from center of sphere, of lower surface—at distance $(R-dR)$, they have latitude coordinates $(\varphi - d\varphi/2, \varphi + d\varphi/2)$ and longitude $(-d\lambda/2, d\lambda/2)$. If to denote the effect of such a prism located at latitude φ, at point $M_0(R_0, \varphi_0 = 90°, \lambda_0 = 0°)$ as $u(M_0)$, and the density values of each prism of this latitude like δ_k, then effect of entire belt will be equal to the sum:

$$U(M_0) = \sum_{k=0}^{N_\lambda - 1} \delta_k u(M_0), \qquad (17.1)$$

where N_λ—number of prisms

Here, under the symbol $u(M_0)$ is meant both the value of the potential and gravity field component. These calculations, as for the case of polar polygons are performed in a Cartesian coordinate system for all latitudinal belts, starting at a latitude of $\varphi_1 = 90° - d\varphi$ to latitude $\varphi_2 = -90° + d\varphi$. Note that due to the symmetry of the location of the belts relative to the equator, calculations performed, for example, for the northern polar point are carried over to the southern.

2.2. We determine effect of latitude belt at latitude φ on all latitudinal belts of the sphere.

First of all, we note that the effect of such a latitudinal belt, consisting of identical approximate prisms, is represented as a circular discrete convolution:

$$U_n = \sum_{k=0}^{N_\lambda-1} \delta_{n-k} u_k = \sum_{k=0}^{N_\lambda-1} \delta_k u_{n-k}, \qquad (17.2)$$

where δ_k—values of approximate prisms densities of belt with longitude φ, u_k—effect of the single prism with latitude φ at the zero meridian to the calculation points at latitude φ_0 ($n = 0, \ldots, N_\lambda-1$; $k = 0, \ldots, N_\lambda-1$) (Fig. 17.3). The effective computation of such a convolution, called the fast discrete convolution, is based on determination of kernel's u_k (effect of a single prism) and convoluted function (discrete density function δ_k) discrete Fourier transforms (DFT). The calculated DFTs are multiplied and then inverse DFT calculates. The computation of the DFT of convoluted functions is carried out using the Fast Fourier Transform (FFT) for calculation of DFT.

For calculation of FFT the number of discrete function values must be equal to the power of two $N = 2^n$, ($n = 2, 3, 4, \ldots$). Since the number of cells N_λ at latitude belt with latitude φ may not be equal to the value N, then to implement the algorithm it is required to extend given series. When calculating the DFT (17.2) the signal is periodic. Therefore, to calculate the circular convolution, it is necessary to periodically continue the discrete density function to N_λ values:

$$\delta_{N_\lambda+i} = \delta_i, \quad i = 0, \ldots, N_\lambda - 1, \quad \delta_{2N_\lambda} = \delta_0, \qquad (17.3)$$

and fill array of the density values by zeroes from the number $2 N_\lambda + 1$ to nearest value of N. To calculate the spectrum (effect of prism with unit density), we place prism to zero meridian at latitude φ, and calculate its effect at latitude φ_0 for

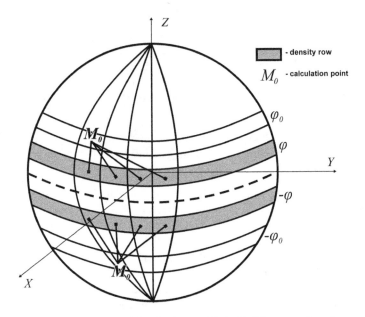

Fig. 17.3 Relationship between calculated points and latitudinal layers on the sphere

longitude $\lambda_{0k} = d\lambda \cdot k$, $k = 0, \ldots, N_\lambda/2$ Further, we mirror these values symmetrically with respect to the point $N/2$:

$$u_{N-i} = u_i, i = 1, \ldots, N_\lambda/2, \qquad (17.4)$$

and the remaining values of this sequence are set equal to zero. It is also necessary to pay attention to the fact that, with the value $k = N_\lambda/2$ the gravity effect $g_{N_\lambda/2}$ or potential $v_{N_\lambda/2}$ of prism is determined, which longitude is 180°, and for, so that this effect is not taken twice into account when calculating the convolution, it is necessary to put the values $u_{N_\lambda/2} = u_{N-N_\lambda/2} = \frac{1}{2} g_{N_\lambda/2}$ or $\frac{1}{2} v_{N_\lambda/2}$. To calculate the DFT of gravity field and its potential, possible to use one discrete transformation. To do this, it is need to create a complex function:

$$V_k = g_k + i v_k, \qquad (17.5)$$

where i—imaginary unit. Since the discrete functions g_k and v_k are even, the DFT of these sequences will contain only real parts. As a result, the actual values of the DFT of the complex series $\widehat{V_k}$ will correspond to the amplitude-frequency characteristic (AFC) of the sequence g_k and the imaginary—amplitude-frequency characteristic of the series v_k.

It should be noted that the gravity effect of rows of spherical rectangular prisms with unit density located at latitudes φ and $-\varphi$, to pints with latitudes φ_0 and $-\varphi_0$ are equal. In this case, the calculation of the effects from such layers is represented as a convolution with the same kernel u. Then function δ_k in Eq. (17.2) can be represented as a complex function:

$$\delta_k = \delta_{Nk} + i \delta_{Sk}, \qquad (17.6)$$

where δ_N and δ_S—the density of the northern and southern belts, respectively. Multiplying the DFT of the function (17.6) with the discrete spectrum of spherical rectangular prism's effect, we obtain the DFT of spherical density row's effect. Based on the Fourier transform additivity property, the summation of effects can be performed in the frequency domain.

2.3. Having obtained the total effect spectra, we perform the inverse discrete Fourier transform for each latitudinal belt and, as a result, we obtain the resultant field.

Consider a model of a sphere that best describes the surface of the Eart. Its radius is 6371 km. We calculate the gravity field and potential of spherical layer with 1 km thickness with a density 1 g/cm³, the upper edge of which will coincide with the surface of the sphere. Effects calculates on the surface of the sphere. In this case, the potential V is 5340.79 m²/s² and gravity field g—83.82 mGal.

At Fig. 17.4 the differences of the calculated potential V and gravity field g values from the theoretical values with different model details are presented. The values of the calculated model are always less than theoretical values. The maximum discrepancy can be seen from the graphs in equatorial latitudes.

17 Calculation of Spherical Layer with Variable Density Gravity ...

It should be noted, that the greatest difference between the theoretical and calculated values of the potential is at polar points. In Fig. 17.4, these points are not marked. The differences of the potential V and the gravity field g from the theoretical values at the polar points are shown in Table 17.1.

Figure 17.5a presents a model of the effective density distribution in a spherical layer with a thickness 1 km on the earth's surface ($R_{up} = 6371$ km,

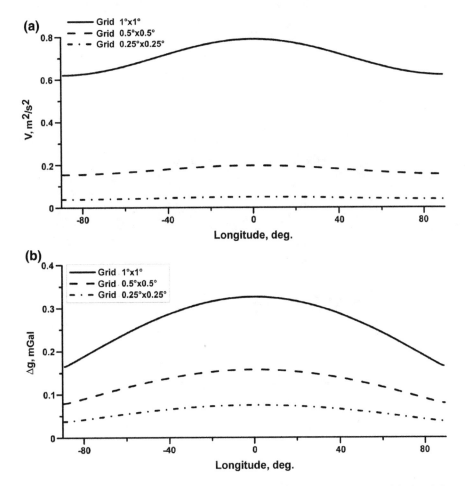

Fig. 17.4 The difference between the theoretical and calculated values of the potential (**a**) and the gravity field (**b**) for the spherical layer model ($R_{up} = 6371$ km, $R_{down} = 6370$ km, $\delta = 1$ g/cm^3)

Table 17.1 Deviations of the potential V and the attractive force g at polar points	Detailing	Difference V, m^2/s^2	Difference g, mGal
	1° × 1°	32.33	0.010
	0.5° × 0.5°	15.48	0.003
	0.25° × 0.25°	7.36	0.002

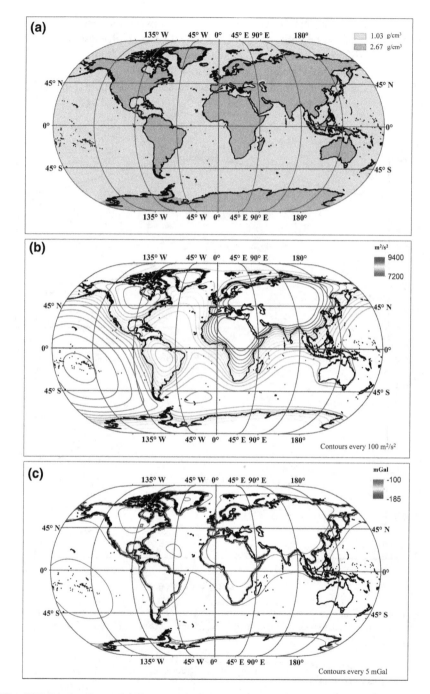

Fig. 17.5 Results of potential V and gravity field g of the spherical layer calculation: **a**—model of spherical layer; **b**—gravity potential; **c**—gravity field

R_{down} = 6370 km). The density in the continents is 2.67 g/cm³, and in the oceans 1.03 g/cm³. Figure 17.5b, c show the results of calculating the potential and the attractive force from a given model to the surface of a sphere.

Main results.

An algorithm is proposed for calculating the gravity potential and the gravity field from a spherical layer with variable density. This algorithm is characterized by high computational speed and high accuracy.

At the same time, it should be noted that the proposed algorithm has certain limitations. One of these limitations is due to the fact that calculations can not be made on land relief, i.e. the algorithm assumes the calculation of effects on the surface of the ocean (on the surface of the sphere).

References

Blokh Yu.I. (2009) Interpretation of gravity and magnetic anomalies / M. MGGRU. 2009. pp. 48–58. (http://sigma3d.com/pdf/books/blokh-2009.pdf). (in Russian).
Bulychev A.A., Gilod D.A., Krivosheya K.V. (2002) Construction of a three-dimensional density model of the ocean's lithosphere along the geoid heights field. // Bulletin of the Moscow University Vol. 4: Geology. 2002. №2. pp. 40–47. (in Russian).
Bulychev A.A., Krivosheya K.V., Melihov V.R., Zal'cman R.V. (1998) Calculation of the anomalous gravitational potential and its derivatives on the sphere. // Bulletin of the Moscow University Vol. 4: Geology. 1998. T.4. № 2. pp. 42–46. (in Russian).
Bychkov S.G., Dolgal' A.S., Simanov A.A. (2015) Calculation of gravity anomalies in high-precision gravimetric surveys. Perm'. UrO RAN. 2015. 142 p.
Kuznetsov K.M., Lygin I.V., Bulychev A.A. (2017) Algorithm of numeric direct gravity calculation of spherical layer with variable density // Geofizika. 2017. №1. pp. 22–27. (in Russian).
Starostenko V.I., Manukyan A.G., Zavorot'ko A.N. (1986) Methods for solving direct problems of gravimetry and magnetometry on spherical planets./ Kiev. Naukova dumka. 1986. 112 p. (in Russian).
Starostenko V.I., Pyatakov Yu.V. (2013) Solution of direct gravimetric problems for spherical approximating bodies. Algorithms. // Izv.Tomsky Polytechnical. Un-ty 2013. Vol. 322. № 1. pp. 28–34 (in Russian).
Strakhov V.N., Lapina M.I. (1986) Direct problems of gravimetry and magnetometry for homogeneous polyhedra. // Geofiz. Journ.1986. Vol. 8. № 6. pp. 20–31. (in Russian).
Hohlova V.V. (2015) Taking into account the sphericity of the Earth when processing gravimetric data. // Geofizika. № 5. 2015. pp. 59–64. (in Russian).
Hellinger S. J (1983). A method for computing the geoid height contribution of three-dimensional bodies within a spherical earth. // Geophysics. 1983. Vol. 48 № 12. p. 1664–1670.
Johnson L. R., Litehiser J. J. (1972) A Method for Computing the Gravitational Attraction of Three-Dimensional Bodies in a Spherical or Ellipsoidal Earth. // Geophysics. 1972. Vol. 77 № 35. p. 6999–7009.

Chapter 18
Possibility of Identification of Modeling in Complex Analysis Geological and Geophysical Data

O. M. Muravina, E. I. Davudova and I. A. Ponomarenko

Abstract The possibilities of the method of group accounting of arguments for the analysis geological and geophysical data are considered. Examples of the use of the method in the procedure for the formation of petrophysical models of sedimentary and crystalline rocks of Voronezh anteclise are given.

Keywords Method of group accounting of arguments · Petrophysics Identification modeling · Voronezh anteclise

Results of modeling by the Group Method of Data Handling (GMDH) are given. The method of group accounting of arguments is a method of inductive modeling, which allows to reveal the interrelations between geological and geophysical attributes. The theoretical foundations of the method were developed by academician A.G. Ivakhnenko in the eighties of the last century. The theory of the method is based on the ideas of synergetics, the main task of which is to identify the laws of the organization and the formation of order in complex systems. According to synergetics in many cases there is an opportunity among the set of interacting factors and variables to identify the most important processes and key factors. In this case, a complex dynamic linear ore nonlinear system can be described by a mathematical model with a small number of variables (Kapitsa et al. 1997).The best model is selected in the process of enumeration from a variety of options according to a certain criterion. This approach is consistent with one of the basic ideas of mathematical geophysics about the existence of a set of feasible solutions and

---

O. M. Muravina (✉) · E. I. Davudova
Voronezh State University, Voronezh, Russia
e-mail: muravina@geol.vsu.ru

E. I. Davudova
e-mail: eldina.davudova.94@mail.ru

I. A. Ponomarenko
Russian State University of Oil and Gas Named After I.M. Gubkin,
Moscow, Russia
e-mail: ponomarenko.i@gubkin.ru

© Springer Nature Switzerland AG 2019
D. Nurgaliev and N. Khairullina (eds.), *Practical and Theoretical Aspects of Geological Interpretation of Gravitational, Magnetic and Electric Fields*, Springer Proceedings in Earth and Environmental Sciences, https://doi.org/10.1007/978-3-319-97670-9_18

constructing of a final solution for this set (Strakhov 2000, Muravina 2012). To organize the generation of multiple models, we use a combinatorial algorithm with a multiple-row analysis of the variants. In Fig. 18.1 shows a scheme for generating models for 4 variable arguments (Muravina and Ponomarenko 2016).

The result of identification is a model equation, the structure and parameters of which are determined during the modeling process. The choice of optimal solutions in the GMDH is carried out using a system of criterial choice. The methods for calculating and applying the criteria as well as the choice of the basis function are determined by the requirements for the desired model. In GMDH, external criteria are applied, for the determination of which the method of dividing experimental data into learning and testing sequences is used. This allows to use the additional (external) information to select the optimal solution.

The most commonly used external criteria are quadratic forms, different functionals. The best model is chosen in accordance with the minimum of the external criterion. The advantage of these criteria is the understandable logic of the requirements for the model chosen. For example, the regularity criterion defines a model with a minimum forecast error, such a criterion can be used in prediction problems in the study of dynamic processes and in interpolation-extrapolation problems in the case of analysis of parameters of stationary geophysical fields. The minimum bias criterion is used in the problems of finding a trend. This criterion allows us to choose the least controversial (unbiased) model on two disjoint sets of input data (Ivakhnenko 1982; Muravina 2012).

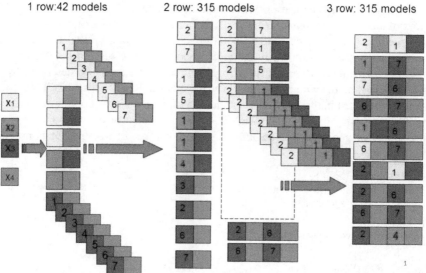

Fig. 18.1 Scheme of model generation for 4 variable-arguments in the combinatorial multi-row algorithm of GMDH. Numbers in rectangles denote variants of model equations

Organization of the structure of the initial data (numerical matrix), their method of parametrization and obtaining results in the form of an identification structural-parametric model of the geological environment or process allows the method to be adapted to existing automated geophysical data processing systems.

The resultant structural-parametric model is the equation of the relationship between geological parameters and various inhomogeneous attributes of geophysical fields, which makes it possible to use the capabilities of the entire complex of geophysical data. Such an equation can be used in the classification analysis when solving the pattern recognition problem. In this case, a certain range of values of the dependent variable may correspond to a certain class of studied objects. If the dependent variable is also a geophysical field parameter or a petrophysical characteristic, the model equation will reflect the relationships between these parameters that can be used in a further complex interpretation.

Let's consider some results of practical application of the considered method at the decision of various problems connected with processing and the analysis of the geologo-geophysical information. Identification modeling by the method of group accounting of arguments was used in the procedure for the formation of petrophysical models of sedimentary and crystalline rocks of the Voronezh anteclise according to the summary petrophysical definitions of the core of wells drilled in the region.

Identification equations allowed to take into account the influence on the petrophysical characteristics of such factors as the depth and lithological type of the rock, to reveal stochastic relationships between different petrophysical parameters. As an example, we give an equation relating the propagation velocity of elastic waves with the density and depth of sampling. The equation was obtained for sedimentary rocks of different ages and lithological type. In Fig. 18.2. graphs of experimental and model values of the velocity of propagation of elastic oscillations in the depth interval from -20 to -10 m are shown.

The results obtained for sedimentary rocks of different ages were used in the formation of a generalized spatial structural model of the sedimentary cover of the region, which is necessary for carrying out a complex interpretation of geophysical fields (Muravina 2013; Muravina and Glaznev 2013, 2014).

When working with petrophysical data relating to the crystalline rocks of the upper part of the basement, equations of identification of stratigraphic and intrusive-metasomatic formations of the Early Precambrian of the Voronezh crystalline massif were obtained. The analysis of petrophysical data made it possible to solve such problems as: elimination of emissions in experimental values, classification of data, establishment of interrelation between petrophysical characteristics (Muravina et al. 2016, 2017).

Consider the results of the application of the GMDH to solve the problem of identifying the petrophysical parameters of the mihailovskaya series. The stratified formations of the mikhaylovskaya series refer to the upper Archean (AR_2mh) and are subdivided into two suites: the lower alexandrovskaya (AR_2al) and the upper lebedianskaya (AR_2lb). The representativeness of a sample of petrophysical characteristics for the rocks of the alexandrovsky suite (petrophysical determinations for

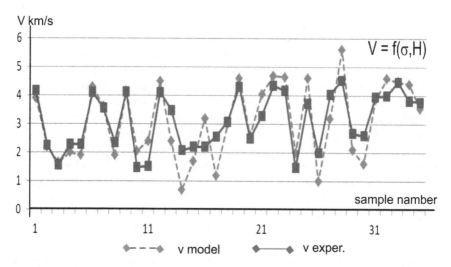

Fig. 18.2 Results of simulation of elastic wave velocity for sedimentary rocks of Voronezh anteclise

1396 samples) made it possible to obtain empirical model dependencies for various petrological rock types in order to identify data referred to the undivided mikhaylovskaya series. The results of identification of the rocks of the undivided mikhaylovskaya series the results are shown in the Fig. 18.3. The model equation obtained for the petrophysical parameters of the plagiomigmatites of the alexandrovskaya suite, relates the rock density with the polarizability and depth of sampling. The reliability of the model is confirmed by the fact that for 82% of the samples the discrepancy between the model experimental values of the density lies in the interval $\pm$ 0.02 g/cm^3. The application of the equation to unseparated rocks of the mikhailovskaya series led to the following results: for 6 of 18 samples of plagiomigmatites of the mikhailovskaya series, the absolute error of the density estimate does not exceed $\pm$ 0.02 g/cm^3. This allows us to classify these samples as rocks of the alexandrovskaya suite (Muravina et al. 2016).

Let us consider an example of the application of GMDH for detecting emissions of experimental data. In Fig. 18.4 shows graphs of the experimental and model values of the velocity of elastic waves for the crystalline shales of the alexandrovskaya suite (AR$_2$al) of the Voronezh crystal massif. Model values were calculated in accordance with the received identification equation relating the speed of elastic waves with the density and depth of sampling.

Preliminarily, the most probable limits of variations of the velocity values were determined from the values of 5% and 95% of quantiles from 4.7 to 6.8 km/s. The standard deviation of the sample was 0.6 km/s. As it appears from the figure, there are three variants of emissions in the data. In the first case (number 1 in the Fig. 18.4), both experimental and model values go beyond the limits of the accepted range of the scatter of data. However, the difference between them does not exceed the standard deviation value, which indicates that these values

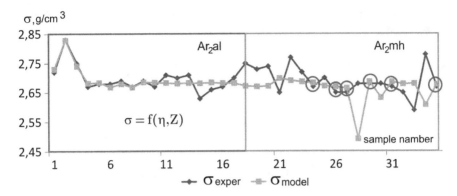

Fig. 18.3 Results of identification analysis of plagiomigiatae of the undivided mikhaylovskaya series

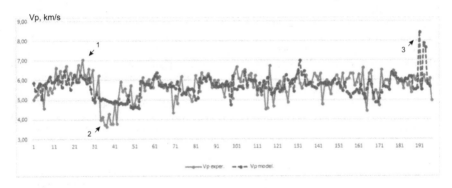

Fig. 18.4 Results of the identification analysis. The numbers indicate the types of data releases. The frame shows the limits of the allowed range of data

correspond to the identification equation and not be considered as emissions. In the second case, a reliable range of experimental data is observed (samples № 36, 39, 40—number 2 in Fig. 18.4), the experimental values of the velocity are lower than the maximum permissible values, which are established by the value of a quantile of 0.05 order. However, model values do not exceed the range limits. In the third case, for the samples № 194 and № 195, there are significant deviations in the model velocity values. The reason for this behavior of the model curve becomes clear when studying the arguments of the identification equation: samples 194 and 195 have abnormal density values uncharacteristic for the shales of the alexandrovskaya suite (Muravina et al. 2017).

A high degree of geophysical study of the region (gravimetry, magnetometry, electrical survey data) and developed petrophysical models allowed the use of identification analysis by the method of group accounting of arguments for complex interpretation of geophysical and petrophysical data (Glaznev et al. 2016; Muravina 2016).

Acknowledgements The main results of the research were obtained in the framework of scientific research under the RFBR grant No. 18-05-00226.

References

Glaznev, V. N, M. V. Mints and O. M. Muravina (2016). Density modeling of the central part of the East European platform. Vestnik KRAUNTS, Series of "Earth Sciences" T. 29, p. 53–63.

Ivakhnenko, A G (1982). Inductive method of self-organization of models of complex systems: Kiev, Nauk. Dumka, 296 p.

Kapitsa, SP, SP Kurdyumov and GG Malinetsky (1997). Synergetics and forecasts of the future: M., Nauka, 39p.

Muravina, O.M. (2012). The method of group accounting of arguments in the analysis of geophysical data. Geophysics, № 6, p. 16–20.

Muravina OM Identification analysis of petrophysical characteristics of sedimentary cover rocks of Voronezh anteclise (2013). Vestnik KRAUNTS, Series of "Earth Sciences" T. 29, p. 20–25.

Muravina, OM, Glaznev VN (2013) Some results of a statistical analysis of the petrophysical parameters of sedimentary cover rocks of the Voronezh anteclise. Collection of materials of the XIV International Conference "Physical-Chemical and Petrophysical Studies in Earth Sciences". Moscow, pp. 190–193.

Muravina, O. M., Glaznev V. N. (2014). Structural and parametric models of petrophysical parameters of the sedimentary cover of the Voronezh anteclise. News of Siberian Branch of the Russian Academy of Sciences , 44, № 1, pp. 81–87.

Muravina, O. M., E.I. Davudova and I.A. Ponomarenko (2016). Separation of the rocks of mikhailovsky series of the Voronezh crystalline massif for petrophysical and spatial parameters. Collection of materials of the XVII International Conference "Physical-Chemical and Petrophysical Studies in Earth Sciences", Moscow, pp. 235–237.

Muravina, O. M., E.I. Davudova and I.A. Ponomarenko (2017). Use of the method of group accounting of the arguments for detecting emissions in experimental petrophysical data. Collection of materials of the XVIII International Conference "Physical-Chemical and Petrophysical Studies in Earth Sciences". Moscow, pp. 200–203.

Muravina, O. M. (2016). Density model of the crust of the Voronezh crystalline massif. Bulletin of VSU, Series Geology, No. 1, pp. 108–114.

Muravina, O. M. and I.A. Ponomarenko (2016). Program implementation of the group account method for arguments in identification modeling of geological-geophysical data. Bulletin of VSU, Series Geology, No. 2, pp. 107–110.

Strakhov, VN, (2000). Geophysics and Mathematics. Methodological foundations of mathematical geophysics. Geophysics, №1, pp. 3–18.

Chapter 19
"Native" Wavelet Transform for Solving Gravimetry Inverse Problem on the Sphere

N. Khairullina (Matveeva), E. Utemov and D. Nurgaliev

Abstract We present a novel algorithm to interpret the geopotential data obtained on the surface of a sphere. Suggested method is based on CWT with so called "native" basis. Computational experiments show that location and depth of synthetic causative sources are uniquely determined by the proposed method. Comparison presented results with seismic data demonstrates a good agreement.

Keywords Gravimetry · Inverse gravity problem · Gravitational potential Sphere · Wavelet transform

Introduction

The determination of the causal potencial sources gravitational and magnetic anomalies obtained at the Earth's surface is the global problem. It doesn't lose relevance currently. Among a large number of well-known approaches for solving this problem we can distinguish the category based on using of the homogeneity properties potential fields of elementary sources (e.g., point mass, dipoles and so on). Further, this category may be devided into two groups. The first group includes the approaches such that they implement a search of causal sources in the time domain (e.g., Euler and Werner deconvolution and their modifications (Werner 1953; Hood 1965; Thompson 1982; Reid et al. 1990). The second group's technique use the time-frequency space of continuous wavelet transform for the same purpose (Moreau et al. 1997, 1999; Sailhac et al. 2000; Gibert and Pessel 2001).

Above mentioned approaches of both groups are useful tools to express-interpretation of the potential anomalies, but they don't give clear solutions in case of high interference of potential fields due to destruction of their homogeneity properties. There is problem in case of recovering deep sources. In our opinion, the approaches from second group is more attractive because it presents several new

N. Khairullina (Matveeva) (✉) · E. Utemov · D. Nurgaliev
Kazan Federal University, Kazan, Russian Federation
e-mail: limonich@mail.ru

features with respect to other methods but eventually the "weak point" approaches of both groups lies in phenomenon of high interference of real geophysical potential fields.

The technique of recovery causative sources are offered in this study in the wavelet domain. (Utemov and Nurgaliev 2005; Matveeva et al. 2014, 2015) This approach allows to obtain satisfactory results of potential fields obtained on the spherical surfaces.

Mathematical Framework

The family of "native" wavelets has the form (Utemov and Nurgaliev 2005; Matveeva et al. 2014, 2015):

$$\psi_{(n)}(h,x) = \frac{2^{n-3}h^{n-2}}{(n-2)!\pi^2 f} V_{(n)}(h,x), n > 1 \quad (19.1)$$

where x is horizontal coordinates, h is vertical coordinate (depth), f is gravitational constant, function $V_{(n)}(x, h)$ is n-th vertical derivative of the gravitational potential of a two-dimensional point mass.

In three-dimensional case this formula turns into

$$\psi_{(n)}(x,y,h) = \frac{2^{n-1}h^{n-2}}{(n-2)!f} V_{(n)}^{3D}(x,y,h), n > 1 \quad (19.2)$$

where superscript «3D» emphasizes the three-dimensional of the model.

The first radial derivative of the gravitational potential of a two-dimensional point mass on 2D-sphere has the follow form

$$V_r(\gamma) = 2f \frac{R - H\cos(\gamma)}{H^2 + R^2 - 2HR\cos(\gamma)} \quad (19.3)$$

where R is radius of the sphere, γ—zenith angle, h is depth of point mass, $H = R-h$. Function (19.2) is even and $2\pi R$-periodic, therefore its Fourier transform is

$$S_{V_r}(\omega) = \frac{2G}{R\pi} \int_0^{2\pi R} \frac{[1 - q\cos(\frac{t}{R})]e^{i\omega t}}{q^2 + 1 - 2q\cos(\frac{t}{R})} dt = \frac{2G}{\pi} \int_0^{2\pi} \frac{[1 - q\cos(y)]e^{iR\omega y}}{q^2 + 1 - 2q\cos(y)} dy, \quad (19.4)$$

here $q = (R-h)/R$. In order to calculate this integral is need the theory of residues. Finally, there was found following ultimate result:

$$S_{V_r}(\omega) = 2fe^{-R|\omega|\ln(\frac{1}{q})} = 2fe^{-h_k|\omega|}, \quad (19.5)$$

where function

$$h_k = -Rln\left(\frac{1}{q}\right) = -Rln\left(1 - \frac{h}{R}\right) \tag{19.6}$$

might be named "seeming" depth. Here we should compare the result (19.6) with formula of Fourier spectrum of gravity anomaly of two-dimensional point source in planar case:

$$S_{V_z}(\omega) = 2Ge^{-h|\omega|} \tag{19.7}$$

Comparison of (19.5) and (19.7) shows that radial derivative of the gravitational potential of a two-dimensional point mass on 2D-sphere and its planar analogue have the same formula of Fourier spectrum taking but different parameters of depth into account formula (19.6).

This result allowed to consider a data obtained on 2D-spherical surface as a data on the plane but with other depth parameters of causative sources, consequently if we perform wavelet transform with "native" basis we will get solution of inverse problem on 2D-sphere as well as on plane.

Approaches that have been offered for 2D spherical case [Eqs. (19.4)–(19.7)] are fit for 3D spherical case except that instead of the fields radial and vertical derivative of gravitational potential we have to use just the field of the gravitational potential. In addition, it is important that the Eq. (19.7) is still in working order in 3D spherical case.

In 3D spherical case we encountered with a few mathematical issues linked with features of integration on the sphere and existence of poles, but eventually we have solved their. As simple example, we demonstrate the results of the recovery two

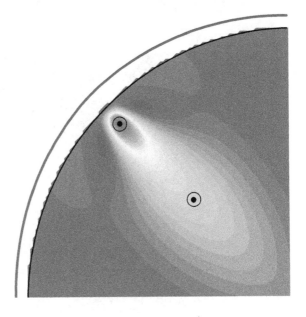

Fig. 19.1 Recovery of location of two 3D point sources one under another. Red line corresponds gravitational potential

causal sources by their total gravitational potential obtained on the sphere. It is shown in Fig. 19.1.

In Table 19.1 we present the numerical values of parameters of initial and reconstructed sources.

Table 19.1 Comparison of the initial and received parameters of causal sources

Longitude, degree		Latitude, degree		Depth, km		Magnitude, 10^5 kg	
Initial	Received	Initial	Received	Initial	Received	Initial	Received
135	135	45	44.95	500	500	1	0.985
135	135	45	45.14	3000	3000	100	100.62

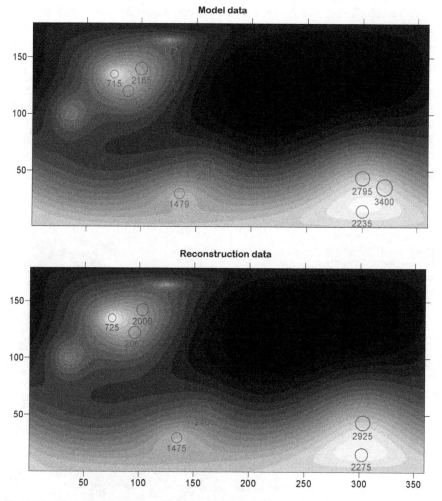

Fig. 19.2 Recovery of location of 10 random 3D point sources

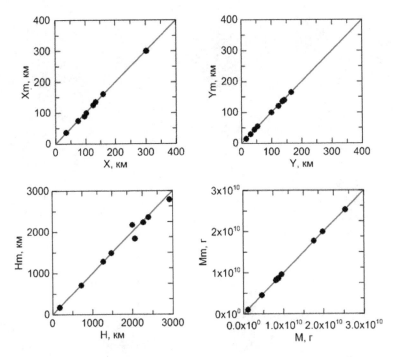

Fig. 19.3 Comparison of the initial and received parameters of 10 random causal sources

Now let's go to more realistic synthetic example. In the experiment we use 10 random sources and place them (Fig. 19.2).

As one can see the positions of the reconstructed sources are well correlated with the ones of the true (model) sources. (Figure 19.3)

Analysis of Gravity Field of the Earth

The results of wavelet transform have value in itself because wavelet-slices with different scale parameters allow to reveal some hidden features of the gravity field, especially in case of our approach because wavelet coefficients have the dimension of density (as a solution of the inverse problem).

We have constructed some different wavelet-slices. For example the slice corresponding to 50 km reveals lithospheric heterogeneity of density. Thus, one can see such well-known geological structures as mid-ocean ridges and transform faults, continental reefs, convergent boundaries, folded regions, hot spots due to mantle plumes (Hawaii, Monarch Ridge, etc.). If scale parameter represent slice at 150 km the inhomogeneities of density of the upper asthenosphere are shown such as convergent boundaries and subduction zone. The mantle-core boundary and the top of outer core are obtained for slice corresponding to 3500 km depth (Fig. 19.4).

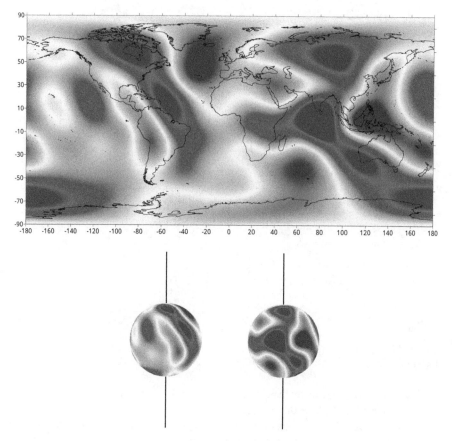

Fig. 19.4 CWT slice corresponding to a depth of 3500 km

It might be assumed that inhomogeneities of density in this slice correspond the thermodynamic state of the substance of the outer core. Interest is the presence of long strips both negative and positive anomalies of density, which form an angle to the axis of rotation of the Earth approximately 20–30°. Note also the huge negative anomaly of density in region of the Indian Ocean.

Conclusion

Our method has several apparent advantages. At first, the "native" wavelet transform allows getting a formal solution of the inverse gravity problem, therefore wavelet transform has clear physical meaning. At second, our approach has pointed easy way to recover of the parameters of the causative sources.

Using this technique, there was determined the distribution of causative sources of the Earth's gravity anomalies. The analysis using our approach allows to investigate plenty of well-known geological structures. For calculating were used gravimetric data of International Centre for Global Earth Models (ICGEM) (http://icgem.gfz-potsdam.de/ICGEM).

References

Gibert D. and Pessel M. (2001), Identification of sources of potential fields with the continuous wavelet transform: Application to self-potential profiles. Geophys. Res. Lett., 28(9), 1863–1866.

Hood P. (1965), Gradient measurements in aeromagnetic surveying: Geophysics, 30, 891–802.

Matveeva N., Utemov E., Nurgaliev D (2015) "Native" wavelet transform for solution inverse problem of gravimetry on the spherical manifold. International Multidisciplinary Scientific GeoConference Surveying Geology and Mining Ecology Management, SGEM, v.3, p. 1067–1074, 2015.

Matveeva N., Utemov E., Nurgaliev D (2014) Solutions of inverse problem of gravimetry on the sphere using "native" wavelet transform. International Multidisciplinary Scientific GeoConference Surveying Geology and Mining Ecology Management, SGEM, v.1, p. 621–628, 2014.

Moreau F., Gibert D., Holschneider M. and Saracco G. (1997), Wavelet analysis of potential fields, Inverse probl., 13, 165–178.

Moreau F., Gibert D., Holschneider M. and Saracco G. (1999), Identification of sources of potential fields with the continuous wavelet transform: Basic theory. J. Geophys. Res., 104 (B3), 5003–5013.

Reid A.B., Allsop J.M., Granser H., Millett A.J. and Somerton I.W. (1990), Magnetic interpretation in three dimensions using Euler deconvolution: Geophysics, 55, 80–91.

Sailhac P., Galdeano A., Gibert D., Moreau F. and Delor C. (2000), Identification of sources of potential fields with the continuous wavelet transform: Complex wavelets and application to aeromagnetic profiles in French Guiana, J. Geophys. Res., 104 (B8), 19455–19475.

Thompson D.T. (1982), EULDPH – A new technique for making computer-assisted depth estimates from magnetic data, Geophysics, 47, 31–37.

Utemov E.V. and Nurgaliev D.K. (2005), Natural Wavelet Transformations of Gravity Data: Theory and Applications, Izvestia Physics of the Solid Earth, 41(4), 88–96.

Werner S. (1953), Interpretation of magnetic anomalies at sheet-like bodies. Sver. Geol. Unders. Ser. C. C. Arsbok, 43 (06).

Part III
Deep Structure Studying

Chapter 20
Earth's Crust Magnetization Model of the Nether-Polar and Polar Urals

N. Fedorova, L. Muravyev and A. Roublev

Abstract We performed a study of the anomalous magnetic field features of the Nether-Polar and Polar sectors of the Urals (48–72° E and 60–68° N). We identified and mapped magnetic anomalies due to the Earth's crust layers. The calculated local magnetic anomalies map allows allocating basic-ultrabasic massifs in the upper parts of the basement within the sedimentary basins. After regional magnetic anomalies interpretation and deep seismic sounding data along profiles, located in the study area we build an Earth's crust structure model. Comparison of the deep structure of the cross-sections produced by independent interpretation methods using seismic and magnetic data has enabled us to extract two layers with different magnetic properties from the consolidated crust. Top layer of the Earth's crust does not make a significant contribution into regional magnetic field and is characterized by a low magnetization (less than 0.3 A/m). Within this layer we identified magnetized local sources. The lower layer has greater crustal magnetization. The magnetic data interpretation shows that the crust's basalt layer magnetization value is 3–4 A/m. The average depth to the top surface of the layer is 18–20 km. The resulting parameters were used for three-dimensional modeling. We considered a model with uniform magnetization directed along the modern geomagnetic field. Thus, we built the upper surface of the magnetized layer, which allowed clarifying mafic layer in the space between the deep seismic sounding profiles. We found that at the Northern, Circumpolar and Polar Urals basalt layer plunged to a considerable depth of 26–30 km.

Keywords Regional magnetic anomaly · Magnetization · The earth's crust Urals region

---

N. Fedorova (✉) · L. Muravyev · A. Roublev
Institute of Geophysics Ural Branch of RAS, Ekaterinburg, Russia
e-mail: nataliavf50@mail.ru

L. Muravyev
e-mail: mlev@igeoph.net

L. Muravyev
Ural Federal University, Ekaterinburg, Russia

© Springer Nature Switzerland AG 2019
D. Nurgaliev and N. Khairullina (eds.), *Practical and Theoretical Aspects of Geological Interpretation of Gravitational, Magnetic and Electric Fields*, Springer Proceedings in Earth and Environmental Sciences, https://doi.org/10.1007/978-3-319-97670-9_20

Introduction

At present, when studying the deep structure of the Earth's crust and upper mantle, geophysicists trends from constructing two-dimensional cross-sections along profiles to creating three-dimensional models of large regions. The initial volumetric model is based on information of the seismic profiles and geotraverses. For the Circumpolar sector of Eurasia within the geographical coordinates of 60–68° N, 48–72° E the cross-sections for ten DSS profiles were constructed by two-dimensional seismic tomography (Ladovskiy et al 2016). On their basis, new maps-schemes of the main boundaries of the Earth's crust have been created: the basement, the Moho and the basalt layer (Fedorova et al. 2017a). Tectonically, the region covers the northeastern part of the East European Platform, the Timan-Pechora Plate, the Polar and Subpolar Urals, and the northern segment of the West Siberian Plate adjoining the east.

In this paper we present the investigation results the anomal magnetic field of this region's lithosphere. With the help of modern computer technologies developed at the Institute of Geophysics of the Ural Branch of the Russian Academy of Sciences (Martyshko et al. 2016), the anomalies from the sources located in the upper layers of the Earth's crust and the regional component of the geomagnetic field are identified, its rectangular components are calculated, and a three-dimensional magnetic model of the "basite" layer is constructed.

Local Anomalies and Regional Component Separation

Most of the study area is covered by sedimentary cover, whose power within the cavities reaches 8–12 km. The observed magnetic field contains contribution from all sources located in the upper lithosphere. To distinguish the anomaly from the layers of the Earth's crust, a method based on high-level recalculations was used (Martyshko and Prutkin 2003). In the large areas study it is necessary to handle large data amounts, resulting in a significant uniprocessor computing time. The developed computer technology based on parallel computing on multiprocessor computer systems significantly reduces the calculation time (Martyshko et al. 2012, 2014). The developed technology application results and map of anomalies from different layers of the Earth's crust for the study area are published in articles (Fedorova et al. 2015, 2017b). These data were used to construct the forecast map of basite-ultrabasite massifs located under sedimentary rocks in the upper part of the basement to a depth of 5 km, (Fedorova and Rublev 2016). Using this technique, anomalies were also identified from magnetized massifs in the deeper basement layers to the depth $H = 20$ km. Figure 20.1 shows an anomalous magnetic field map and anomalies created by local sources in the upper layer to a depth of 5 km, in a layer from 5 to 20 km, and the regional component.

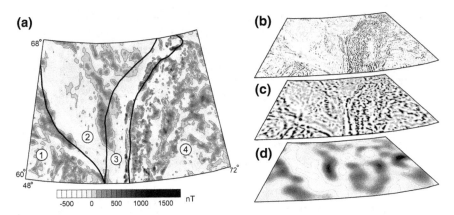

Fig. 20.1 Anomalous magnetic field map (**a**). The anomalies separation from sources in different layers of the Earth's crust: local anomalies from the upper layer to a depth of 5 km (**b**); anomalies from sources located in the layer at a depth of 5—20 km (in); regional anomalies (**d**). Maps are shown in perspective projection. The tectonic structures boundaries are shown by black lines: 1—East European platform, 2—Timano-Pecherskaya plate, 3—Ural fold system, 4—West Siberian plate

The long-wave regional anomalies intensity varies from −250 to +250 nT. In the western part of the map (Fig. 20.1), a large negative anomaly is the most notable feature, it occupies half of the Timan-Pecherskaya plate territory. Its transverse dimension reaches 350 km, and the intensity is −250 nT. In the southern part of the area, this anomaly merges with a chain of negative anomalies extending within the Ural fold system. Despite the fact that within the Ural Mountains there are crystalline rocks (gabbro, serpentinites, diorites), which is located close to the Earth's surface and create an intense local positive anomalies, the long-wave part of the magnetic field consists of negative anomalies. The Ural Mountains stretch more than 2300 km and is divided into South, Middle, North, Polar and Subpolar sector. Previously, negative regional anomalies were identified over the Southern, Middle and Northern ranges of the Ural Mountains. As can be seen on the map (Fig. 20.1), the presence of negative regional anomalies of the magnetic field is typical for the Polar and Nether-Polar sectors of the Urals, which undoubtedly indicates a similarity in the deep structure features of all sectors of the Ural orogen.

The main deep rocks magnetization carriers are minerals of the titanomagnetite series, and primarily magnetite. Therefore, the lower limit of the magnetoactive layer of the lithosphere can be limited by the Curie isotherm 580° C or along the Moho boundary, in those regions where the temperature at the base of the Earth's crust is lower than the magnetite's Curie temperature. It is difficult to estimate the magnetoactive layer thickness by geothermal data because of the fairly rare and uneven network of heat flow measurements, and the lack of data in the northern part of the territory. Due to ambiguity, the geothermy inverse problems solution obtaines significantly different estimates of the temperature conditions in the Earth's crust of the Ural region. The difference in estimates reach 20–30 km (Fedorova and

Kolmogorova 2013). The Earth's crust thickness varies in large ranges from 35 to 57 km, with an average value of about 40 km. Moreover, a significant increase in the Earth's crust thickness within the Ural fold system is marked by negative regional anomalies. Most likely, rocks near the base of the Earth's crust have weak magnetic properties.

For Nether-Polar Eurasian sector statistical studies have shown that the regional magnetic anomalies cannot be explained by changes in the basement topography and the Moho, as well as variations in the consolidated crust and basalt layer thickness. The correlation coefficient between regional magnetic anomalies intensity and variations in the crust boundaries relief reaches the maximum value for the basalt layer surface. It is shown that in the study area the lower limit of the magnetically active layer of the lithosphere is located significantly higher Moho and the main contribution to regional magnetic anomalies introduces the top surface of the basalt layer (Fedorova et al. 2017a).

Long-Wave Magnetic Anomalies Sources Modeling

Interpretation of magnetic anomalies, as a rule, is based on models consisting of spatially limited magnetized blocks. The background magnetization influence of the Earth's crust layers is not taken into account. Therefore, the interpretation of negative magnetic anomalies often causes difficulties. When using a layered model, negative anomalies can easily be explained by a reduction in the power of the magnetized layer.

Studies have shown that the upper layer of the crystalline Earth's crust is characterized by a low average magnetization and does not make a significant contribution to the regional magnetic field. It is possible to neglect the background magnetization within the this layer limits. The lower basaltic layer of the crust is magnetized significantly more strongly. Based on the two-dimensional modeling results, we estimated this layer average magnetization $\approx$ 3 A/m (Fedorova and Kolmogorova 2013). The upper boundary of the layer coincides quite well with the 6.5 km/s velocity level position in the corresponding seismic sections, and it's average depth is 20 km. We used these results for the three-dimensional interpretation of the regional component of the magnetic field and the basal layer relief reconstruction. To solve the inverse magnetometry problem, we developed a program wich based on the modified local corrections method (Martyshko et al. 2010).

The method is based on the assumption that the change in the field value at a certain point is most affected by the change in the part of the surface closest to a given point, which is the boundary between two layers with different physical properties. This iterative method does not use non-linear minimization, which allows you to quickly solve large-scale problems. So, the initial field, set on 100×100 pixels grid, is reconstructed with a relative low error (up to 0.01%) for 300 iterations and the calculation process takes several minutes.

The preliminary interpretation results of the Timan-Pechora region regional anomalies are given in the paper (Fedorova et al. 2013). The depth to the magnetized lower layer of the crust varies from 14 to 30 km. The result obtained is in good agreement with 2D modeling results for DSS profiles. In the Timano-Pechora regional negative magnetic anomaly zone, the magnetized layer boundary in the lower crust is submerged up to 24–30 km, and up to 15–16 km rises on the northeastern edge of the Russian plate. The maximum rise to 14 km occurs within the Pre-Ural trough.

To calculate the vertical component of the anomalous magnetic field Z from it's modulus ΔT_a distribution we developed the algorithm based on the approximation of data ΔT_a as a set of fields from singular sources. This procedure allows us to calculate the vertical component of the magnetic field Z from the selected source distribution. As a model source, a set of rods uniformly magnetized along its axis was used. The algorithm is implemented using parallel computing technology on the NVidia graphics processor (Byzov et al. 2016). For the region, the vertical component Z is determined and reduction to pole is performed: the values of Z_v for the vertical magnetization of all sources are calculated. Despite the fact that in the circumpolar region the geomagnetic field direction is close to vertical (inclination 74–80°), however, the discrepancies between the anomalies ΔT_a and Z_v are significant. Z_v anomalies epicenters are shifted to the north relative to the epicenters of the ΔT_a anomalies, the distance sometimes reaching 20–40 km.

Results and Discussion

For the Nether-Polar and Polar sectors of the Ural region, the result of modeling the magnetized layer is given for a rectangular section of 800 × 800 km. The basaltic layer surface, calculated for the magnetization of 3 A/m, is shown in Fig. 20.2.

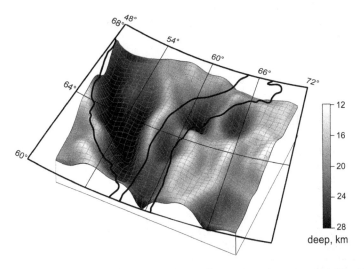

Fig. 20.2 Model of the upper boundary surface of the lower magnetized layer of the Earth's crust; tectonic structures boundaries are shown on a horizontal surface at a depth of 20 km

The surface relief varies from 14 to 30 km. On the eastern margin of the Timano-Pechora plate within the Predural edge of the trough there is an ascent up to 14 km. Under the Ural mountains there is a sharp immersion of the border. The dive has the form of an extended ravine and reaches a depth of 26 km in the Subpolar Urals and 30 km in the Polar Urals. In the lithosphere of Western Siberia, there are more smooth changes in the relief than in the Urals, and the depth varies from 14 to 27 km.

Conclusion

Regional magnetic anomalies contain important information on the deep structure of the Earth's crust. The most contrasting lithosphere layers in the according to the elastic physical properties, the velocities of seismic waves and density, are the boundary between the sedimentary layer and the crystalline basement, as well as the Moho boundary separating the Earth's crust and the upper mantle. Inside the crust, changes in these parameters during the transition from layers containing predominantly acid rocks to layers consisting of basic and ultrabasic rocks may be insignificant. In this case, the contrast of the magnetic properties can reach large values. Regional magnetic field anomalies interpretation makes it possible to obtain the magnetized layer boundary relief inside the crust. For the circumpolar sector of Eurasia on the cross-sections for the DSS profiles, the magnetized layer boundary coincides quite well with 6.5 km/s velocity level position. The obtained results give a spatial location of the lower high-speed boundary and more dense basite layer of the Earth's crust.

Negative regional magnetic anomalies have been identified along the entire Central Ural anticlinorium. As a result of their interpretation in the Subpolar and Polar Urals, it was found that in the Earth's crust there is a significant thickness reduction of the magnetized basal layer and the dipping of its upper surface to a depth of 26–30 km.

References

Byzov D., Muravyev L. and Fedorova N. (2016). The Approximation of Anomalous Magnetic Field by Array of Magnetized Rods. AIP Conference Proceedings 1863(1):560051. https://doi.org/10.1063/1.4992734.

Fedorova N.V., Kolmogorova V.V. (2013). Models of distribution of the magnetization and the longitudinal wave velocity of the Earth's crust in Timan-Pechora and Northern Ural regions. Lithosphere (5), 160–169 (in Russian).

Fedorova N.V., Kolmogorova V.V., Roublev A.L., Tsidaev A.G. (2013). The magnetic model of the North-Eastern region of Europe. Geophysical Research, 14 (2), 25 – 37 (in Russian).

Fedorova N.V., Martyshko P.S., Gemaidinov D.V., Rublev A.L. (2015). Computer technology for highlighting the magnetic anomalies from the deep layers of the Earth's crust. XIVth International Conference – Geoinformatics: Theoretical and Applied Aspects. 11-14 May 2015, Kiev, Ukraine. https://doi.org/10.3997/2214-4609.201412426.

Fedorova N.V., Muravyev L.A., Kolmogorova V.V. (2017a) Statistical estimates of the relationship of regional magnetic anomalies with the seismic boundaries of the Urals region nether polar sector. Ural'skij geofizicesij vestnik (1), 47–51. (in Russian).

Fedorova N.V., Rublev A.L. (2016). Ultramafic-mafic massif mapping in the upper horizons of the basement polar sector of Eurasia. 16th International Multidisciplinary Scientific GeoConference SGEM 2016, Conference Proceedings, June 28 - July 6, 2016, Book1 Vol. 3, 703–710 https://doi.org/10.5593/sgem2016/b13/s05.089.

Fedorova N.V., Rublev A.L., Muravyev L.A., Kolmogorova V.V. (2017b). Magnetic anomalies and model of the magnetization in the Earth's crust of circumpolar and polar the sectors of Ural region. Geofizicheskiy zhurnal, 39(1). 111–122. https://doi.org/10.24028/gzh.0203-3100.v39i1.2017.94014.

Ladovskiy I.V., Martyshko P.S., Fedorova N.V., Kolmogorova V.V. (2016) XP constructing three-dimensional density-velocity model on velocity sections DSS. Ural'skij geofizicesij vestnik (2), 108–119 (in Russian).

Martyshko P.S., Fedorova N.V., Akimova E.N., Gemaidinov D.V. (2014). Studying the Structural Features of the Lithospheric Magnetic and Gravity Fields with the Use of Parallel Algorithms Izvestiya. Physics of the Solid Earth, 50 (4), 508–513. https://doi.org/10.1134/s10693513 14040090.

Martyshko P.S., Fedorova N.V., Gemaidinov D.V. (2012). Study of the anomalous magnetic field structure in the Ural region using parallel algorithms. Doklady Earth Sciences, 446 (1), 1102–1104. https://doi.org/10.1134/s1028334x12090127.

Martyshko P.S., Ladovsky I.V., Fedorova N.V., Tsidaev A.G., Byzov D.D. (2016) Theory and methods of complex interpretation of geophysical data. Ekaterinburg, 94 p.

Martyshko P.S., Prutkin I.L. (2003). Technology for separating the gravity sources by the depth, Geofizicheskiy zhurnal, 25 (3), 159–168 (in Russian).

Martyshko P.S., Rublev A.L., Pyankov V.A. (2010). Using local corrections technique to solve structural magnetometry problems. Geophysics (4), 3–8 (in Russian).

Chapter 21
Computer Modeling of Lateral Influence of the Ladoga Anomaly (Janisjarvy Fault Zone) on the AMT Sounding Results

A. A. Skorokhodov and A. A. Zhamaletdinov

Abstract In the northern Ladoga Lake area in 2015 and 2017 the audio-magnetotelluric (AMT) soundings were performed. The main task of AMT was to study the deep structure of the high conductive anomalies discovered before by the di-rect current (DC) profiling. A special AMT experiment has been made over the Janisjarvi anomaly. This zone of anomalous conductivity is situated on the contact between the Proterozoic and the Archaean rocks. Several AMT sounding sites were situated symmetrically on the both sides of anomaly. Data processing al-lowed to make a conclusion about the quasi-two-dimensionality of the medium under investigation. The computer 2D modeling has been made with the use of grid method. Results of modeling showed that Janisjarvi conductive zone has sub vertical position. Its continuation to the depth is restricted in the range of 0.5 km. The influence of Janisjarvi conductive zone creates effects of fictitious conductive layers on results of AMT soundings. The depth of fictitious conductive layers increase from units to 10–20 km depending to increase of the distance between the center of the Janisjarvi conductive zone and AMT sounding sites. It means that all AMT soundings should be made with taking into account the influence from the steeply dipping uppermost conductive layers.

Keywords Audiomagnetotellurics · Lateral influence · Conductive zone Simulation

---

A. A. Skorokhodov (✉) · A. A. Zhamaletdinov
Geological Institute of the Kola Scientific Centre of RAS, Apatity, Russia
e-mail: sammicne@yandex.ru

A. A. Zhamaletdinov
e-mail: abd.zham@mail.ru

A. A. Zhamaletdinov
SPbF IZMIRAN, St. Petersburg, Russia

© Springer Nature Switzerland AG 2019
D. Nurgaliev and N. Khairullina (eds.), *Practical and Theoretical Aspects of Geological Interpretation of Gravitational, Magnetic and Electric Fields*, Springer Proceedings in Earth and Environmental Sciences, https://doi.org/10.1007/978-3-319-97670-9_21

Introduction

In the Northern Ladoga area in 2015 and in 2017, a DC profiling and audio-magnetotelluric sounding (AMT) were carried out for to study the nature of the Ladoga deep anomaly of electrical conductivity. DC profiling has been made with a step 500 m by a 180-km profile that crossed the Ladoga anomaly (Fig. 21.1a). The average depth penetration of DC profiling was 150–200 m. Two most intensive anomalies of electrical conductivity were identified. One of them is situated near to Elisenvaara village (Grand anomaly of 7 km visible thickness) and another one near to Suistamo village (Janisjarvi anomaly of 200 m thickness). Apparent resistivity over both anomalies decreases up to 1 Ωm and less. In this presentation a special AMT experiment is depicted that has been done over the Janisjarvi anomaly. Six AMT sounding sites were situated symmetrically on the both sides of the anomaly (Fig. 21.1b).

Observed Data

The measurements of AMT-field on the Janisjarvi site (Fig. 21.1b) were made directly over the anomaly (point 1), as well as on both sides of the anomaly at distances of 200, 800 and 1600 m (points 2–6). The work was performed with the use of VMTU-10 equipment developed by VEGA LTD. Data processing was carried out using the 3-component algorithm (Semenov 1985). An analysis of the

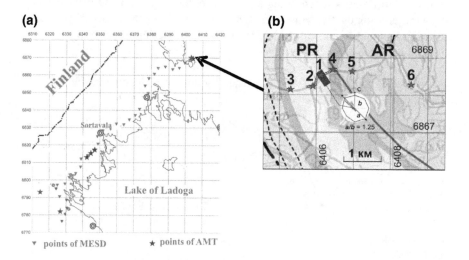

Fig. 21.1 Scheme of AMT measurements over the Janisjarvi anomaly. *a*—location of DC measurements (triangles) and AMT soundings (asterisks) on the profile Sevstjanovo-Suinjarvi; *b*—Location of detailed AMT soundings 1–6 situated symmetrically across Janisjarvi anomaly

polar diagrams and skew factor estimation showed that the medium has an elongated structure in the northwestern direction and can be offered as conditionally two-dimensional medium. In this case, the MT field is considered in view of 2 independent (TM and TE) modes.

TM-mode appears when the electric field is directed across the stretch of the structures. It has the galvanic nature of the distortions and is affected by near-surface anomalies mainly of high resistivity. The second, TE mode exists when the electric field is directed along the structures. It has the inductive nature of the distortions. It is sensitive to deep conducting structures.

Results of measurements are presented in the form of curves of apparent resistivity and phases of impedance (Fig. 21.2).

It can be seen from the figure that apparent resistivity curves have symmetrical shape relative to the center of anomaly. The level of curves increases with increase of distance between the center of anomaly and the AMT sites location. That is evident on both modes (Fig. 21.1a).

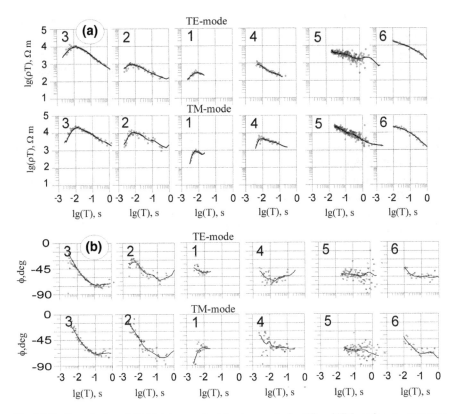

Fig. 21.2 Compilation of AMT data over Janisjarvi anomaly for TE and TM modes. *a*—apparent resistivity, *b*—phase of impedance

Simulated Data

The simulation was carried out using a software package developed by Vardanyants (1979), which calculates the field components at grid nodes using the finite difference method. The problem of constructing a priori model was reduced to constructing a normal section and of anomalous object. A normal section was created by asymptotic analysis of the effective resistivity curve at a distance of 1600 m. The resistivity of the upper layer is accepted equal to the value of highest apparent resistivity of the "normal" curve. Asymptote analysis of the normal curve gave a two-layered model with conductive layer 100 Ω m at the depth of 10 km (Fig. 21.3).

The anomalous body was specified taking into account of the DC profiling data. So the thickness and the depth of the upper edge were set. The value of the resistivity (ρ) and the depth of the lower edge (z) were offered as variable parameters (Fig. 21.4).

The structure was considered as vertical due to the bilateral behavior of the observed apparent resistivity curves with respect to the anomaly Since the depth z and the resistivity are related to each other by the condition of equivalence with respect to the parameter of the electrical conductivity (within the framework of our problem), we will consider further work by investigating the dependence of the response of the TE mode, more sensitive to high conductivity, on the parameter z.

It is seen from the Fig. 21.4a that at the most distant points the modeling data are in rather good agreement with the measured data. But when observed sites approache to the anomaly, the low frequency (long period) parts of modeling curves does not coincide with measured ones. But their level decreases symmetrically. With the increase of conductive body extension up to the depth Z to 400 m and 1000 m (Fig. 21.4b, c), the divergence between measured and modeled curves increases at all points. Moreover, subsidence of the low-frequency part is observed, and the further from the anomaly, the stronger it is. If these points are considered separately, it can be seen that there is the image intermediate horizontal conductive layer which depth varies, depending on the distance to the anomaly. The depth to the conducting base also increases with the increase of the distance to anomalous body. Further, with increasing Z (Fig. 21.4c), these affects are amplified, especially on the great distances from anomaly.

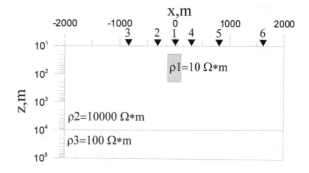

Fig. 21.3 A priori model of Yanisjarvi fault zone

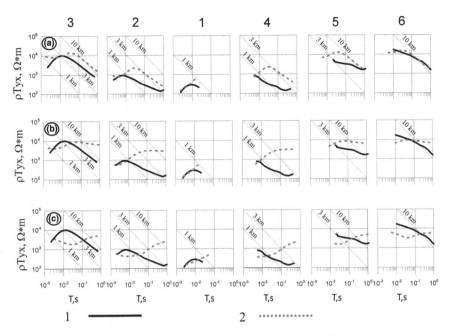

Fig. 21.4 Comparison of measured (1) and simulated (2) curves of apparent resistivity (TE-mode). *a*—The depth of lower edge z = 200 m. *b*—The depth of lower edge z = 400 m, *c*—the depth of lower edge z = 1000 m

From here we can conclude that the Yenisyavri anomaly does not have a connection with the deep Ladoga conductivity zone, and apparently is limited in depth to the first hundreds of meters (300–500 m). The displacement of the apparent resistivity curves is caused by the lateral influence of the shallow conductor. This is a classic case of the manifestation of a low resistivity anomaly of, described by Berdichevsky and Dmitriev (2009).

Conclusions

The simulation results allowed us to establish that the Yanisyarvi anomaly continues to the depth of the first hundred meters and, apparently, it doesn't connect with deep conducting structure of Ladoga anomaly. The decrease in the level of the apparent resistivity curves as we approach the anomaly is caused by the lateral influence of the conductor, and this effect is more pronounced in the TE mode. Although, there are several theoretical methods those allow to separate these effects from measured data, but we must thoroughly select the AMT-sounding sites location with the use of geological data and DC electrical profiling. In other case formal interpretation of AMT sounding curves in condition of Ladoga anomaly can create an erroneous intermediate conductive layers at the depth of from units to dozens kilometers.

Acknowledgements The authors are grateful to Mikhail Birulja, the employee of "VEGA" Ltd, for operative measurements and troubleshooting of magnetotelluric station VMTU-10.

References

Semenov V, Yu. (1985) Processing of magnetotelluric sounding data. Nedra, M. p. 25–72.
Vardanyants I.L. (1979) Calculations by the grid method of magnetotelluric fields over two-dimensional inhomogeneous media. // Questions of geophysics№28, L. p. 40–50 (part 2).
Berdichevsky M.N. Dmitriev V.I., (2009) Models and methods of magnetotelluric. M.: Nauchnyy Mir, p. 566–571.

Chapter 22
Application of Frequency-Resonance Method of Satellite Images Processing for the Oil and Gas Potential Assessment of "Onisiforos West-1" Well Drilling Site in the Mediterranean Sea

S. Levashov, N. Yakymchuk, I. Korchagin and D. Bozhezha

Abstract The potentialities of the direct-prospecting method of satellite images frequency-resonance processing is demonstrated by the results of their practical application for monitoring over drilling site of the "Onisiforos West-1" prospecting well in the eastern Mediterranean (Block 11, Cyprus offshore). In the area of the well location, an anomalous zone "Gas-1" was mapped, within witch the estimation of the maximum value of fluids pressure in reservoir is 53.5 MPa. The area of the anomalous zone along the isoline of 0 MPa is approximately 42 km^2, and along the isoline of 50 MPa it is about 29 km^2. The channel of vertical fluid migration was detected in the fault zone, adjacent to the anomaly from east. Estimation of reservoir pressure values at the central point of the channel was about 80 MPa. By the geological cross-section scanning in the depth interval of 1700–6000 m seven anomalous polarized layers (APLs) of the "gas" type, promising to detect gas deposits, were identified in the cross-section. During the scanning, the APLs depths and thicknesses, fluids pressure and also porosity were estimated. The most promising for gas detection are two search intervals: 1800–2250 m; 4050–4250 m. The results of operatively conducted experimental studies of reconnaissance character increase, as a whole, the probability of gas accumulations (deposits) detection in the "Onisiforos West-1" well.

S. Levashov (✉) · N. Yakymchuk · D. Bozhezha
Institute of Applied Problems of Ecology, Geophysics and Geochemistry,
Kyiv, Ukraine
e-mail: geoprom@ukr.net

I. Korchagin
Institute of Geophysics, NAS Ukraine, Kyiv, Ukraine
e-mail: korchagin.i.n@gmail.com

Introduction

In connection with a significant drop in oil prices, the problem of accelerating and optimizing the process of prospecting and exploration of industrial hydrocarbon accumulations in traditional and non-traditional reservoirs has become quite urgent in the oil and gas sector of the world economy. Low oil prices force large and small oil and gas companies to cut both expenses (costs) in the search and exploration for commercial hydrocarbon accumulations, and at the time of introduction of discovered and explored deposits into development.

Under the circumstances, mobile (low-cost) direct-prospecting technologies can be claimed and used more actively at various stages of the oil and gas prospecting process.

Below, the potentialities of the direct-prospecting methods are demonstrated by the results of their practical application for monitoring of the local drilling site of the "Onisiforos West-1" prospecting well in the eastern Mediterranean (Block 11, Cyprus offshore).

The main purpose of the operatively conducted work at the prospecting well site is an additional demonstration (on a concrete example) of the operability and potential capabilities of mobile and direct-prospecting technology of frequency-resonance processing and interpretation (decoding) of remote sensing data (satellite images).

Method of Research

Mobile technology of frequency-resonance processing and decoding of remote sensing data was used during experimental investigation conducting. Distinctive features of the technology are described in numerous publications, including given in Levashov et al. (2010, 2011, 2012), examples of practical applications are given in Levashov et al. (2010, 2011, 2012, 2015a, b, 2016, 2017a, b), Yakymchuk et al. (2015).

For many years, the technology has been actively used for prospecting and exploration of ore and combustible minerals. In article (Levashov et al. 2015a) the results of mobile methods application on the sites of prospecting wells drilling are analyzed. Articles (Levashov et al. 2016, 2017a) present the results of the technology application for the detection and localization of channels for the deep fluids (hydrocarbons) vertical migration in various regions of the world (on onshore and offshore).

Additional information about the mobile technology used can be found at site [http://www.geoprom.com.ua/index.php/en/]. Here is a video that shows the features of the work at various stages of research, as well as a presentation with numerous results of practical application of mobile direct-prospecting methods.

In this region (deep-water part of the eastern Mediterranean), the authors conducted demonstration studies in 2015 and 2016. In September 2015, a frequency-resonance processing of a satellite image of area of first prospecting well location in the Zohr gas field was performed. The scale of image processing is 1:150,000; this is a reconnaissance mode of search operations by direct-prospecting methods. The results of studies at this stage are presented in paper (Levashov et al. 2015b).

In March 2016, during the search for channels of deep fluids vertical migration in different regions of the world, frequency-resonance processing of two local sites within Block 9 (Shorouk) on the Egyptian offshore and Block 11 on the offshore of Cyprus was performed. Scale of these areas images processing was 1: 50,000, detailed. The channels of deep fluids vertical migration were detected within these sites. Materials of research at this stage are published in Levashov et al. (2016, 2017a).

Initial Data

The following data were used during the works conducting:

(A) Information messages on various Internet sites on the preparation for well drilling, including those listed in the list of reference (Energy…; Drilling…).
(B) Materials of earlier conducted studies in this region (Levashov et al. 2015b, 2016, 2017a).
(C) Approximate location of the "Onisiforos West-1" prospecting well. It is established on the coordinates of the drilling vessel location, which are regularly updated on the site (Vessel…).

Processing Results

The frequency-resonance processing of a satellite image of the local site of "Onisiforos West-1" prospecting well location was conducted at August 18–22, 2017.

For the experimental studies, a satellite image of the sea area was prepared at a scale of 1: 80,000 (Fig. 22.1). This processing scale is not detailed.

At the initial stage of the work, a zone of tectonic fracture of the north-north-western strike was detected and traced in the surveyed area. The "Onisiforos West-1" well is located in the immediate vicinity of this fracture (Fig. 22.1).

In the area of the well location, an anomalous zone "Gas-1" was discovered and mapped, which adjoins to the fault zone from the west. Within the limits of the detected "Gas-1" anomaly the estimation of the maximum value of fluids pressure in reservoir is 53.5 MPa. The area of the anomalous zone along the isoline of 0 MPa is approximately 42 km^2, and along the isoline of 50 MPa it is about 29 km^2.

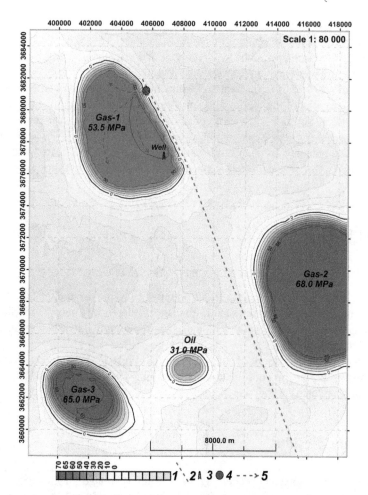

Fig. 22.1 Map of anomalous zones of the "Gas-Oil" type at the area of the "Onisiforos West-1" search well drilling in the Mediterranean. *1*—scale of maximum values of reservoir pressure, MPa; *2*—tectonic fracture; *3*—well position; *4*—vertical channel of gas migration; *5*—gas migration paths

Within the contour of the detected anomaly in the depth interval of 1700–6000 m, the anomalous responses were recorded in the following intervals of reservoir pressures: (1) 18.50–22.90 MPa; (2) 27.22–28.30 MPa; (3) 37.89–39.30 MPa; (4) 42.8–43.70 MPa; (5) 51.36–53.95 MPa.

Two more anomalous zones of the "gas" type ("Gas-2" and "Gas-3") and one small anomalous zone of the "oil" type ("Oil") were found in the southern half of the surveyed area (Fig. 22.1).

Using the technique of detection and localization of channels of deep fluids vertical migration (Levashov et al. 2016, 2017a), searches for such channels in the contours of the anomalous zone "Gas-1" were carried out. As a result, the channel of vertical fluid migration was detected in the fault zone, adjacent to the anomalous zone from the east (Fig. 22.1). Approximate coordinates of the channel position are following: X = 3681,141.20, Y = 405,538.18. Estimation of reservoir pressure values at the central point of channel was about 80 MPa.

Note that in this region, the channels of deep fluids vertical migration were also detected and localized in the contours of two mapped anomalous zones (Levashov et al. 2016, 2017a).

Scanning Results

In the vicinity of the prospecting well in the depth interval of 1700–6000 m, the geological cross-section was scanned in order to determine the depths and thicknesses of the anomalous polarized layers (APLs) of the "gas" type. In the course of the scanning, the fluids pressure and also porosity in the APLs of "gas" type were additionally estimated. The difference between the reservoir pressure in the APLs and the conventional hydrostatic pressure at this depth (parameter D) was also calculated. Based on the scanning results, seven APLs of the "gas" type, promising to detect gas deposits, were identified in the cross-section. The parameters of the selected APLs of "gas" type are as follows: (**1**) 1825–1830 m (the depths of the roof and the base of the layer), 5 m (layer thickness), 20.35 MPa (fluids pressure estimation in the reservoir), +2.08 (parameter D), 13% (APL porosity); (**2**) 1920–1930 m, 10 m, 20.6 MPa, +1.35, 13%; (**3**) 2030–2040 m, 10 m, 21.3 MPa, +0.95, 18%; (**4**) 2110–2120 m, 20 m, 22.4 MPa, +1.25, 20%; (**5**) **2188–2210 m, 22 m, 22.8 MPa, +0.85, 22%**; (**6**) 4070–4080 m, 10 m, 41.3 MPa, +2.25, 13%; (**7**) 4220–4226 m, 6 m, 43 MPa, +0.77, 15%. The scanning results are shown in Table 22.1 and in Fig. 22.2.

The most promising for gas detection are two search intervals: (1) 1800–2250 m; (2) 4050–4250 m.

Possible Additional Investigations

1. To improve the accuracy of the experimental studies results, it is advisable to carry out the processing of a satellite image of the "Gas-1" anomalous zone location on a larger scale.
2. When processing the image of the anomaly area on a larger scale, the detected by scanning at the drilling site of a well APLs of a "gas" type can be traced over the area of entire "Gas-1" anomaly. Scanning materials by area will allow to calculate the volumes of forecasted reservoirs and estimate, in the first approximation, the forecasted gas resources in them.

Table 22.1 The results of vertical scanning in the area of the "Onisiforos West-1" well drilling in the Mediterranean sea (scanning interval: 1700–6000 m)

No.	Layer	Roof (m)	Sole (m)	Thickness (m)	Pressure (MPa)	D	Porosity (%)	Note
1	Gas	1729	1740	11	17.3	0.0		
1a	Water + Gas	1740	1754	14				
2	Gas	1825	1830	5	20.35	+2.08	13	Search
2a	Water	1830	1844	14				
3	Gas	1920	1930	10	20.6	+1.35	13	Search
3a	Water + Gas	1930	1939	9				
4	Gas	2030	2040	10	21.3	+0.95	18	Search
5	Gas	2110	2120	10	22.4	+1.25	20	Search
5a	Water	2120	2121	1				
6	Gas	**2188**	**2210**	**22**	22.8	+0.85	22	Perspective Search!!!!
6a	Water	2210	2212	2				
7	Water + Gas	2225	2232	7				
8	Gas	2759	2770	11	27.65	0.0	14	
8a	Water	2770	2774	4				
9	Water + Gas	2820	2835	15				
10	Water + Gas	3400	3423	23				
11	Water + Gas	3590	3609	19				
12	Water + Gas	3779	3783	4				
13	Gas	4070	4080	10	43.1	+2.25	13	Search
13a	Water	4080	4090	10				
14	Gas	4220	4226	6	43.0	+0.77	15	Search
14a	Water	4226	4232	6				
15	Water + Gas	4760	4783	23				
16	Water + Gas	5230	5241	11				

Processing Results of Minous Structure Area (Conophagos et al.)

The satellite image of the area location was processed on a scale of 1:150,000 (Fig. 22.3). This is a reconnaissance mode for images processing. The preliminary, frequency-resonance processing of the image was performed on August 20–21, 2017.

A relatively large anomalous zone of the "Oil & Gas" type was detected and mapped on the surveyed area. In Fig. 22.3, the anomalous zone is represented in isolines of fluids pressure in reservoir. The maximum pressure value is 25.5 MPa.

Anomalous responses at resonance frequencies of the gas were detected in the contours of the detected anomaly (within the isoline of 0 MPa). In addition, anomalous responses at resonant frequencies of oil are recorded in the contour of 20 MPa isoline.

By additional processing of the image in the depth interval 0–5000 m two intervals of reservoir pressures were established: (1) 12.22–13.1 MPa; (2) 23.44–25.5 MPa.

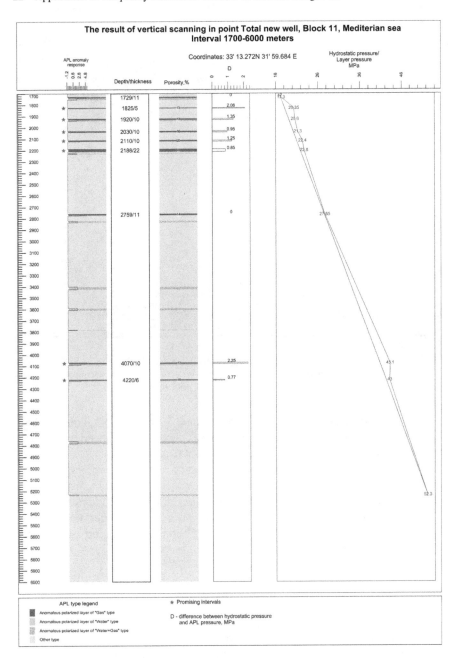

Fig. 22.2 Results of vertical scanning at the drilling point of the search well "Onisiforos West-1" in the Mediterranean (Block 11 on the Cyprus offshore). Interval of scanning: 1700–6000 m

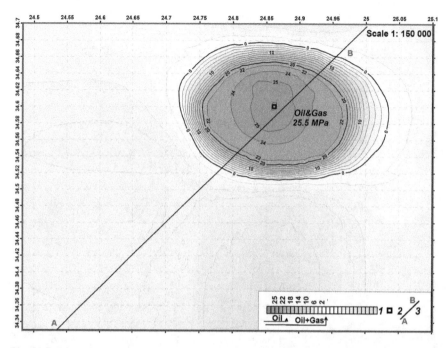

Fig. 22.3 Map of anomalous zone of the "Oil & Gas" type on the offshore to south of Crete. *1*—scale of maximum values of reservoir pressure, MPa; *2*—point of the anomalous zone maximum; coordinates: 24.86175, 34.60043°; *3*—position of the seismic profile

The area of the detected anomaly: (a) along the 0 MPa contour—about 426 km^2; (b) along an isoline of 20 MPa—about 213 km^2. Within the mapped anomaly the channel of deep fluids vertical migration was not detected.

The operatively obtained results of frequency-resonance processing of the satellite image of MINOAS seismic profile (structure) location testify to the expediency of performing the detailed search operations on this site (area).

At the initial stage of this site further study, it is advisable to perform a satellite image processing of the location of the detected anomalous zone on a larger scale (on 1:50,000 and larger).

Conclusions

1. Within the surveyed search area in the Mediterranean Sea the works of demonstration character have been carried out, not detailed. As a result, **an anomaly of the "gas deposit" type was discovered and mapped in the area of the "Onisiforos West-1" well drilling!** The area of the anomalous zone along the isoline of 0 MPa is of the order of 42 km^2, and along the isoline of 50 MPa it is approximately 29 km^2.

2. By the scanning of the vertical cross-section in the depth interval of 1700–6000 m in the area of the prospecting well location seven anomalous polarized layers (APLs) of the "gas" type were identified, promising to detect gas deposits. The depths and thicknesses of the APLs of "gas" type have been determined, and reservoir pressures and porosity have been estimated in the intervals of their location.
3. Within the contours of the mapped anomaly, **a channel of deep fluids vertical migration was detected and localized.** Estimation of reservoir pressure values at the central point of the channel was about 80 MPa.
4. The results of operatively conducted experimental studies of reconnaissance character increase, as a whole, the probability of gas accumulations (deposits) detection in the "Onisiforos West-1" well. However, the minimal volume of research performed does not allow authors to unequivocally conclude that these accumulations will be in commercial volumes or not. Detailed studies in the contours of the detected "Gas-1" anomaly on a larger scale can contribute to the formation of more concrete (certain) conclusions on this problem.
5. The received materials will be used in the future to demonstrate (advertise) the operability and potential capabilities of mobile and direct-prospecting technology of frequency-resonance processing and decoding of remote sensing data.

References

Conophagos E., Lygeros N., Foscolos A. PGS shows a Giant target-reserve south of Crete. http://www.lygeros.org/articles.php?n=21015&l=en

Levashov S.P., Yakymchuk N.A., Korchagin I.N. (2010) New possibilities of the oil-and-gas prospects operative estimation of exploratory areas, difficult of access and remote territories, license blocks. Geoinformatika, 3, 22–43. (in Russian)

Levashov S.P., Yakymchuk N.A., Korchagin I.N. (2011) Assessment of relative value of the reservoir pressure of fluids: results of the experiments and prospects of practical applications. Geoinformatika, 2, 19–35. (in Russian)

Levashov S.P., Yakymchuk N.A., Korchagin I.N. (2012) Frequency-resonance principle, mobile geoelectric technology: a new paradigm of Geophysical Investigation. Geophysical Journal, 34, 4, 167–176. (in Russian)

Levashov S.P., Yakymchuk N.A., Korchagin I.N., Bozhezha D.N. (2015a) Mobile technologies of direct prospecting for oil and gas: feasibility of their additional application in selecting sites of well drilling. Geoinformatika, 3, 5–30 [in Russian]

Levashov S.P., Yakymchuk N.A., Korchagin I.N., Bozhezha D.N. (2015b) Operative assessment of hydrocarbon potential of area in region of Zohr gas field on the Egypt offshore in the Mediterranean Sea by the frequency-resonance method of remote sensing data processing and interpretation. Geoinformatika, 4, 5–16. (in Russian).

Levashov S.P., Yakymchuk N.A., Korchagin I.N., Bozhezha D.N., Prylukov V.V. (2016) Mobile direct–prospecting technology: facts of the channels detection and localization of the fluids vertical migration - additional evidence for deep hydrocarbon synthesis. Geoinformatika, 2, 5–23 (in Russian)

Levashov, S.P., Yakymchuk, N.A., Korchagin, I.N. and Bozhezha, D.N. (2017a), Application of mobile and direct-prospecting technology of remote sensing data frequency-resonance processing for the vertical channels of deep fluids migration detection. NCGT Journal, v. 5, no. 1, March 2017, p. 48–91. www.ncgt.org

Levashov Sergey, Yakymchuk Nikolay, and Korchagin Ignat. (2017b), On the Possibility of Using Mobile and Direct -Prospecting Geophysical Technologies to Assess the Prospects of Oil -Gas Content in Deep Horizons. Oil and Gas Exploration: Methods and Application. Said Gaci and Olga Hachay Editors. April 2017, American Geophysical Union. p. 209–236.

Yakymchuk, N. A., Levashov, S. P., Korchagin, I. N., & Bozhezha, D. N. (2015, March 23). Mobile Technology of Frequency-Resonance Processing and Interpretation of Remote Sensing Data: The Results of Application in Different Region of Barents Sea. Offshore Technology Conference. https://doi.org/10.4043/25578-ms. https://www.onepetro.org/conference-paper/OTC-25578-MS

Energy…http://www.cyprusprofile.com/en/articles/view/energy-giants-upbeat-over-block-11-prospects

Drilling … http://www.leptosestates.com/news/leptos-cyprus-news/Drilling-Ship

Vessel … http://www.marinetraffic.com/en/ais/details/ships/shipid:419249/mmsi:356736000/ imo:9372523/vessel:WEST_CAPELLA#HPBV6rLfRvkTh11y.99

Chapter 23
Evolution of Ideas on the Nature and Structure of Ladoga Anomaly of Electrical Conductivity

A. A. Zhamaletdinov, I. I. Rokityansky and E. Yu. Sokolova

Abstract Several stages of the Ladoga conductivity anomaly study are discussed. The Ladoga anomaly was discovered in the late 70s by means of magnetovariational observations by research team of the Institute of geophysics from Kyyiv (Ukraine). At the next step, in 80s the anomaly was studied by magnetotelluric method. The third stage of the deep studies was realized in 2013–2015 by means of integrated magnetotelluric and magnetovariational profiling. Each of three stages were completed by construction of specific geoelectrical models of the anomalous area up to the depth of 30–40 km. In the fourth stage, in 2015 and 2017 the studies were performed by means of DC electrical profiling with the use of multielectrode installations in complex with AMT soundings. According to results of the fourth stage, an unambiguous conclusion was drawn on the connection of the upper part of the Ladoga anomaly (till the depth of 1–2 km) with electronically-conducting sulfide-carbonaceous rocks. The structure and nature of the deeper part of the Ladoga anomaly requires further investigations on the base of integration of geological and complex geophysical methods.

Keywords Ladoga anomaly · Electrical conductivity · Magnetotellurics DC profiling

---

A. A. Zhamaletdinov (✉)
SpbF IZMIRAN, St. Petersburg, Russia
e-mail: abd.zham@mail.ru

A. A. Zhamaletdinov
Geological Institute of the Kola Science Centre of RAS, Apatity, Russia

I. I. Rokityansky
Institute of Geophysics of NANU, Kiev, Ukraine
e-mail: rokityansky@gmail.com

E. Yu. Sokolova
Institute of Physics of the Earth RAS, Moscow, Russia
e-mail: sokol_l@mail.ru

E. Yu. Sokolova
All-Russian Research Geological Oil Institute, Moscow, Russia

Introduction

Ladoga anomaly of electrical conductivity is a subject of research for more than three decades. On the Russian territory, Ladoga anomaly is traced along the northwestern and southeastern coasts of Lake Ladoga. It is a part of Ladoga-Bothnian anomalous zone (LBA) that stretches in the northwestern direction for more than 1000 km in Finland and Sweden (Fig. 23.1).

The Northern Ladoga electrical conductivity anomaly is confined within the system of long-lived faults, oriented in the northwest direction and located in the region of the junction of two major tectonic units—Karelian (Archaean) and Svecofennian (Proterozoic). The anomalous behavior of magnetotelluric field in the Northern Ladoga area was discovered firstly by N.V. Lazareva on the profile Lahdenpohja-Sortavala-Lake Kaitonijarvi of a total length 150 km. According to the data of magnetotelluric soundings (MTS) in the range of periods from 10 to 500 s, abrupt changes of apparent resistivity values from 10 Ωm to 600,000 Ωm were recorded. N.V. Lazareva had connected the reason for the drastic changes with the influence of tectonics. In the following years the main interest was aroused by the question of the deep structure of the Ladoga anomaly. The researches were conducted in several stages which are considered below in chronological order.

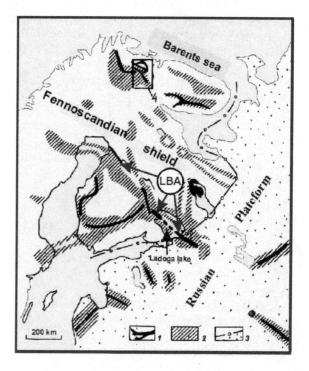

Fig. 23.1 Fennoscandian Ladoga-bothnian anomaly (LBA) of electrical conductivity. 1—axis of high conductivity (1000 Sm and more), 2—middle conductivity (10–100 Sm), 3—the boundary of Russian platform. 4—Pechenga area

Magnetovariational Study

The first model of the Ladoga anomaly deep structure was proposed by results of magnetovariational profiling (MVP), performed by Institute of Geophysics of Ukrainian Academy of Sciences (Rokityansky et al. 1981). The position of measuring points of MVP and conductive body location in the plan is shown in Fig. 23.2a.

The asymmetric behavior of the horizontal component of the anomalous magnetic field (Fig. 23.2b) was supposed as possible interpretation factor. On this basis it was suggested that Ladoga anomaly can have the northeastern dipping at an angle of about 30–45°. The center of the anomalous body was estimated at a depth of 10 km. The integral conductivity $G = \sigma \cdot S = 10^8 \, m/\Omega$, where σ-electrical specific conductivity in $1/\Omega m$ and S is the square of the cross section of the anomalous body in m^2.

Magnetotelluric and Audiomagnetotellurics Sounding

The next model of the deep structure of Ladoga anomaly has been compiled on the base of results of audiomagnetotelluric and magnetotelluric soundings (MTS and AMTS), conducted in 80s of the last century by Saint Petersburg (former Leningrad) Mining Institute (SPbMI) and Saint Petersburg (former Leningrad) State University (SPbGU). By means of numerical simulation of MTS and AMTS data the two-dimensional model has been constructed in Vasin et al. (1993). In the anomaly cross-section the model demonstrates a wide area of low resistivity, extending to a depth of 30–40 km and not having a certain dip angle. The model is given on the Fig. 23.3.

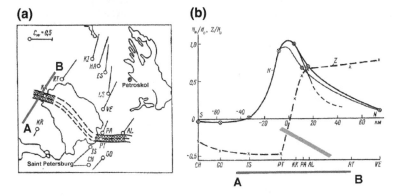

Fig. 23.2 Ladoga anomaly. *a*—location in plan; *b*—cross section and graphs of MVP data

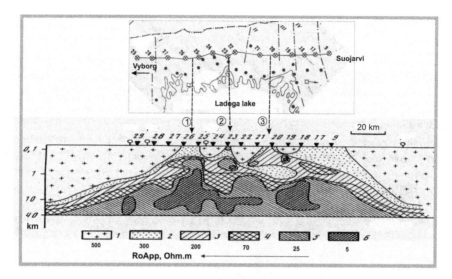

Fig. 23.3 2D model of the Ladoga anomaly by results of AMT-MT data. Legend—signs 1–6 show resistivity from 500 till 5 Ωm

According to this model the width of the Ladoga anomaly is approximately 60 km at the surface, and 100–120 km at a depth of 30 km. The resistivity of the anomaly decreases from 200–300 Ωm at the surface to 5–25 Ωm at a depth of 30 km. Three epicenters of high conductivity are marked on the Fig. 23.3 by arrows. Arrow 1 coincides with I. Rokityansky anomaly near to Elisenvaara village (Fig. 23.2). The similar model of the deep electrical conductivity of Northern Ladoga anomaly was obtained independently on the profile Gdov-Spasskaya guba by means of MTS-AMTS researches implemented in complex with the deep seismic soundings (Sviridenko et al. 2017). Low-frequency magnetotelluric soundings conducted by the group of Kovtun et al. (2004) on the territory of the Russian platform, revealed the LBA continuation to the south of Lake Ladoga by more than 100 km under the sedimentary cover at depths of 3–10 km (Fig. 23.3).

Integrated MT-MV Soundings

The third model of the Ladoga anomaly structure was derived on the results of synchronous MT-MV soundings performed by working group "Ladoga" (Sokolova et al. 2016, 2017).

The interest to Ladoga anomaly has recently arisen again in response to actual needs of Precambrian geological studies. That happened due to increased possibilities of modern magnetotelluric and magnetovariational deep sounding methods: digital synchronous recording, GPS positioning and synchronization, advanced processing and inversion techniques developed and introduced into MTS practice by the end of the first decade of XXI century. The MT group of SPbGU (A. A. Kovtun) has invited MT researches from Moscow (Moscow State University) to combine efforts and to initiate new stage of Ladoga anomaly instrumental investigations in the frames of "Ladoga" WG collaboration. "Nord-West" Ltd has provided invaluable assistance in the implementation of this idea and the large MT/MV sounding project has been organized.

The observations were implemented by the 200 km long Vyborg-Suojarvi profile in 2013–2014 with the use of "Phoenix" and "LEMI" stations. The study have resulted in 50 broad-band and 9 long period MT/MV soundings with synchronous recording in remote bases (Fig. 23.4).

Several modern processing techniques and software provided by equipment manufacturers and remote reference and multi-RR schemes with magnetic and/or electric remotes were used to suppress EM noises. The data of geomagnetic observatories Nurmijarvi, Mekkrijarvi, Suwalki and permanent geomagnetic observations of St. Petersburg Branch of IZMIRAN in lake Krasnoe were used for long period synchronization of data obtained by different stations. The resulting local (impedance Z, tipper Wz) and inter-stations (horizontal magnetic tensor M) transfer functions have been estimated in broad band of periods (0.003–2048 s). Their invariant analyses has defined general azimuth (45–50° NE) and dimensionality (quasi-2D with local 3D distortions) of the data and thus approved application of 2D interpretational approach (Sokolova et al. 2016).

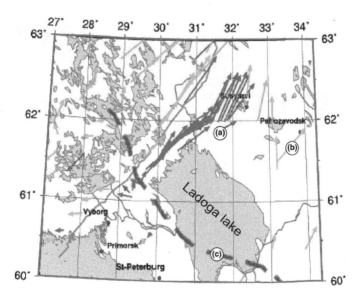

Fig. 23.4 Location and results of MT-MV investigations, made by the "Ladoga" WG. Letters in circles: *a*—real induction vectors at period 1024 s (after Sokolova et al. 2016); *b*—real induction vectors at 1800 s; *c*—proposed axis of the deep crustal conductor

The inversions based on the robust code (Varentsov 2007) have produced the model of Ladoga anomaly, which demonstrates, that it is caused not by a unique anomalous object but a complicated ensemble of conductive features of different structural identity (Fig. 23.5a) (Sokolova and Ladoga 2017).

Results of this work revealed the general South-Western fall of the underlying conductive structures that form the Ladoga anomaly, and allowed to spend the geotectonic interpretation of the section of electrical conductivity of the crust taking into account the regional results given by reflection seismic and by potential geophysics (Mints et al. 2017).

The high resolution of the cross-section has permitted to carry out the informative tectonic interpretation (Fig. 23.5b). At mid-crustal levels the conductive structures are generally characterized by distinct S-W dipping. Presumably they correspond to thrust zones, developed along graphite-bearing slippery surfaces of supracrustal Palaeoproterozoic formations during their accretion/thrusting upon the SW border of Karelian Craton in the late Palaeoproterozoic. At the upper levels they are steepening and connecting to the major fault zones recognized at the surface, including activated neotectonic ones bordering Ladoga-Bothnian tectonic zone. Probably at these levels the conductance of the layers is increased by

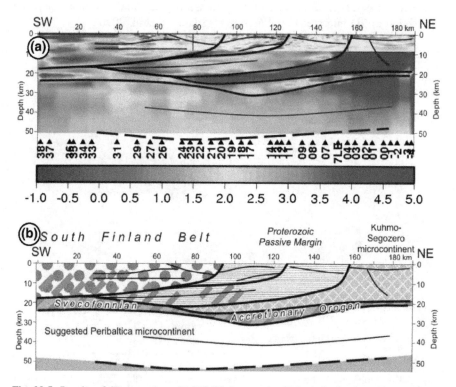

Fig. 23.5 Results of 2D inversion of MT-MV data on the Vyborg-Suojarvi profile (**a**) and its geological interpretation (**b**)

mineralized aqueous fluids (due to meteoric water) gathered in weakened fault zones. In the upper crust (5–7 km) of NE part of the profile bowl-formed association of conductive features is imaged, which describes the structure of Raahe-Ladoga suture zone across its ∼50-km width.

In the SW part of the resistivity cross-section significant inflation of the deepest conductive layer (about 15–25 km depth) was found which produces the very long-period Ladoga anomaly originally discovered by pioneering MVS surveys. One can assume that this extremum of the conductivity is caused by deeply metamorphosed complexes of South-Finland Granulite-Gneiss Belt which could be similar to exhumed formations of Lapland Granulite Belt and include crystalline graphite. The formations of the coarse- and medium-flaky graphite exposed at the famous Ikhal'sky deposit, located in the nappe of the South-Finland Belt, probably represent these deep-seated fabric. However at NE segment of the profile, in pericratonic zone, the enhance upper-crustal conductivity is connected with often exposed graphite- and sulphide-bearing sedimentary-volcanic Kalevian and Ladoga series of lower metamorphic stages.

Electrical DC Profiling in Combination with AMT

One of the weak points in the study of the Ladoga anomaly, hindering the elucidation of its nature, is the absence of more or less complete information about the conductivity of rocks coming close to the surface. This is observed both in Russia and in foreign parts of LBA (Fig. 23.1). Existing data of electrical prospecting is fragmental and do not provide information about the conductivity rocks across the width of the Ladoga anomaly. At the same time, the specificity of the geoelectrical conditions of the Baltic shield requires consideration of the electrical conductivity of rocks near the location of the receiver line when performing deep soundings with natural and controlled sources (Zhamaletdinov 1990). The main reasons for the lack of extended electrical studies in the Northern Ladoga area was difficult terrain conditions and extreme tortuosity of the coastline of Lake Ladoga and adjacent roads. In order to fill up the gap the DC electric profiling in combination with AMT-MTS has been made by the joint team of Geological Institute of the Kola science center of RAS and Institute of Geology of the Karelian research center in 2015 and in 2017 (Fig. 23.6).

Two of the most contrasting electrical conductivity anomalies have been discovered near the villages Elisenvaara and Suistamo (Fig. 23.6). Both anomalies are of extremely low values of apparent resistivity (units and tenths of ohmmeters), testifying to their electronic conductive nature. They are related to the presence of sulfide-carbonaceous rocks. The Western anomaly, near Elisenvaara (Grand anomaly) has a visible width of 7 km. It coincides with epicenter of MVP anomaly (Fig. 23.2a) and with the Western epicenter of AMT-MTS anomaly marked by arrow number 1 on Figs. 23.3 and 23.6.

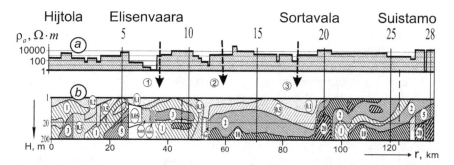

Fig. 23.6 Results of electrical DC profiling by MESD method *a*—graph of the specific resistivity according to results of 1D inversion; *b*—the 2D section according to MESD

The effect of the influence of conductive zone on the AMT data is demonstrated on the Fig. 23.7 on the example of Janisyarvi anomaly situated near to Suistamo village (site 28 on the Fig. 23.6).

The thickness of the conductive layer is 0.2 km. With the increase of distances between conductor and AMT sites to the West from anomaly (on Proterozoic rocks) and to the East from anomaly (on Archaean rocks) the level of apparent resistivity curves happens to be higher and higher. Accordingly, the depth till the roof of hypothetical (seeming) conductive layer became more and more, changing from 0.3 km at the center till 5–10 km and more at distance of 1–2 km from anomaly. The experiment on the Fig. 23.7 shows that the question of the nature and structure of the deep Ladoga anomaly more fully can be solved only by a detailed study of AMT-MTS and MVP in a wide range of frequencies and in close correlation with the DC profiling. Digits in circles—distances between AMT site and the center of conductor to the West (−0.2, −0.8 km) and to the East (0.2, 0.8, 1.6 km). The central AMT curve is marked by 0.

This is especially evident in the Grand anomaly (30–37 km on the Fig. 23.6) appearing in the complex of supercrustal rocks. Its detailed study shows that electron-conductive objects appearing on the day surface can not be described within the framework of a two-dimensional approach. Moreover, within the greater

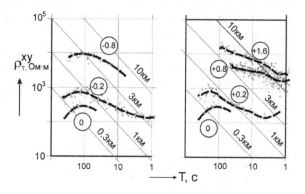

Fig. 23.7 The influence of electron-conductive zone on results of AMT sounding on the example of Janisyarvi fault zone

part of the Northern Ladoga they are distributed sporadically in the supracrustal sequences flanking the dome-shaped structure granites and gneisses (Sviridenko et al. 2017). In these conditions when processing results of the deep MT-AMT soundings requires careful, reasoned invariant analysis of the MT/MV transfer functions, the selection of material for interpretation. Not in vain in Vasin et al. (1993) of the 29 curves AMT-MTS selected only 15 different "best consistent" when building two-dimensional numerical model of the structure of the Ladoga zone.

Conclusion

After this overview, we can conclude that the question on the deep structure and nature of the Ladoga anomaly can't be resolved with the use of the deep electromagnetic soundings only. The further research should integrate joint analysis of the near-surface DC profiling and the deep electromagnetic soundings in complex with seismic, geothermic and potential fields data, physical rock properties and should be finalized by geological syntheses. The graphite hypotheses of Ladoga anomaly nature, supported by many authors, should be comprehensively verified and estimated taking into account the hypothesis of Klabukov (2006), assuming two-level structure of the conductivity anomalies in the whole Ladoga-Bothnian zone (electronically conductive rocks at the surface and fluid zones at a depth of 10–40 km)

Acknowledgements The authors are grateful to the ccolleagues from IG KarRC of RAS, GI KolRC of RAS, geomagnetic observatories Nurmijarvi, Mekkrijarvi, Suwalki, Krasnoe as well as the Russian Foundation for Basic Research (RFBR) for supporting projects No. 15-05-01214, 16-05-00543, 16-05-00975, 18-05-00528.

References

Klabukov B.N. (2006). Possibilities of petroelectrics in the study of the earth's crust of Karelia. // Geology and minerals of Karelia. Issue. 9. Petrozavodsk: KarRC RAS, 2006. P. 127–134.
Kovtun A.A., Vardanyants I.L., Legenkova N.P., Smirnov M.Yu., Uspensky N.I. (2004). Features of the structure of the Karelian region according to geoelectric researches. // Deep structure and seismicity of the Karelo-Kola region (ed. N.V. Sharov). Petrozavodsk: Kar.SC RAS, 2004. P.102–130.
Mints, M.V., Sokolova, E.Yu., LADOGA Working Group, (2017). Volumetric model of the deep structure of the Svecofennian accretionary orogen based on data from CMP seismic profiling, MT sounding and density modeling. Proceedings of the Karelian Research Center of the Russian Academy of Sciences, Precambrian Geology. No. 11 (in Russian). https://doi.org/10.17076/geo656
Rokityansky I.I., Kulik S.N., Rokityanskaya D.A. (1981) Ladoga anomaly of electrical conductivity // Geofiz. Journal. Ukrainian Academy of Sciences. 1981. Volume 3. №2. Pp. 97–99.

Sokolova E.Yu., Golubtsova NS, Kovtun AA, Kulikov VA, Lozovsky IN, Pushkarev P.Yu., Rokityansky II, Taran Ya.V., Yakovlev A. G. (2016). Results of synchronous magnetotelluric and magnetovariational soundings in the Ladoga anomaly of electrical conductivity // Geophysika. №1. P. 48–64 (in Russian).

Sokolova E.Yu., LADOGA WG. (2017). Synchronous MT/MV sounding experiment across Lake Ladoga Conductivity Anomaly. // In: Deep structure and geodynamics of Lake Ladoga region. Petrozavodsk, Inst. of Geology of the KRC RAS. P. 204–2014.

Sviridenko L.P., Isanina E.V., Sharov N.V. (2017). The deep structure of volcanoplutonism and tectonics of the Ladoga area. // Proceedings of the Karelian Research Center of the Russian Academy of Sciences No. 2. 2017. P. 73–85. DOI:10.17076 / geo336

Varentsov Iv.M. (2007). Joint robust inversion of magnetotelluric and magnetovariational data. Electromagnetic sounding of the Earth's interior. Methods in geochemistry and geophysics. V. 40. / Ed. Spichak V.V. Elsevier. 2007. P. 189–222.

Vasin N.D., Kovtun A.A., Popov M.K. (1993) Ladoga anomaly of electrical conductivity. // In kN.: The structure of the lithosphere of the Baltic Shield (edited by NV Sharov). Moscow. VINITI, 1993. p. 69–71.

Zhamaletdinov A.A. (1990) Model of the electrical conductivity of the lithosphere from results of studies with controlled sources of the field (Baltic Shield, Russian Platform). // L. Science., 1990. 159 p. (monograph in Russian).

Chapter 24
The Use of Gravimetry for Studying Shelf of the North Barents Basin

M. Chadaev, V. Kostitsyn, V. Gershanok, R. Iblaminov,
G. Prostolupov and M. Tarantin

Abstract The continental shelf and slope of the Arctic seas contain many hydrocarbons and mineral deposits. An important object of shelf crust study is a Mohorovicic (Moho) discontinuity as a principle factor in formation of the structural and fault features. The paper considers the feasibility of gravimetry for studying inhomogeneities of the structure of the crust. On the basis of the VECTOR technology, a new technological scheme "Extension of gravitational anomaly lower half-space, using seismic profile as a baseline" was developed and implemented. As a result, direct relationships between the anomalies of Δg and Moho depths were established. Conclusion about the feasibility of applying the transformations of the field in the VECTOR system for the study of the Earth's crust on the shelf and continental transition zone was made.

Keywords Basin · Shelf · Earth's crust · Moho · Gravimetry

Possibility of the use of gravimetry in studying inhomogeneities of the structure of the Earth's crust had previously been repeatedly proven and is undisputed. This gives prerequisites for successful application of remote geophysical methods, in particular the gravimetry, for the study of the continental shelf of the Arctic seas near the northern borders of the Russian Federation, containing rich hydrocarbon and mineral deposits (Petrov et al. 2016). An important object of shelf crust study is a Mohorovicic (Moho) discontinuity as a principle factor in formation of the structural and fault features, contributing in formation and placement of mineral deposits.

M. Chadaev · G. Prostolupov (✉) · M. Tarantin
Mining Institute UB RAS, Perm, Russian Federation
e-mail: genagravik@gmail.com

V. Kostitsyn · V. Gershanok · R. Iblaminov
Perm State University, Perm, Russian Federation

Computer system VECTOR developed in the Mining Institute UB RAS (Prostolupov et al. 2006; Chadaev et al. 2014; Chadaev et al. 2016) was used as a main operating tool for interpretation of gravity data. Produced in the system 3D transform of the field can be considered as a quasi-density model of medium. Horizontal slices and vertical cuts were used for analysis, integration, and interpretation of geological data.

The main features of the structure of the Arctic shelf, continent and their juncture are presented at the horizontal slice of a three-dimensional gravity chart built into the system VECTOR (Fig. 24.1). Area of study comprises the waters of the Arctic Ocean with coastal seas and part of land of the Russian Federation, including more detail territory of the northwestern part of the Barents Basin. On the map, the contour line (purple color) with the absolute sea-bed level-200 m taken as a morphological boundary, and the contours of the local uplifts within the Pechora and Russian Plates (blue) are drawn.

Anomaly field is clearly differentiated on tramline extended positive anomalies, associated with ridges and uplifts, and negative ones, associated with basins and depressions. Many varying in size and intensity anomalies create an anomalous gravitational field, which differs from the fields of the continent. Ridges and basins are fairly regular, continuing into the Polar region.

Profiles a, b, c, d and e were formed on the map offshore the North Barents Basin, to the west of the northern edge of the Novaya Zemlya Archipelago (Fig. 24.2). Profile b coinciding with seismic profile MOV 200,707 was used as reference. It also corresponds to the density model built earlier using the GRAV-3D program. Density was defined taking into account the seismic velocity (Zhemchugova et al. 2001).

On these five lines, the vertical cuts of 3D gravitational field were formed in VECTOR system. Obtained cuts on profiles a, b, c, d and e provide extended visualization of anomalouse space surrounding the reference profile b (Fig. 24.3).

Positive (1–5) and negative (6–10) anomalies are highlighted on the reference cross-section (Fig. 24.3). Location of anomalies, their intensity, size, and configuration were used for the analysis and forecast of the Moho discontinuity surface between profiles.

On the reference cross-section, the position of the boundary of the Moho is shown (Grad et al. 2009). Smallest depth to the Moho surface is about 33 km, and maximal depth is up to 35 km. Depth distribution of the poles P_I representing variety of equivalent sources (blue dots) was built using the program Polus2d (Prostolupov and Tarantin 2013). Poles P_I locate mainly in the basin regions.

Positive relief forms of the Moho discontinuity and shelf uplifts match mainly the positive anomalies of the gravitational field transformed in the VECTOR system. So on the profile a, in the depth range of 300–500 km, the positive anomaly coincides with the elevated Moho structure, which amplitude is about 2 km. The

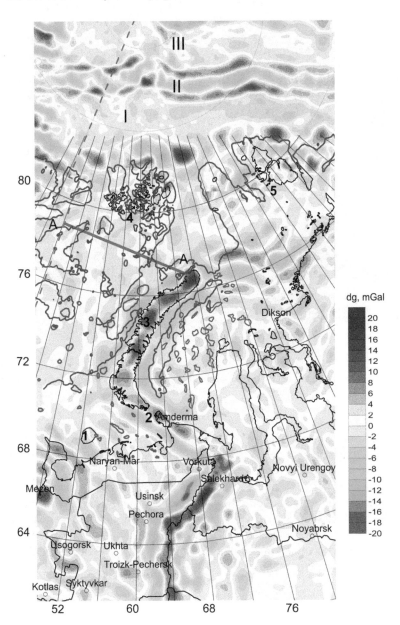

Fig. 24.1 Horizontal slice of the gravitational field transform dg obtained in the system VECTOR. Shelf and the continent: I-Nansen Basin, II-Gakkel Ridge, III-Amundsen Basin. Islands: 1-Kolguev, 2-Vaygach; 3-arch. Novaya Zemlya, 4-arch. Severnaya Zemlya. Blue shows the contours of the uplift. Red line shows the position of the seismic profile (Piskaryov et al. 2016). Contour (purple) marked −200 m refers to the sea-bed level. The contours of the local uplifts are depicted in blue

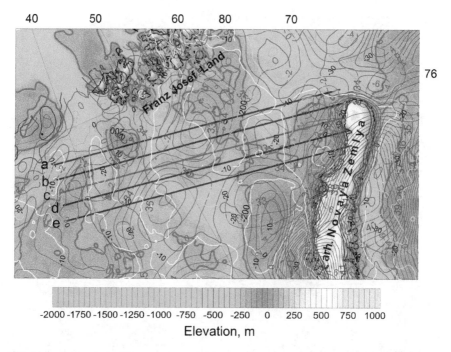

Fig. 24.2 Profiles layout (lines a, b, c, d and e) on the North Barents Basin. Contours: gravity anomalies Δg (white color); anomaly field from VECTOR system (dark-blue); Moho depth section (blue); isobath −200 m (purple)

negative form with an amplitude of 1.5 km is observed on profile d. It corresponds to the relative negative gravity anomaly. Raised structures of bottom related to dikes can be used as indicators of the Moho boundary uplifts.

The shelf bottom has uneven surface. There are elevations and depressions. Assumed position of the edge of the shelf on the −200 m isobath is marked with dotted lines (Fig. 24.3). On the vertical cuts, the limits of the shelf are marked with "diamond" (on the top of each cut). Positive anomalies in the system VECTOR (1–5) are usually confined to the high elevation areas of the seabed and the Moho discontinuity surface.

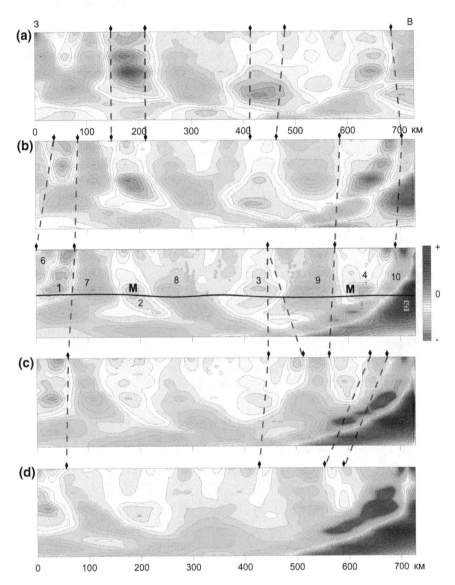

Fig. 24.3 Vertical cuts obtained in the VECTOR system at profiles, shown in Fig. 24.2. The "diamonds" show the edge of shelf location at −200 m isobath

Conclusions

A new technique to create 3D diagrams of the gravitational field transforms in the VECTOR system and related horizontal slices and vertical cuts allows primarily carry out express analysis of large territories. Usage of the seismic data significantly expands the opportunities of cross-section detailization aimed at better examining the Moho boundary and consolidated crust. Experimental data allowed identifying additional elements of the Earth's crust structure outside the seismic profile. It is possible to deploy the anomaly intensity decay to trace continuation of field sources and to make an assumption about the continuation of the continental forms into the shelf territory. It was found that positive gravity field anomalies usually correspond to the higher elevation of the Moho surface and shelf bottom.

References

Zhemchugova, V.A., et al. (2001). Lower Paleozoic of the Pechora oil and gas bearing basin (structure, formation conditions, oil resources). M.: Publishing House of the Academy of mining Sciences. 110 p. (in Russian)

Petrov O.V., et al (2016). Large and unique mineral deposits in the Arctic region. Razvedka i okhrana nedr 12, 8–11. (in Russian)

Piskaryov A.I. et al (2016) Structure of the eastern side of the North Barents Basin and the prospects of discovery of major oilfields in the region.. Razvedka i okhrana nedr 1, 44–48. (in Russian)

Prostolupov G.V. et al (2006) About interpretation the gravitational and magnetic fields based on transformation of horizontal gradients in the system VECTOR. Fizika Zemli 6, 90–96. (in Russian)

Prostolupov G.V. and M.V. Tarantin (2013) Polar transformation of gravitational potential derivative. Geofizika 2, 13–18. (in Russian)

Chadaev M.S. et al (2014) Crustal structure according to gravimetry and magnetometry. Perm. PSU. 95 p. (in Russian)

Chadaev M.S. et al (2016) Parametric relationships between geophysical and geochemical fields in applied geology. Perm. PGNIU. 100 p. (in Russian)

Grad, M., T. Tiira and ESC Working Group (2009). The Moho depth map of the European Plate. Geophysical Journal International 176: 279–292. https://doi.org/10.1111/j.1365-246x.2008.03919.x

Chapter 25
An Iterative Solution of the 2-D Non-Linear Magnetic Inversion Problem with Particular Attention to the Anisotropy of Magnetic Susceptibility of Rocks

A. B. Raevsky, V. V. Balagansky, O. V. Rundkvist and S. V. Mudruk

Abstract A new method of the solution of the 2-D non-linear magnetic inverse problem is considered as a tool for reconstructing the shape of large synforms that contain strongly magnetic layers. It is based on an interpretation model consisting of two components. The first is a 2-D geometrical model which imitates a synform cross-section normal to the synform strike. The second is a set of elementary layers making up a fold model. A specific feature of the layers is their magnetic susceptibility values that are different in directions parallel and normal to planar shape fabrics of rocks. That is why of particular interest has been a consideration of the impact of the magnetic susceptibility anisotropy of rocks in a measured magnetic field. It has been shown that it can be significant and should be taken into account at solving the magnetic inversion.

Keywords Magnetic field · Anisotropy · Inverse problem · Sheath fold

Introduction

The quantitative interpretation of the magnetic survey is crucial for reconstructing the shape of near-surface structures from the magnetic inversion. In this paper we provide an advanced version of the algorithm based on the solution of the 2-D non-linear magnetic inverse problem and described in Raevskii (2008). This method has been used for reconstructing the shape of the large-scale Serpovidny synformal fold in the Keivy terrane of the northeastern Baltic shield which has a sheath shape from orientation data (Mudruk et al. 2013). A specific feature of this

A. B. Raevsky · V. V. Balagansky · O. V. Rundkvist (✉) · S. V. Mudruk
Geological Institute of the Kola Science Centre of the Russian Academy of Sciences, Apatity, Russian Federation
e-mail: rundkvist@yahoo.com

© Springer Nature Switzerland AG 2019
D. Nurgaliev and N. Khairullina (eds.), *Practical and Theoretical Aspects of Geological Interpretation of Gravitational, Magnetic and Electric Fields*, Springer Proceedings in Earth and Environmental Sciences, https://doi.org/10.1007/978-3-319-97670-9_25

fold is that strongly magnetic layers have different values of their magnetic susceptibility. Therefore, the impact of the magnetic susceptibility anisotropy of rocks in a measured magnetic field should be considered among other factors that are decisive for the solution of the 2-D magnetic inverse problem.

General Information

The land magnetic survey was performed for the western part of the Serpovidny fold on grids 25 × 5 and 50 × 5 m. In its eastern part, measurements were done only along four profiles with a step of 20 m. All profiles are oriented orthogonally to the fold strike. The magnetometer receiver was located 0.5 m above the ground surface. The measurement uncertainty in areas of low and high gradient magnetic field was ±5 nT and ±20 nT, respectively. The magnetic field values varied in an interval of 7000 nT. The accepted normal magnetic field parameters corresponded to those of the model IGRF 11 (Maus et al. 2010).

To consistently model the Serpovidny fold, we have used eight parameters: x_1, x_a, x_n, β_1, β_n, H_1, H_q, H_n (Fig. 25.1a). These parameters allow approximating synformal folds of different morphological types by varying these parameters. Zones of the maximum magnetic field gradient are accepted to indicate contacts between rock units that are strongly and weakly magnetic.

Measurements of magnetic susceptibility in 97 rock samples showed that longitudinal κ_t (in the plane orthogonal to planar shape fabric) exceed transverse κ_n (in the plane parallel to planar shape fabric). According to measured anisotropy values (Fig. 25.1b), we may divide rocks into two groups based on factor $f = \kappa_n/\kappa_t$: weakly anisotropic ($f > 0.5$) and highly anisotropic ($f < 0.5$) (Fig. 25.1c). It is obvious that rocks with f values about 0.4 and 0.7–0.8 have a considerable influence on amplitude and the magnetization vector inclination angle in each elementary

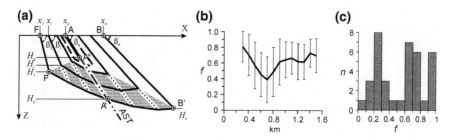

Fig. 25.1 a 2-D geometric model visualizing an orthogonal section through the Serpovidny fold core (AST, axial plane trace). F, B. F′, B′, reference points defining the fold shape. β_1, β_i, β_a, β_n, dip angle (β_1, β_a, β_n, measured; β_a, weighted average of β_1 and β_n); x_1, x_i, x_a, x_n, horizontal coordinate; H_1, H_q, H_i, H_a, H_n, vertical coordinate (depth; H_a, function of coordinates of H_1 and H_n). **b** Variations and **c** histogram of the magnetic anisotropy factor f through the generalized cross-section of the Serpovidny fold core

layer of the fold. Therefore, magnetic anisotropy parameters of model synform elementary layers should be included into an equation that describes an observed anomalous field. For each elementary layer we consider longitudinal magnetic susceptibility and magnetic anisotropy factor to be constant. The lower part of the synform is approximated by two objects with magnetic properties identical to its limbs (Fig. 25.1a). Values of κ_t and f in the lower parts of the synform separated by the axial plane (i.e. parts of the hinge) are equated to the respective parameters of the upper part.

To solve the 2-D forward magnetic problem we need to calculate components of the induced magnetization vector for all elementary layers (J) in a vertical plane orthogonal to the fold strike. An effect of magnetic anisotropy should be incorporated in the equation, therefore we set the following parameters: $|B_0|$, modulus of normal magnetic field B_0; α, inclination of vector B_0; Θ, magnetic azimuth of fold strike; $\bar{\beta}_i = (\beta_i + \beta_{i+1})/2$, average dip angle of elementary layer (i); k_i, longitudinal magnetic susceptibility of elementary layer (i); f_i, magnetic anisotropy factor of elementary layer (i); B_θ, the normal field vector projection on the vertical plane orthogonal to fold strike; α_θ, inclination of vector B_θ. Thus, modulus $|J_i|$ and inclination angle φ_i of magnetization vector J_i (Fig. 25.1) can be defined for each elementary layer by formulas

$$|J_i| = \kappa_i |B_\theta| \sqrt{\cos^2 \gamma_i + f_i^2 \sin^2 \gamma_i}, \quad \varphi_i = \bar{\beta}_i - \delta_i.$$

where

$$|B_\theta| = |B_0| \sqrt{\sin^2 \theta \cos^2 \alpha + \sin^2 \alpha},$$

$$\alpha_\theta = \arcsin\left(\frac{\sin \alpha}{\sqrt{\sin^2 \theta \cos^2 \alpha + \sin^2 \alpha}}\right)$$

$$\delta_i = \arctan(f_i \tan \gamma_i), \gamma_i = \bar{\beta}_i - \alpha_\theta.$$

A reconstructed shape of the original synform represents a synform cross-section which varies in dependence on reference points' coordinates given for synform. The magnetic susceptibility of each layer is set to be uniform (Fig. 25.2c). Solving the 2-D forward magnetic problem for the synform, we can estimate an impact of the magnetic anisotropy on an anomalous magnetic field (Fig. 25.2d). According to tests, the influence of magnetic anisotropy may change the magnetic field value by up to 500 nT even at boundaries of magnetic anomalies. Therefore, the magnetic anisotropy of rocks is crucial for the solution of the 2-D non-linear magnetic inversion.

Furthermore, average values of factor f calculated from measurements along the generalized profile can significantly vary across the fold strike (Fig. 25.1c). Hence, the solution must also include the relationships between factor f and horizontal coordinate x_i of elementary layer i. To show factor f variations along the profile, we use a smoothing approximation with cubic splines based on three parameters: f_1 at

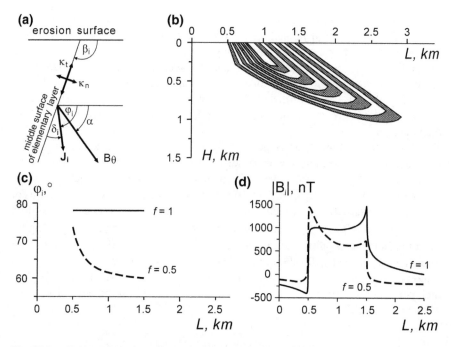

Fig. 25.2 **a** Relationships between magnetization vector J_i and geomagnetic induction vector B_Θ in an elementary layer dipping at angle β_i and displaying the magnetic susceptibility anisotropy. **b** Geometry of a synform imitating the Serpovidny fold core (cross-section). **c–d** Variations of values of the magnetization vector inclination φ_i (**c**) and magnetic induction modulus (**d**) across a test fold at anisotropy factor values of 1 and 0.5

x_1, f_a at x_a and f_n at x_n. Combining these parameters with horizontal coordinate x_i, we can define factor f for each elementary layer and generate functions of f either monotone or having extremums at x_a.

After setting the magnetic anisotropy factor and geometrical parameters defining magnetization of each elementary layer i, the longitudinal magnetic susceptibility k_i remains the only unknown parameter. This parameter is a linear factor in equations of components of anomalous magnetic field. Therefore, it can be determined by the solution to the linear magnetic inverse problem with respect to the longitudinal magnetic susceptibility.

Iterative Linearization Method and Other Approaches

To solve the magnetic inverse problem with respect to magnetic susceptibility, we apply a linearization procedure that is based on the iterative method proposed in (Raevskii 2008) instead of observed values of magnetic induction B_{obs}. Applying the quadratic form $\tau = B^2 - B_0^2$ where B_0 is the normal magnetic field modulus we have got an equation which for the 2-D magnetic field at point j is represented by a recurrent formula

$$\tau_{\text{mod}}^{(l+1)}(j) \approx Z_a^{(l+1)}(j)\left[Z_a^{(l)}(j) + 2Z_0(j)\right] + H_a^{(l+1)}(j)\left[H_a^{(l)}(j) + 2H_0(j)\cos A\right] \quad (25.1)$$

where Z_0, H_0 and Z_a, H_a are respectively normal and anomalous field components, and A is the magnetic azimuth of a plane orthogonal to the synform strike.

Considering anomalous field components as linear operators and taking into account magnetic susceptibility values of elementary layers $k(i)$ we can express Eq. (25.1) as an iterative process on the l-th step:

$$Z_a(j) = \Sigma_i k(i) a(i,j) \quad \text{and} \quad H_a(j) = \Sigma_i k(i) b(i,j), \quad (25.2)$$

$$a(i,j) = |B_\theta| \sqrt{\cos^2 \gamma_i + f_i^2 \sin^2 \gamma_i} [W_{xz}(i,j)\cos\gamma_i + W_{zz}(i,j)\sin\gamma_i] \quad (25.3)$$

and

$$b(i,j) = |B_\theta| \sqrt{\cos^2 \gamma_i + f_i^2 \sin^2 \gamma_i} [W_{xz}(i,j)\sin\gamma_i - W_{zz}(i,j)\cos\gamma_i]. \quad (25.4)$$

In these formulas W_{xz} and W_{zz} are geometric factors of 2-D quadrangular prism gravitational potential derivatives approximating elementary layers (Mudretsova, 1981).

Anomalous field components $Z_a^{(l)}(j)$ and $H_a^{(l)}(j)$ can be presented for the next iteration in form (25.2) and incorporate into Eq. (25.1), we can write a recurrent formula which is a linear system with respect to magnetic susceptibility $k(i)^{(l+1)}$ and is written as

$$\sum_i \kappa(i)^{(l+1)} c(i,j)^{(l)} = \tau_{obs}(j) \quad (25.5)$$

where

$$c(i,j)^{(l)} = a(i,j)\left[Z_a^{(l)}(j) + 2Z_0(j)\right] + b(i,j)\left[H_a^{(l)}(j) + 2H_0(j)\cos A\right].$$

In the given expression parameter $c(i,j)^{(l)}$ is defined by Eqs. (25.2), (25.3), and (25.4), using values of anomalous field components from previous iteration (25.5). Finally, implementing of the misfit minimization problem into recurrent form (25.1) permits evaluating the quality of a solution of the magnetic inverse problem. It is the base for choosing the optimal synform model

$$\|\tau_{obs} - \tau_{\text{mod}}\| = \min, \quad \delta = \|B_{obs} - B_{\text{mod}}\| = \min.$$

The proposed method is based on an interpretation model that incorporates both the geometry and the anomalous magnetic field properties. For the further operations we represent their geometrical parameters (Fig. 25.1), the magnetic anisotropy factor and coordinates of left and right background noise compensators as components of parametric vector $\mathbf{P} = \{p_i, i = 1,\ldots,13\}$. For each of its components we set an interval of values that are constrained by components of vector $\mathbf{L} = \{p_i(\min), p_i(\max); i = 1,\ldots,13\}$ where $p_i(\min) \leq p_i \leq p_i(\max)$. Then we set magnetic susceptibility vector $\mathbf{K} = \{\kappa_i, i = 1,\ldots, 2M+4\}$ for all model synform elements (M, number of layers in a fold limb).

The most acceptable variant of an interpretation model k is thus formed by components of vector $\mathbf{L}$. Consequently, the model magnetic field can be defined as a function B_{mod} of two parametric vectors: vector $\mathbf{P}(k)$ of predetermined geometrical synform parameters and vector of magnetic susceptibility values $\mathbf{K}(k)$. The ground surface relief should also be taken into account:

$$B_{\mathrm{mod}}(k) = B_{\mathrm{mod}}[\mathbf{P}(k), \mathbf{K}(k)].$$

Interpreting measured magnetic induction B_{obs} has to result in finding optimal vector $\mathbf{P}_{opt}$ in a set of vectors $P(k, k = 1,\ldots,m)$, that were obtained taking into account constraints for vector L. Each vector $P(k)$ is paired with congruent vector $K(k)$ constrained by the minimum misfit between measured and model fields:

$$\delta(k) = \|[B_{obs} - B_{\mathrm{mod}}(k)] \cdot W\| = \min[\mathbf{K}(k)]. \tag{25.6}$$

Weight function W determined by evaluation a local noise in each observation point describes the quality of B_{obs} measurements. Evaluating the results, we choose one vector that best suits the minimum misfit condition:

$$\delta(\min) = \left\|[B_{obs} - B_{\mathrm{mod}}(k_{opt})] \cdot W\right\| = \min[\mathbf{P}(k), \mathbf{K}(k)].$$

To diminish the effect of measured field interferences, we consider vector $\mathbf{P}_{opt}$, which components are average values of parametric vector $P(k^*)$ that satisfy the condition $\delta(k^*) \leq \varepsilon$. To define the vector, we perform a synthetic sampling by the iterative approach.

In conclusion, the presented algorithm includes the following operations. At first, we define vector $L(n)$ with components to satisfy the condition: $p_i(n, \min)$ and $p_i(n, \max)$. Using a random-number generator with the uniform distribution, we then limit vector $P(k, k = 1,\ldots, N(n))$ to a range of allowed values as $p_i(k)$ for each k at $P(k)$. The next operation is a solution of an inverse problem with respect to magnetic susceptibility of elementary layers (vector $K(k)$). After that we select 2–3% variants with the least misfit $\delta(k^*)$ for the next iteration $L(n+1)$ from a set of predetermined variants $N(n)$. Then we calculate weighted means $\bar{p}_i^n$ and congruent standard deviations D_i^n based on inverse square of the misfit and parameters of model variants. Finally, we set a constraint vector $L(n+1)$ for the next iteration according to the following conditions:

$$p_i(n+1, \min) = \bar{p}_i^n - \frac{3}{2} D_i^n \quad \text{and} \quad p_i(n+1, \max) = \bar{p}_i^n + \frac{3}{2} D_i^n.$$

This cycle of operations is repeated a given number of times or until stable results are achieved. Performed numerical modelling showed that both the misfit and average values of model parameters become stable after six or seven iterations. The obtained results have been statistically verified by implementing the iterative procedure with the varying number of elementary layers and different initial parameters.

Test Models and the Reconstructed Serpovidny Fold Shape

To assess the validity of the approach we carried out a number of computational experiments of solving the inverse problem for multiple synform models whose cross-sections reflect possible cross-section of the Serpovidny fold. Two of the models are shown in Fig. 25.3a, b. They consist of 30 elementary layers whose magnetic susceptibility values were randomly generated in the range from 0 to 50×10^{-3} SI (Fig. 25.3c–f). Magnetization vector inclination angles of the upper model part layers are given in Fig. 25.3g, h. Magnetic field values were calculated with respect to the real conditions of the ground survey and complicated by random pulse interferences and a background noise. Six to seven iterations were enough for achieving reasonable results (Fig. 25.3i, j).

Numerical test results displayed that the precision of the geometrical fold parameters is not accurate when background noise level is high. If the background noise intensity is, however, comparatively low, the geometry of the fold can be determined satisfactorily. Also, evaluating the intensity of local interferences in Eq. (25.6), makes the solution significantly more accurate and stable. Furthermore, we do not need the high precision for identification of all fold parameters since determining only critical features of the fold are much more important e.g. whether it is inclined, upright, or overturned or at what depth its trough is located. These data along with the fold contour on the ground surface make possible to determine the hinge line curvature. Numerous tests also demonstrated very reasonable results when strongly magnetic rocks were folded.

Using the proposed approach, we reconstructed the shape of the Serpovidny fold core which is strongly magnetic. A 3-D model based on eight cross-sections reconstructed through the solution of the 2-D non-linear magnetic inverse problem is consistent with a structural model based on field observations according to which the Serpovidny fold is a sheath isoclinal synform whose core is pitched (Mudruk et al. 2013). In all parts of the core, except from fold closures, rocks dip north-northeastward, with the calculated dip angles being 30–50° in the northern limb and 50–70° in the southern limb. The trough of the strongly magnetic inner part of the core reaches the depth of about 2 km whereas the length of this part along the sheath length is equal to ~6.5 km. As the core includes a weakly

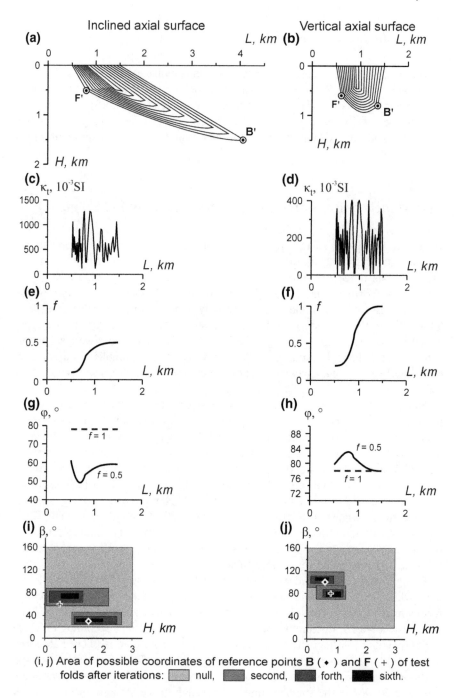

(i, j) Area of possible coordinates of reference points **B** (♦) and **F** (+) of test folds after iterations: null, second, forth, sixth.

◂**Fig. 25.3 a, b** Cross-sections of test folds. **c–h** Variations of magnetic field parameters across test fields: **c, d** longitudinal magnetic susceptibility κ_t, **e, f** magnetic anisotropy factor f and **g, h** inclination of the magnetization vector φ at the normal magnetic inclination (78°) in the study area (67° 45′ N, 37° 00′ E). **i, j** Plots showing the dependence of reference points **B** and **F** possible coordinates from the number of iterations

magnetic unit, the total length may reach 10 km. Together with weakly magnetic limbs of the entire Serpovidny fold dimensions of the entire fold may exceed two or three tens kilometers. Its axial plane becomes gentler (30–40°) at deeper levels.

Conclusions

1. A new iterative algorithm of the solution of the 2-D non-linear magnetic inverse problem is developed. It permits to reconstruct the shape of poorly exposed large-scale synformal folds which contain strongly magnetic layers.
2. The impact of the magnetic susceptibility anisotropy of rocks in a measured magnetic field has been investigated. It was demonstrated that it may be significant and should be taken into account at solving the magnetic inversion.
3. The given approach provides stable results of the determination of reference fold points' coordinates. The accuracy of these coordinates is sufficient to confidently answer a number of questions important for the geologist. Inter alia, whether the synformal fold is straight or inclined and even overturned, or what depth its trough line reaches. In turn, the data along with the fold outline on the ground surface gives an opportunity to evaluate the hinge line curvature.

Acknowledgements The study was finalized in the frame of research project № 0231–2015–0004 in the Geological Institute of the Kola Science Centre of the Russian Academy of Sciences with the financial support of the Russian Foundation for Basic Research (project 16–05–01031A).

References

Maus, S., Manoj, C., Rauberg, J., Michaelis, I., and Lühr, H. (2010). NOAA/NGDC candidate models for the 11th generation International Geomagnetic Reference Field and the concurrent release of the 6th generation Pomme magnetic model. Earth, Planets Sp. 62: 729–735. http://dx.doi.org/10.5047/eps.2010.07.006

Mudretsova, E.A. (1981). Direct problem of gravity prospecting for bodies of the correct geometric shape. In: Mudretsova, E.A. (ed.). Gravity prospecting (geophysicist reference book). Hedra Publishing, Moscow, pp. 173–197. (In Russian).

Mudruk, S.V., Balagansky, V.V., Gorbunov, and I.A., Raevsky, A.B. (2013). Alpine-type tectonics in the Paleoproterozoic Lapland-Kola Orogen. Geotectonics 47: 251–265. http://dx.doi.org/10.1134/S0016852113040055

Raevskii, A.B. (2008). An iterative algorithm for magnetic induction modulus inversion. Izv. Phys. Solid Earth 44: 548–554. http://dx.doi.org/10.1134/S1069351308070057

Chapter 26
On 2D Inversion of MTS Data in Tobol-Ishim Interfluve of Western Siberia

N. V. Baglaenko, V. P. Borisova, Iv. M. Varentsov, T. A. Vasilieva and E. B. Fainberg

Abstract The results of 2D inversion of amplitude and phase effective curves from magnetotelluric soundings in the Tobol-Ishim interfluve of Western Siberia are presented. These results confirm the existence of conductive regional faults of submeridional strike and a conductive asthenospheric layer at depths of 70–80 km. The revealed deep conductive anomalies may indicate peculiarities of the oil and gas generation regimes.

Keywords Magnetotellurics · Tobol-Ishim interfluve · Conductive layer

In the article (Borisova et al. 2013) the analysis and 1D inversion of the amplitude and phase curves of magnetotelluric (MT) soundings made by the Tyumen Geological department in 1980–1981 was carried out, followed by the construction of a two-dimensional geoelectric model by the "stitching" of 1D sections obtained. A series of regional faults of submeridional strike have been identified in the geoelectric section of the Tobol-Ishim interfluve. In addition, at depths of 55–70 km, the roof of a conducting horizon is traced, whose resistivity is tens of Ωm, and two extents of an abnormal rise of the roof of the conductive horizon are identified (Fig. 26.1), one of which corresponds to the Kiselevsky fault (at depth of 60–65 km), another—the onboard zone of the Ishim branch of the Triassic rift system of the West Siberian Plate (at depth of 55–60 km). It has been supposed that the Ishim geoelectric anomaly corresponding to the geothermal anomaly and the features of the related deep geological structure is due to the element of the mantle-crustal fluid paleosystem. It is critically important to verify these results within the class of a two-dimensional (2D) inversion.

For the construction of a 2D model of the studied area, the profiles 16 and 20 of 108 and 57 km lengths were chosen. The MTS data measured along these profiles,

---

N. V. Baglaenko · V. P. Borisova · Iv. M. Varentsov
T. A. Vasilieva · E. B. Fainberg (✉)
GeoElectroMagnetic Research Centre, Schmidt Institute of Physics
of the Earth Russian Academy of Sciences, Moscow, Russia
e-mail: edfain@yandex.ru

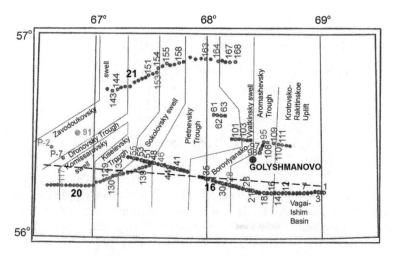

Fig. 26.1 The layout of profiles, MTS points and geological structures

namely the effective (square determinant) amplitude-phase curves of the apparent resistivity along the profile 16 (Fig. 26.2) and the curves in the measurement directions (close to meridians and parallels) along the profile 20, look heterogeneous.

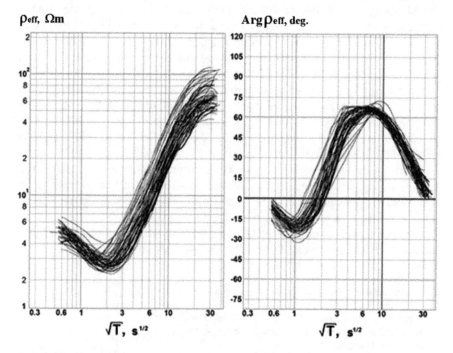

Fig. 26.2 Effective amplitude and phase apparent resistivity curves of MT soundings, profile 16

Since there were greater amount of effective data (profile 16), a variant of their joint inversion along combined profiles was chosen. On this way, the bimodal data at the profile 20 were recalculated into the values of the effective impedance phases Arg Z_{eff} and the apparent resistivities ρ_{eff}. All sites from profiles 20 and 16 were projected using the GMT tool (Wessel and Smith 2007) to a common profile (the dotted line in Fig. 26.1) passing close to both original profiles across the strike of dominant regional structures. The inversion was carried out using the code (Varentsov 2002, 2015a), developed to construct models in the class of 2D piecewise-continuous media. This code allows selecting the distribution of electrical resistivity in 2D models from profile observations of various components of EM fields and their transfer functions for several periods. Geoelectric section is represented as a superposition of the background block geoelectric structure and superimposed rectangular windows "scanning" the anomalous structure. Resistivity of individual blocks of the background structure can also be optimized in the course of solving the inverse problem. The code uses modern schemes of stable optimization of Tikhonov's nonlinear functionals, effective finite-difference modeling procedures, and robust metrics of minimized data misfits. As a result, it reliably treats multicomponent data ensembles, including effective impedance data (Varentsov et al. 2013).

MT data were limited to a period range of 0.25–1000 s and the full impedance tensor was not available at many MT sites and for some periods in this range. Missing Arg Z_{eff} and ρ_{eff} data elements in the range of 400–1000 s were filled with the interpolation from neighboring sites and got sufficiently increased error estimates. The final data set included data from the profile 20 in the 15–400 s range, and from the profile 16 in the 15–1000 s range.

The impedance phases are less sensitive to the distorting effect of near-surface inhomogeneities than the amplitudes. Moreover, the phase curves begin to feel the effect of the deep structures at earlier periods than the amplitude curves. Taking into account these factors, inversion weights for the phase data were set twice larger than for apparent resistivities at periods greater than 36 s. The background ("normal") geoelectric section was taken as a horizontally-layered structure, and a superimposed rectangular "scanning" window for the anomalous structure was set under the observation sites from the ground surface to a depth of 160 km. Resistivities of the starting inversion model were selected taking into account the results of one-dimensional inversion of MT soundings at the Askino-Tyukalinsk geotraverse (Dyakonova et al. 2008).

The results of the inversion are presented in Fig. 26.3a–g in two variants. Figure 26.3a shows the model obtained from the data at the middle profile (dashed line in Fig. 26.1) for the period interval 15–1000 s. The accuracy of the inversion can be estimated from the Fig. 26.3b–g: the misfits do not exceed first percent for apparent resistivity and first degrees for the phase in the whole period range along the whole profile extent, i.e. lie within the experimental data error level.

The obtained 2D model generally confirms the main conclusions from 1D analysis and inversion (Borisova et al. 2013). Comparing the disclosed anomalies of electrical conductivity with the location of the main geological structures presented

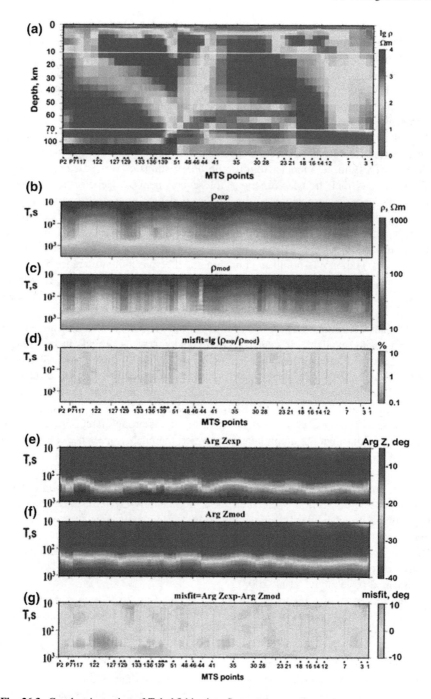

Fig. 26.3 Geoelectric section of Tobol-Ishim interfluve

in Figs. 26.1 and 26.3a, we can state that the vertical conductive zones in the area of the MT sites 055, 044-041, 023-012 are very close to Kiselevsky, Pletnevsky, Aromashevsky (Turgay) faults. MT sites 007-001 at the right side of the profile indicate the presence of a wide conductive zone in the upper part of the section, coinciding with the rift zone of the West Siberian Triassic system (Kostyuchenko 1992). The layer of increased conductivity at a depth of ~ 55–65 km is clearly distinguished. The steep dip of this layer in the western part of the profile has not yet been explained and is the subject of further research. To refine and improve the obtained results, which are important for studying the deep structure of the region, new combined magnetotelluric and magnetovariational soundings with modern equipment and data processing and interpretation procedures capable to treat simultaneous observation systems (Varentsov 2015a, b) are desirable.

References

Borisova V.P., T.A. Vasilyeva, S.L. Kostyuchenko, N.V. Narskii, B.K. Sysoev, E.B. Fainberg, A.G. Charushin and A.L. Sheinkman (2013). On the deep electrical conductivity of the lithosphere of the Tobol-Ishim interfluve, Western Siberia. Izv., Phys. Solid Earth. V. 49(3).

Dyakonova A.G., K.S. Ivanov, O.V. Surina, P.F. Astafyev, V.S. Vishnev and A.D. Konoplin (2008). The structure of the tectonosphere of the Urals and the West Siberian Platform according to electromagnetic data. DAN. Series: Geophysics. V. 423(5). pp. 685–688.

Kostyuchenko S. L. (1992). The deep structure of the Earth crust of West Siberian plate by the results of complex geological and geophysical research. PhD Thesis in geological and mineralogical sciences. M.: MSU. 247 p.

Varentsov Iv.M. (2002). A general approach to the magnetotelluric data inversion in a piecewise-continuous medium. Izv. Phys. Solid Earth. V. 38(11). pp. 913–934.

Varentsov Iv.M. (2015a). Methods of joint robust inversion in MT and MV studies with application to synthetic datasets. Electromagnetic sounding of the Earth's interior: theory, modeling, practice (Ed, V.V. Spichak). Elsevier. pp. 191–229.

Varentsov Iv.M. (2015b). Arrays of simultaneous electromagnetic soundings: design, data processing, analysis, and inversion. Electromagnetic sounding of the Earth's interior: theory, modeling, practice (Ed, V.V. Spichak). Elsevier. pp. 271–299.

Varentsov Iv.M., V.A. Kulikov, A.G. Yakovlev and D.V. Yakovlev (2013). Possibilities of magnetotelluric methods in geophysical exploration for ore minerals. Izv., Phys. Solid Earth. V. 49(3). pp. 309–328.

Wessel P. and W.H.F. Smith (2007). The generic mapping tools. Technical reference and cookbook, v. 4.2. http://gmt.soest.hawaii.edu.

Chapter 27
Non-hydrostatic Stresses Under the Local Structures on Mars

A. Batov, T. Gudkova and V. Zharkov

Abstract To evaluate the stress field in the Martian interior a static-state approach is applied. We use trial interior structure model having 150–300 km thick lithosphere overlying a low rigidity layer, which partly lost elastic properties. Calculations of stresses are performed with spatial resolution a 1×1 arc-deg spherical grid and down to 1000 km depth. Stress estimates are calculated in the interiors of the planet under local topography structures, these areas are of interest to reveal the zones of possible marsquakes sources. Large non-hydrostatic stresses under Hellas Planitia, Argyre Planitia, Mare Acidalia, Arcadia Planitia and canyon Valles Marineris may lead to relatively increased seismic activity for these regions.

Keywords Mars · Topography · Gravity · Stress field

Introduction

Studying of stress field in the Martian interior is of importance for the seismic exploration of Mars. Discovery Program mission InSight (Interior Exploration using Seismic Investigations, Geodesy and Heat Transport) will place a single geophysical lander with a seismometer in Elysium Planitia on Mars to study its deep interior (Banerdt et al. 2013), as well as the international project of Russian Space Agensy and European Space Agency suggesting seismic sounding of Mars is under preparation (Manukin et al. 2016). The estimates of global seismicity are reported in (Knapmeyer et al. 2006).

Theoretical stress modelling studies attempting to understand the sequence of events and mechanisms responsible for the specific features of the Martian surface have been performed starting from the Viking-era (e.g., Banerdt et al. 1982; Sleep

---

A. Batov (✉)
Institute of Control Sciences of RAS, Moscow, Russia
e-mail: batov@ipu.ru

T. Gudkova · V. Zharkov
Schmidt Institute of Physics of the Earth RAS, Moscow, Russia

and Phillips 1985; Banerdt and Golombek 2000; Arkani-Hamed 2000; Zhong and Roberts 2003; Belleguic et al. 2005; Dimitrova et al. 2006). Recently, subcrustal stress field on Mars was computed by Tenzer et al. (2015).

Joint analysis of gravity and topography data let us some knowledge on stress field in the crust and in the lithosphere of Mars. Estimates of the stress distribution in the lithosphere of Mars in frame of the static method reported in (Zharkov et al. 1991; Koshlyakov and Zharkov 1993) were based on the gravity field data complete to spherical harmonic degree and order 50. Since then, the data on gravity and topography have been consistently expanded and improved.

Information on Martian topography comes from the high resolution MOLA data (Smith et al. 2001), which are expanded into spherical harmonics. Two recent spherical harmonic expansions of the Martian gravity field complete to degree and order 120 are MRO120D by Konoplive et al. (2016) and GMM-3 by Genova et al. (2016) (the models are available at https://urldefense.proofpoint.com/v2/url?u=http-3A__pds-2Dgeosciences.wustl.edu&d=DwIDaQ&c=vh6FgFnduejNhPPD0fl_yRaSfZy8CWbWnIf4XJhSqx8&r=Sso2B5iThJQwICrxPF1WAiVf3MsSKjSKbbiLUEa68aqF5T3kgwJnrbhDgmVskN9J&m=af3uRN2Jujnacry0NX9rm5PcI8pJPoIQuLeSL1U6MYs&s=HfbgRS2nIHIbMFFq1pPMYiIxBdBqYHlBbsg6jkMsBXo&e=).

The recent progress in developing the gravity and topographic models of Mars (Smith et al. 2001; Konopliv et al. 2016; Genova et al. 2016) allowed us to study the stress field in detail and to recalculate previous estimates of static stresses (Gudkova et al. 2017).

The purpose of this paper is to reveal areas of large shear and extension-compression stresses in the lithosphere of Mars as possible marsquakes' sources. We perform numerical calculations of non-hydrostatic stresses (extension- compression stresses and maximum shear stresses) in Mars for a trial interior structure model having 150–300 km thick lithosphere overlying a low rigidity layer, which partly lost elastic properties, with a 1 × 1 arc-deg spherical grid and down to 1000 km depth.

Method

The Numerical simulation is based on a static approach (the loading factors technique or the Green's functions method) (Marchenkov et al. 1984; Marchenkov and Zharkov 1989; Zharkov et al. 1986). According to this method a planet is modeled as an elastic, self-gravitational spherical body. It is assumed, that deformations and stresses which obey Hooke's law are caused by the pressure of relief on the surface of the planet and anomalous density $\delta\rho(r, \theta, \varphi)$, distributed by a certain way in the crust and the mantle.

The anomalous density field is represented in the form of weighted thin layers positioned at different characteristic depths. Imposing the anomalous density waves (ADW) on the surface or in the interior leads to the deformation of the planet interior and the distortion of the surface and the boundary interfaces. In spherical coordinates (r, θ, φ), it is convenient to expand the distribution of anomalous density into spherical harmonics:

$$\delta\rho(r,\theta,\varphi) = \sum_{i,n,m} R_{inm}(r)Y_{inm}(\theta,\varphi) = \sum_{i=1}^{2}\sum_{n=2}^{\infty}\sum_{m=0}^{n} R_{i,n,m}(r)Y_{i,n,m}(\theta,\varphi), \quad (27.1)$$

where R_{inm}—expansion coefficients (amplitudes),

$$Y_{inm}(\theta,\varphi) = P_{nm}(\cos\theta)\begin{cases} \cos(m\varphi), i=1 \\ \sin(m\varphi), i=2 \end{cases},$$

$P_{nm}(x)$ are the associated Legendre polynomials and are taken here as fully normalized such that

$$P_{nm}(x) = \left(\frac{2(n-m)!(2n+1)}{(n+m)!}\right)^{1/2} P_n^m(x), m \neq 0;$$

$$P_{nm}(x) = \left(\frac{(n-m)!(2n+1)}{(n+m)!}\right)^{1/2} P_n^m(x), m = 0 \quad (27.2)$$

$$P_{nm}(x) = (1-x^2)^{m/2}\frac{d}{dx^m}(P_n(x)).$$

With the addition of ADW to a planet it goes to a new state of elastic equilibrium, that is, it "adjusts" to the ADW. So, the problem is reduced to the determination of Green's response function for the case of a single ADW located at some depth level.

The gravitational field at the surface of a planet from such a spherical layer is expressed as

$$\Delta V = 4\pi GR \sum_{i,n,m} \left(\frac{r}{R}\right)^{n+2} \frac{R_{i,n,m}(r)}{(2n+1)} Y_{i,n,m}(\theta,\varphi) \quad (27.3)$$

The anomalous density layer acts as a load on the planet, its interior undergoes deformations. To account additional perturbations of the potential due to the global deformation of the planet, the coefficient $(1 + k_n(r))$ is introduced:

$$\Delta V = 4\pi GR \sum_{i,n,m} \left(\frac{r}{R}\right)^{n+2} \frac{R_{i,n,m}(r)(1+k_n(r))}{(2n+1)} Y_{i,n,m}(\theta,\varphi), \quad (27.4)$$

$k_n(r)$ is the load number for the n-th harmonic of the anomalous density wave located at radius r, $1 + k_n(r)$ defines the total change in the gravitational potential on the planetary surface.

Load Love numbers $h_n(r)$ which are used to describe the deformation under the action of the load are introduced

$$D(\theta, \phi) = \frac{4\pi GR}{g} \sum_{i,n,m} \left(\frac{r}{R}\right)^{n+2} \frac{R_{i,n,m}(r)h_n(r)}{(2n+1)} Y_{i,n,m}(\theta, \varphi). \quad (27.5)$$

Load coefficients for deeply buried density anomalies $k_n(r)$ and $h_n(r)$ of n-th harmonic of ADW located at radius r define the total change in the gravitational potential on the surface of the planet and deformation of the planetary surface under the action of load, respectively. It is clear that a unique inversion of ADW from the data on the gravitational field of a planet is not possible, an infinite number of mass distributions can give rise to the same potential on the bounding surface. A unique solution is possible only if addition conditions are imposed, for example, by assuming that there are two levels of concentration of anomalous density in Mars—on its surface and at the crust-mantle boundary. The amplitudes of ADW are selected so that the anomalous gravitational field can be reproduced.

Let the topographic load as an equivalent infinitely weighted thin layer be at the reference surface R, while the rest of the anomalous mass as an equivalent infinitely weighted thin layer being at the crust—mantle boundary R_1. Spherical expansion coefficients of the anomalous density waves on the surface $R^1_{i,n,m}(\theta, \varphi)$ and at the crust-mantle boundary $R^2_{i,n,m}(\theta, \varphi)$ are related to spherical expansion coefficients of the anomalous gravitational field C_{ginm} and the Martian topography C_{ginm} by equations (Zharkov et al. 1991):

$$C_{ginm} = \frac{R^1_{inm}}{R\rho_0} \frac{3(1+k_n(R))}{(2n+1)} + \frac{R^2_{inm}}{R\rho_0} \frac{3(1+k_n(R_1))}{(2n+1)} \left(\frac{R_1}{R}\right)^{n+2}, \quad (27.6a)$$

$$C_{tinm} = \frac{R^1_{inm}}{R\rho_c} + \frac{R^1_{inm}}{R\rho_0} \frac{3(1+h_n(R))}{(2n+1)} + \frac{R^2_{inm}}{R\rho_0} \frac{3(1+h_n(R_1))}{(2n+1)} \left(\frac{R_1}{R}\right)^{n+2}, \quad (27.6b)$$

where ρ_0 is the mean density of Mars, ρ_c is the density of the crust, R is the radius of Mars, R_1 is the radius of the crust-mantle boundary. In Eq. (27.6b), the first term describes the contribution to the relief directly from the surface loading, and the second and the third terms describe the deformations by the actions of $R^1_{i,n,m}(\theta, \varphi)$ and $R^2_{i,n,m}(\theta, \varphi)$, respectively. These Eq. (27.6a) establish a unique relationship between coefficients C_{gnm}, C_{tnm} and $R^1_{i,n,m}(\theta, \varphi), R^2_{i,n,m}(\theta, \varphi)$.

First, the components of complete stress tensor σ_{ij} are calculated for every point (r, θ, ϕ) at given depth, longitude and latitude. Then this tensor is reduced to the diagonal form with the principal stresses $\sigma_3 \leq \sigma_2 \leq \sigma_1$. The extension-compression stresses p and maximum shear stresses τ are defined as $p = (\sigma_1 + \sigma_2 + \sigma_3)/3$ and $\tau = \max|\sigma_i - \sigma_k|/2$, $(i, k = 1, 2, 3; i \neq k)$.

Stresses Estimates

As a benchmark real model for the planetary interior we use a trial model of Mars M_50 from (Zharkov et al. 2017), which satisfies currently available geophysical and geochemical data. The mean density of the crust is 2900 kg m^{-3}, the thickness of the crust is 50 km, the density contrast at the crust-mantle boundary is 360 kg m^{-3}.

The definition of the "topography" needs the choice of a reference surface. Mars departs from hydrostatic equilibrium to significant extent. To avoid uncontrollable stresses and deformations in the mantle of the planet due to the significant deviation of Mars from hydrostatic equilibrium state, we take an outer surface of the hydrostatic equilibrium form of the planet, as a reference surface for topography and gravity field of Mars (Zharkov et al. 2009; Zharkov and Gudkova 2016). Parameters s_2, s_4 and gravitational moments J_2, J_4 of the equilibrium spheroid are listed in the Table 27.1. Only nonequilibrium components of gravity and topography fields are considered: they are obtained by subtracting the equilibrium components from the observed external field.

For stress field calculations the rheological cross section of Mars is of importance, but it is not well known. The Green function values depend on the internal structure, particularly on its density and elastic parameters (rheology). The lithosphere thickness of Mars remains presently in debate. Mars deviates much more strongly from the hydrostatic equilibrium than the Earth. We suggest that the average thickness of the Martian elastic lithosphere should exceed that of the Earth's continental lithosphere (Zharkov and Gudkova 2016). Best estimates for the present day elastic thickness are above 150 km (Plesa et al. 2016).

Below we consider models with an elastic, 150–300 km thick lithosphere, overlying a sublithosphere low rigidity layer—a weakened layer, which partly lost elastic properties (shear modulus is ten times lower). Weakened layer is assumed to extend down to the core-mantle boundary. The source of gravity anomalies is assumed to be the topographic loading and density anomalies at crust-mantle boundary. The method, described in Section "Method", allows one to calculate all stresses (extension-compression stresses and shear stresses) at different depths.

Figure 27.1 shows the details of the depth distribution of stresses magnitude beneath such geological structures as Olympus Mons, Hellas, Argyre, Mare Arcadia and Valles Marineris.

We assume that large shear stresses in the zones of large extensional stresses are the most likely areas for marsquakes' sources. Such regions are found beneath Hellas, Argyre, Acidalia Planitia, Arcadia Planitia and Valles Marineris, both the maximum shear stresses and the extensional stresses are quite large, up to 20–40 MPa for extensional stresses in the lithosphere, and about 20 MPa for shear

Table 27.1 Parameters of a trial interior structure model

Model	ρ_{crust} (kg/m^3)	l_{crust} (km)	r_{core} (km)	I/MR^2	k_2^S	$\rho_{crust}/\rho_{mantle}$	$-s_2$, 10^{-3}	s_4, 10^{-6}	J_2^0, 10^{-3}	$-J_4^0$, 10^{-6}
M_50	2900	50	1821	0.3639	0.162	3.00/3.36	3.338	9.374	1.800	7.634

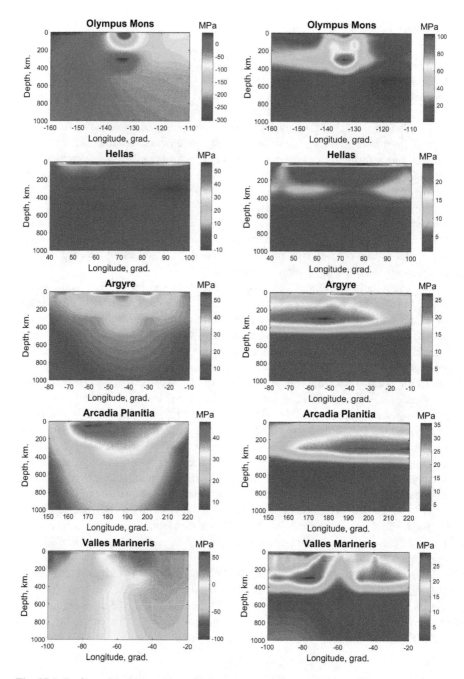

Fig. 27.1 Profiles of maximum shear (right) and extention-compression (left) stresses for the interior structure model M_50 with 300 km thick lithosphere overlying a low rigidity layer which partly lost elastic properties. The profiles are given along the latitude

stresses. For these features, the shear stresses in the crust are comparable in magnitude with those in the Tharsis region. Hellas and Argyre are characterized by rather large non-hydrostatic stresses, which reach their maximum level in the crust and sharply diminish below 50 km depth level of compensation. It might be explained by consideration of these basins as mascons. Hellas is the deepest depression on Mars, while positive gravity anomaly being observed in this region (Zharkov and Gudkova 2016). Argyre shows a slight amount of down warping indicating a weak mascon (Banerdt and Golombek 2000). On the whole Hellas and Argyre are isostatically compensated impact basins, some overcompensation may be related to the accumulation of erosion products on the bottom of the basin, which leads to the formation of a mascon type structure and to an increase of the level of non-hydrostatic stresses in the crust. Moreover, Argyre is in the vicinity of the Tharsis rise, and an additional deformation of the lithosphere under the weight of it can occur. Beneath Valles Marineris and Mare Acidalia, Mare Arcadia stresses particularly leak from the crust into the lithosphere, that can withstand them, and a smaller part of stresses filters into the mantle.

The Valles Marineris has a rather complicated profile of stress state with the areas of compression and extension, the extensional stresses reach 55 MPa, while the shear stresses being about 20–25 MPa. The specific stress distribution beneath the Valles Marineris could be explained by its formation, which is assigned to regional extensional tectonism (Banerdt et al. 1992).

Conclusion

We have applied the static-state approach (Marchenkov et al. 1984; Zharkov et al. 1986, 1991) to evaluate stress field in Mars up to a spherical harmonic degree 90, which corresponds to a spherical resolution of MRO120D gravity model by Konoplive et al. (2016). The solution is computed over a global grid of 10 × 10, assuming various depths.

We have identified those zones in which maximum shear stresses occur along the extensional stresses in the crust and lithosphere of Mars, and non-hydrostatic stresses are quite large, on the order of tens of MPa. From the point of view of seismicity, we assume these regions could be the candidates for potential marsquakes' sources.

Of particular importance are the areas beneath the impact ring basins, mascons Hellas Planitia and Argyre Planitia; the areas adjacent to the Tharsis rise Mare Acidalia and Arcadia Planitia; and huge canyon Valles Marineris. Beneath these regions, there are zones where extensional stresses occur simultaneously with significant shear stresses. Large non-hydrostatic stresses may lead to relatively increased seismic activity for these regions.

Acknowledgements The work was supported by the Russian Foundation for Basic Research and Program RAN 28.

References

Arkani-Hamed, J. (2000). Strength of Martian lithosphere beneath large volcanoes. J. Geophys. Res. 105, E11, 26713–26732.

Banerdt, W.B., Phillips, R.J., Sleep, N.H., Saunders, R.S. (1982). Thick shell tectonics of one plate planets: application to Mars. J. Geophys. Res. 87, B12, 9723–9734.

Banerdt, W., Golombek, M.P., Tanaka, K.L. (1992). Stress and tectonics on Mars. Mars, 1, 249–297.

Banerdt, W.B., Golombek, M.P. (2000). Tectonics of the Tharsis region of Mars: insights from MGS topography and gravity. In Proceedings of the 31st Lunar and Planetary Science Conference. 2038. pdf.

Banerdt, W.B., Smrekar, S., Lognonné, P., Spohn, T., Asmar, S.W., Banfield, D., Boschi, L., Christensen, U., Dehant, V., Folkner, W., Giardini,D., Goetze, W., Golombek, M., Grott, M., Hudson, T., Johnson, C., Kargl, G., Kobayashi, N., Maki, J., Mimoun, D., Mocquet, A., Morgan, P., Panning, M., Pike, W.T., Tromp, J., van Zoest, T., Weber, R., Wieczorek, M.A., Garcia, R., Hurst, K. (2013). InSight: a discovery mission to explore the interior of Mars. In Proceedings of the 44th Lunar and Planetary Science Conference, p. 1915.

Belleguic, V., Lognonné, P., Wiezorek, M. (2005). Constraints on the Martian lithosphere from gravity and topography data J. Geophys. Res. 110. E11005. https://doi.org/10.1029/2005je002437.

Dimitrova, L.L., Holt, W.E., Haines, A.J., Schultz, R.A. (2006). Toward understanding the history and mechanisms of Martian faulting: The contribution of gravitational potential energy. Geophys. Res. Lett. 33, L08202, https://doi.org/10.1029/2005gl025307.

Genova, A., Goossens, S., Lemoine, F.G., Mazarico, E., Neumann, G.A., Smith, D.E., Zuber, M. T. (2016). Seasonal and static gravity field of Mars from MGS, Mars Odyssey and MRO radio science. Icarus 272, 228–245.

Gudkova, T.V., Batov, A.V., Zharkov, V.N. (2017). Model estimates of non-hydrostatic stresses in the Martian crust and mantle: 1. Two-level model. Solar Syst. Res. 51 (6), 457–478.

Knapmeyer M., Oberst J., Hauber E., Wählisch M., Deuchler C., Wagner R. (2006). Working models for spatial distribution and level of Mars' seismicity. J. Geophys. Res. 111, E11006, https://doi.org/10.1029/2006je002708.

Konopliv, A.S., Park, R.S., Folkner, W.M. (2016). An improved JPL Mars gravity field and Orientation from Mars orbiter and lander tracking data. Icarus 274, 253–260.

Koshlyakov, E.M., Zharkov, V.N. (1993). On gravity field of Mars. Sol. Syst. Res. 27 (2), 12–21.

Manukin, A. B., Kalinnikov, I. I., Kalyuzhny, A. V., Andreev, O. N. (2016). High-sensitivity three-axis seismic accelerometer for measurements at the spacecraft and the planets of the solar system. In Proceedings of Solar System conference 7ms3, IKI RAN.

Marchenkov, K.I., Lyubimov, V.M., Zharkov, V.N. (1984). Calculation of load factors for deeply buried density anomalies. Doklady Earth Science Sections 279, 14–16.

Marchenkov, K.I., Zharkov, V.N. (1989). Stresses in the Venus crust and the topography of the mantle boundary. Sol. Astron. Lett. 16 (1), 77–81.

Plesa, A.-C., Grott, M., Tosi, N., Breuer, D., Spohn, T., Wieczorek, M. (2016). How large are presently heat flux variations across the surface of Mars? J. Geophys. Res. Planets, 121, 12, 2386–2403, https://doi.org/10.1002/2016je005126.

Sleep, N.H., Phillips, R.J. (1985). Gravity and lithospheric stress on the terrestrial planets with References to the Tharsis region of Mars. J. Geophys. Res., 90, B6, 4469–4490.

Smith, D.E., Zuber, M.T., Frey, H.V., Garvin, J.B., Head, J.W., Muhleman, D.O., Pettengill, G.H., Phillips, R.J., Solomon, S.C., Zwally, H.J., Banerdt, W.B., Duxbury, T.C., Golombek, M.P., Lemoine, F.G., Neumann, G.A., Rowlands, D.D., Aharonson, O., Ford, P.G., Ivanov, A.B., Johnson, C.L., McGavern, P.J., Abshire, J.B., Afzal, and R.S., Sun, X., (2001). Mars Orbiter Laser Altimeter: Experimental summary after the first year of global mapping of Mars. J. Geophys. Res., 106 (E10): 23689–23722.

Tenzer, R., Eshagh, M., Jin, S. (2015). Martian sub-crustal stress from gravity and topographic Models. Earth and Planetary Science Letters. 425, 84–92.

Zharkov, V.N. Marchenkov, K.I., Lyubimov, V.M. (1986). On long-waves shear stresses in the lithosphere and the mantle of Venus. Sol. Syst. Res., 20, 202–211.

Zharkov, V.N., Koshlyakov, E.M., Marchenkov, K.I. (1991). Composition, structure and gravitational field of Mars. Sol. Syst. Res. 25, 515–547.

Zharkov, V.N., Gudkova, T.V., Molodensky, S.M. (2009). On models of Mars' interior and amplitudes of forced nutations. 1. The effects of deviation of Mars from its equilibrium state on the flattening of the core-mantle boundary. Phys. Earth Planet. Inter. 172, 324–334.

Zharkov, V.N., Gudkova, T.V. (2016). On model structure of gravity field of Mars. Sol. Syst. Res. 50, 250–267.

Zharkov, V.N., Gudkova, T.V., Batov, A.V. (2017). On estimating the dissipative factor of the Martian interior. Sol. Syst. Res. 51, 6, 479–490.

Zhong S., Roberts J.H. On the support of the Tharsis Rise on Mars//Earth Planet. Sci. Lett. (2003). V. 214. P. 1–9.

Chapter 28
Deep Fluid Systems of Fennoscandia Greenstone Belts

A. A. Petrova, Yu. A. Kopytenko and M. S. Petrishchev

Abstract A study of the basement of Fennoscandia on magnetic and gravity anomalies has been carried out. There are detected magnetic and density features of the structure of the greenstone belts of the Early Precambrian crust of the Baltic Shield as a result of interpretation of CHAMP and GRACE satellite data and near surface magnetic anomalies. Magnetic and density sections have been constructed through the mega blocks of Fennoscandia with the age of the Earth's crust ranging from 3.2 to 1.6 billion years that underwent different stages of metasomatosis. There are detected magnetoactive layer in the basement of the Early Precambrian crust ("magnetite zone") and fluid systems at depths of 12–15 and 25–35 km that are the source of the formation of rich deposits of Fennoscandia greenstone belts. Based on the concept of the distribution of fluid supply flows, there are selected several areas promising for electromagnetic monitoring that can help to detail the features of the basement conductivity of the Baltic Shield.

Keywords Magnetite zone · Fluid systems · Gold

Introduction

The goal of this paper is to study the deep structure of the Fennoscandia region from geophysical data for solving a fundamental problem related to the nature of significant concentrations of ore substances and placers in a given area.

One of the important factors in the formation of gold deposits is the carrying out of high-quality concentrates of relics of the Precambrian basement by fluid flows of mineralized aqueous solutions supplied from deep-focus fluid systems. Migration of

---

A. A. Petrova (✉) · Yu. A. Kopytenko · M. S. Petrishchev
St. Petersburg Branch of Pushkov Institute of Terrestrial Magnetism,
Ionosphere and Radio Wave Propagation of the Russian Academy of Sciences,
St. Petersburg, Russia
e-mail: petrischev@gmail.com

deep fluid plays a huge role in the location of mineral deposits and determines the location of gold ore objects.

In this regard, special attention is paid to the following tasks:

- the allocation of blocks of the pre-Riphean basement as possible sources of formation of rich gold deposits;
- mapping of channels of thermofluid flows, carrying out the carrying out of high-quality concentrates, supplied from deep-focus fluid systems;
- detection of the location of deep fluid systems;
- searching of buried troughs, which are the ore-generating structures that have their own geochemical specialization and metallogeny.

There are creating favorable conditions for the formation of the primary deposits and the accumulation of large gold-bearing placers near such zones of increased fracturing in the places of approaches to the surface of fluid flows. The rich gold deposits of the study area are the result of the carrying out of high-quality Archean concentrates by fluid flows under the pulsating influence of deep-focus fluid mantle systems.

Investigation of the Basement by Magnetic and Gravity Anomalies

Investigations of the deep structure of the Early Precambrian Earth's crust of Fennoscandia have been made on the basis of near-surface air and hydromagnetic measurements, magnetic anomalies from CHAMP satellite and gravity anomalies in the Fay reduction obtained using the GRACE altimetry (International Centre; Kaban and Reigber 2005; Korhonen et al. 2007; Kopytenko and Petrova 2014, 2016; Mayer-Gürr et al. 2014).

The visualization of heterogeneities in the Early Precambrian crust of the Baltic Shield in the form of geophysical sections made it possible to clarify the concept of the deep crustal structure with an age of 3.2–1.6 billion years, which underwent different stages of metasomatosis (Fig. 28.1). Magnetic and density sections are performed by the method of spectral-spatial representation of fields that are convertible into the deep sections (Petrova 1976, 2015; Kopytenko and Petrova 2016; Hahn et al. 1976; Petrishchev et al. 2014; Litvinova et al. 2014).

Ore-generating processes determine the regularity of the location of the magnetite layer in the Early Precambrian Earth's crust. Studies have shown that it is allocated along the lower boundary of the zone of regional granitization along the boundary of the granite and granulite-basite layers. Deposits presented by the iron ore formation of the Archean magnetite quartzites form as a result of the granitization process of magnetite formation. Magnetic sections crossing the megablocks of the Baltic shield made it possible to visualize a magnetoactive horizon underlying the bottom of the granite layer in the foundation of the Early Precambrian Earth's crust (Fig. 28.2).

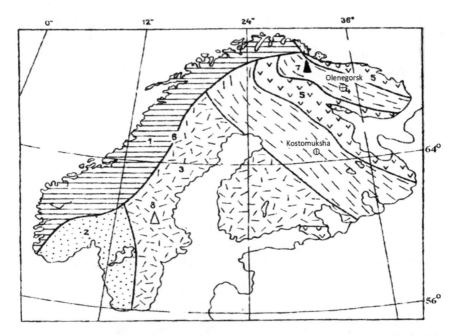

Fig. 28.1 Age characteristics of the Earth's crust of the Baltic Shield: 1—caledonites (PZ); 2—dalclandides (R); beetrofenides (PR_1^1-PR_1^2); 4—Karelides (AR_1^1-PR_1^2); 5—Laplandides and White Seabelomorides (AR_1^1-PR_1^2)

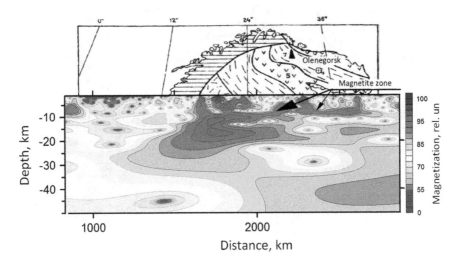

Fig. 28.2 Magnetite zone in the magnetic section of the basement

Mineral deposits in the Early Precambrian Earth's crust are formed in greenstone belts where they are manifested in the form of deposits of ferruginous quartzites of sedimentary-volcanogenic genesis. Large-scale deposits of Kostomuksha and Olenegorsk are known on the Baltic Shield (Fig. 28.1). The iron content in the ore bodies of these deposits of ferruginous quartzites is 20–45%, in the weathering crusts of ore bodies—up to 55–70%.

Anomalies of the magnetic field is played the main role in the investigation of iron ores of the Kostomuksha deposit. This is due to the high magnetization of magnetite-bearing ferruginous quartzites, as well as the contrast of the magnetic susceptibility of the investigated ferruginous-siliceous rocks and their host complexes. In addition, magnetite quartzites and magnetite schists have a high density reaching 3.5–3.7 g/cm^3, so the interpretation of the anomalies of gravity supplements magnetometric studies, allowing a more accurate trace of iron ore horizons.

We have also interpreted magnetic anomalies and gravity anomalies for tracking of iron ore horizons at depth. We have completed deep magnetic and density sections through Olenegorskoe (Figs. 28.1 and 28.3) and Kostomuksha (Fig. 28.1 and 28.4) iron ore deposits.

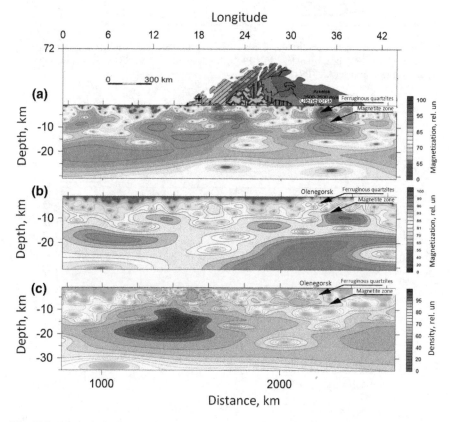

Fig. 28.3 Magnetic (**a, b**) and density (**c**) section through the Olenegorsk iron ore deposit

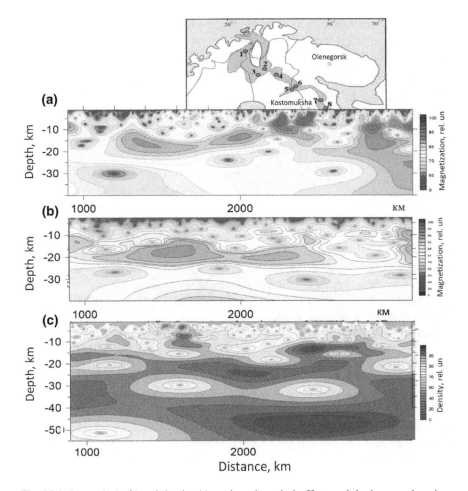

Fig. 28.4 Magnetic (**a, b**) and density (**c**) sections through the Kostomuksha iron ore deposit

Visualization of the correlation of density and magnetic inhomogeneities on the deep sections provides the basis for the formation of a new concept of the geodynamic regularities of the development of greenstone belts. Analysis of density sections has been showed that the greenstone belt of Fennoscandia tends toward blocks with a decompressed mantle. Interpretation of deep magnetic and density sections of the Earth's crust has been allowed to test the hypothesis of the genesis of deposits of archean ferruginous quartzites emerging on the surface from the magnetite zones of the granite layerbasement. It was confirmed by the deep sections that the "magnetite zone" can be a source of formation of rich deposits of ferruginous quartzites in the tectonic structures of greenstone belts (Figs. 28.3 and 28.4).

Magnetite zones are the source of the formation of rich iron deposits and noblemetal ores. Sharing the anomalies of the vertical and horizontal components of the Earth's magnetic field has been allowed to delineate the location of the magnetite zones, which is of great importance for the search for promising deposits.

Identification of Fluid Systems and Zones of Fluid-Magmatic Metasomatosis in the Earth's Crust of the Greenstone Belt of Fennoscandia

There are known the occurrence of gold-platinum-metal mineralization in ferruginous quartzites in the greenstone belts of Fennoscandia (Kostomuksha ore region 2015). Geological and structural features of gold ore occurence of the Kostomuksha region treat the genesis of ores formed as a result of regeneration and redeposition of ore matter as hydrothermal. We have obtained the ideas on the deep features of the gold ore zone based on the interpretation of the anomalies of the vertical and horizontal components of the Earth's magnetic field. Analysis of density and magnetic sections has been allowed to identify the main features of the deep structure of these zones and to reveal specific features of gold ore mineralization for detection of new promising areas.

One of the most important factors in the formation of gold deposits are deep fluid flows of mineralized aqueous solutions supplied from depths of more than 15 km by deep focus tectonic zones from high-quality Archean concentrates. On the density and magnetic sections, the migration paths of the deep fluid streams are reflected in the form of weakly magnetic supply channels of reduced density (Fig. 28.4). Fluid-supply channels are arising under the influence of an ascending heat flow and fluids rising along faults from great depths and diverging in the upper part of the crust into separate streams. There is a directed transformation of rocks inside the channels, changing their physical properties, including magnetic and density.

The migration routes of fluidized hot streams are traced on geomagnetic and density sections in the form of through weakly magnetic tracks of reduced density. Fluid-saturated stream are rising from the depths of 15–30 km, dividing into separate streams near the day surface. Analyzing the ways of fluid ascent to the surface, it can be assumed that the bulk of the fluidized flow of the Kostomuksha deposit rises, mainly from the south-west direction, from a depth of more than 20–25 km (Fig. 28.4). Since the magnetization of geological formations is more dependent on temperature than other parameters of the medium, such as porosity, fracture and fluid saturation, the migration paths of fluids are most clearly traced on magnetic sections (Fig. 28.4a). Figure 28.2 shows that in the Kostomuksha region the fluid system appears in the vertical (Fig. 28.4a) and horizontal (Fig. 28.4b) components and is located in the depth range from 15 to 28 km. It underlies the magnetite layer, which is the source of the iron ore deposit.

At the magnetic and density sections to the East of Kostomuksha, the migration paths of mineralized streams from the fluid system are reflected in the form of weakly magnetic supply channels of reduced density (Fig. 28.4). They are confined to the gold ore troughs (1–8) of greenstone structures (Fig. 28.4a) and capture the northern part of another powerful fluid system south of Belomorsk. There it manifests itself in a magnetic section in the form of three lenses located in the depth interval from 15 to 35 km.

The conducting channel can be clearly seen in the magnetic and density sections. It reflects the trajectory of streams of mineralized fluids, which gradually form deposits with gold ore and copper mineralization in the greenstone structures of the Shombozero, Lekhtinsky and Chirko-Kemsky troughs (Kostomuksha ore region 2015).

Fluid streams, interacting with metamorphosed strata, lead to the formation of gold ore mineralization in the structures of the greenstone belts. Visualization of deep sections is allowed to estimate the depth of occurrence and reveal the location in the Earth's crust of fluid systems and the zone of fluid-magmatic processing of Archean rocks by streams of mineralized fluids forming ore deposits of the structures of the greenstone belt of Fennoscandia.

The ore fields and deposits of Precambrian gold in all known cases are spatially associated with extended systems of deep faults, and the faults themselves usually play the role of fluid and ore-supplying channels. We have revealed the fluid-conducting pathways of the outlet of mantle streams based on the results of an analysis of the distribution of magnetic and density inhomogeneities in deep sections in the basement of the greenstone belt Fennoscandia (Figs. 28.3 and 28.4).

The decomposition of sulphide minerals is happen at one of the last stages of the hypergene-metasomatic process and the release of previously-contained noble metals occur in them. An additional increase in gold in the ore deposition zone is introduced by hydrothermal solutions during the activation of riftogenic structures. At the same time, platinum-metal ore mineralization tends to high-temperature processes, and the formation of gold—to lower-temperature processes.

The development of a geological-genetic model for the formation of noble metal mineralization in ferruginous quartzites will make it possible to determine the role of endogenous processes in its formation. It is possible that in the tectonic structures of greenstone belts the "magnetite zone" is the source of the formation of deposits of ferruginous quartzites and the accompanying noble metal mineralization.

The "magnetite zones" of the continents are of exceptional interest for studying the deep structure of the Earth's lithosphere. Manifested in the anomalies of the magnetic field at temperatures below 560 °C, they indicate the thermal regime of the formations of the ancient crust of the continents and trace the position of the deep boundary of the Curie surface. The study of the Early Precambrian crust forming the basement of ancient continents allows us to approach the evaluation of its deep structure at a real level and show that regional magnetic anomalies reflect the influence of "magnetite zones" occurring at depths of more than 10–15 km along the boundary of granite and granulite-basite layers.

Discussion and Conclusions

We have studied the features of the structure of the ancient blocks of the Doriphean basement, deep fluid systems and fluid supply channels. Hydrothermal study of ancient Precambrian complexes in areas of fluid-magmatic activations manifests itself in the form of deep minima of magnetic anomalies. For example, the traces of fluid-magmatic activations are clearly manifested in the minima of the magnetic anomalies of the horizontal component of the Earth's magnetic field, emphasizing the hydrothermal genesis of gold ore objects. The zones of fluid magmatic activations and intense fluid flow are manifested in the minima of the magnetic field. Magnetic anomalies of the Earth's magnetic field module made it possible to identify zones of deep thermofluid processing that affect younger rocks.

Ways of migration of heat fluxes, fluids and the location of fluid systems can be clearly seen on deep sections in the form of channels and lenses of reduced density and magnetization. The location of fluid channels and terrestrial systems can be estimated from magnetic and density sections. In the lower crust and in the mantle, where the rocks are in a demagnetized state due to high temperatures (more than 560 °C), the location of deep-focus fluid systems is estimated from density depth sections.

Faults are reflected in the form of block boundaries that differ in physical characteristics. Inside the weakened layers that are located in the fluid channels are appeared the weakly-magnetic low density lenses. The trajectory of a deep fault that passes through the Earth's crust to the mantle is traced through a chain of low density lenses. The migration routes of mineralized streams are manifested as weakly magnetic feed channels and lenses of reduced density.

Our studies have been shown that injective dislocations associated with the penetration of deep matter into the Earth's crust led to the decomposition of its matter in vertical fault zones. As a result, powerful subvertical ore-bearing magma-metasomatic columns with a length of 3–8 km arose. Favorable conditions for the continuous circulation of ore-bearing solutions and subsequent ore deposition are created due to the presence of resistive horizons overlapping large ore-bearing highly permeable near-fault zones at depths of 2.5–3 km.

Vertical thermofluid flow, rising along the fault zones of the basement, creates favorable conditions for the formation of large ore-bearing columns and above them —promising gold ore sites. Prospective areas are located above the gold columns in the ore sites. Magmatic processes were accompanied by metasomatic phenomena with the formation of ore-metasomatic columns with a length of up to 8 km in magmatic-fluid systems of the near-fault zones.

Deep density sections showed that the endogenous processes and the mining of the examined gold ore deposits are most likely due to the prolonged impact of multiple stages of tectonic magmatic activation of deep sources—thermal and fluid —on rocks enclosed in zones of subvertical faults of the upper crust.

Thus, we have presented the new ideas of the stratification and vertical crumbling of the Earth's crust, taking into account the features of the influence of factors of depths. It is based on the usage of original technology of interpreting the complex of geophysical data—module anomalous magnetic field, the horizontal component of the magnetic field and gravity field anomalies. The application of the original technology reveals new possibilities for studying the deep structure of the lithosphere and the search for criteria for the localization of mineralization. This has made it possible to clarify the prospects of gold mining in the area and recommend exploratory work on potentially ore-bearing areas above the outlets of vertical channels.

References

International Centre for Global Earth Models (ICGEM). URL: http://icgem.gfz-potsdam.de.

Hahn A., Kind E. G. and Mishra D. C., 1976. Depth estimation of magnetic sources by means of Fourier amplitude spectra, *Geophys. Prospect.* 24, 287–308.

Kaban M.K., Reigber C. (2005) New possibilities for gravity modeling using data of the CHAMP and GRACE satellites, *IZVESTIYA-PHYSICS OF THE SOLID EARTH*. 41 (11), 950–957.

Kopytenko Yu.A., Petrova A.A. (2014). New age magnetic maps for the marine magnetic navigation/Magnitnye karty novogo pokoleniya dlya celei morskoi magnitnoi navigacii. Proc. of the XII All-Russian Conference "Applied Technologies of Hydroacoustics and Hydrophysics GA-2014". Nestor – History, St. Petersburg. 258–261. (in Russian).

Kopytenko Y. A., Petrova A. A. (2016) The results of the development and application component model of the magnetic field of the Earth for the benefit of magnetic mapping and geophysics. *Fundamental and Applied Geophysics*, 9, 2, 88–106 (in Russian).

Korhonen J.V., Fairhead J. D., Hamoudi M., Hemant K., Lesur V., Mandea M., Maus S., Purucker M., Ravat D., Sazonova T. and Thébault E. (2007) Magnetic Anomaly Map of the World, Equatorial scale 1:50 000 000. Map published by the Commission for the Geological Map of the World, supported by UNESCO, 1st Edition, GTK, Helsinki.

Kostomuksha ore region (geology, deep structure and mineralogy) (2015) Kostomukshinskii rudnyi raion (geologia, glubinnoe stroenie i minerageniya) [Eds. Gorkovets V.Ya., Sharov N. V.] Petrozavodsk, Karelian Research Center of the Russian Academy of Sciences. 323 p. (in Russian).

Litvinova T., Petrova A. and M. Petrishchev. (2014) Electromagnetic imaging of lithosphere permeable zones//*Geophysical Research Abstracts*, Vol. 16, EGU General Assembly, Vienna, Austria. EGU2014–8938.

Mayer-Gürr T., Zehentner N., Klinger B., Kvas A. (2014) ITSG-Grace2014: a new GRACE gravity field release computed in Graz; Potsdam

Petrishchev M.S., Petrova A.A. and Yu.A. Kopytenko (2014). Deep structure of thermal zones based on the results of the integration of geophysical fields/Glubinnoe stroenie termalnyh zon po rezultatam kompleksirovaniya geofizicheskih polei//*Proc. 38th session of the Uspensky international scientific seminar "Problems of the theory and practice of geological interpretation of geophysical fields"*. Perm, GI UrB RAS. 24–28. (in Russian).

Petrova A.A. (1976) Method of spectral-correlation analysis of anomalous geomagnetic field/ Metodika spektralno-korrelyacionnogo analiza anomalnogo geomagnitnogo polya. PhD thesis. Moscow, IZMIRAN. - 25 p. (in Russian).

Petrova A.A., (2015) Digital maps of the components of the magnetic field induction vector/ Cifrovye karty komponent vektora indukcii magntinogo polya. Proc. of IZMIRAN, Moscow. 412–416. (in Russian).

Chapter 29
On Deep Electroconductivity of Tobol-Ishim Interfluve

V. P. Borisova, T. A. Vasilieva, S. L. Kostuchenko and E. B. Fainberg

Abstract The results of 1D and 2.5D modeling and inversion of amplitude and phase curves of magnetotelluric sounding in the Tobol-Ishim interfluve of Western Siberia are presented. Cross sections of the electrical conductivity of the crust and asthenosphere are constructed as well. It is suggested that there is a conductive layer lying at a depth of ~ 60–70 km. A decrease in the electrical resistivity of the overlying crustal layer from 1000–3000 to ~ 300 Ω-m which may be a sign of an increase in the conductivity of the earth's crust is noted.

Keywords Magnetotellurics · Western Siberia · Conductive layer

Prospecting magnetotelluric soundings in the southwest of the West Siberian Plate (WSP) have been carried out by the explorers of the Tyumen Geological Administration in 1980–1981 on one of the oil and gas promising areas of the Tobol-Ishim interfluve (Fig. 29.1).

The area of the MT survey is confined to the joint of the Caledonian folded Kazakhstan system (the western side of the Vagay-Ishim depression) and the Hercynian folded Ural system (the eastern side of the Tobolsk anticlinorium) and is located in the inter-rift area between the Nizhnetavdinsk and Ishim branches of the Triassic rift system (Fig. 29.2). MT profiles are located mainly in the zone between two wedge-jointing deep faults (marginal seams) of the northeast strike—the Middle Paleozoic East Urals and Early Paleozoic Central Turgay—and cross a series of Middle-Upper Paleozoic intra-structural faults also in the northeast direction. These include the well-pronounced Kiselevsky fault and the less extended Dronovsky fault. The faults are traced by features in the behavior of geophysical fields, crushing zones, chains of intrusions of the main and ultrabasic compositions.

V. P. Borisova · T. A. Vasilieva · E. B. Fainberg (✉)
Geoelectromagnetic Research Centre—Schmidt Institute of Physics of the Earth,
Moscow, Russia
e-mail: edfain@yandex.ru

S. L. Kostuchenko
Limited Liability Company "Rosgeology", Moscow, Russia

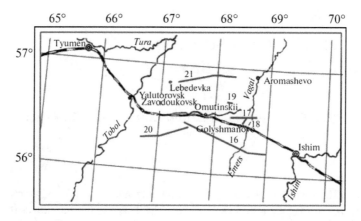

Fig. 29.1 Scheme of the MT survey area, ▬▬ profiles and their numbers

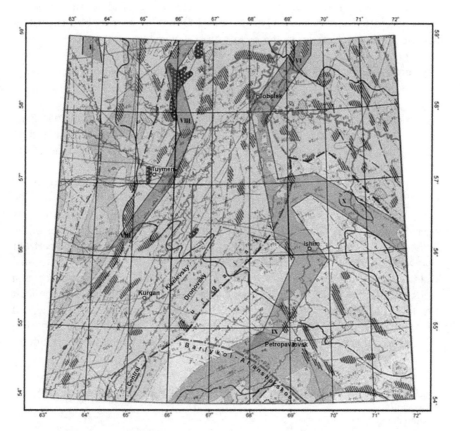

Fig. 29.2 Riftogenic zones of the West Siberian Triassic Rift System: VIII-Nizhnetavdinsk, IX-Ishimsk, X-Tyukalinsk. ☐ MT survey area. ▬▬ MT profiles

Boundary and intrastructural comagmatic faults are considered as coromanthic faults (Barykin and Bochkarev 1993; Fault map 1980).

The MT survey profiles crossed a number of structures of the Mesozoic-Cenozoic cover of the northeastern strike, namely, Zavodoukovsky, Komissarovsky, Sokolovsky, Borovlyansko-Vyatkinsky swells, Dronovsky, Kiselevsky, Pletnyevsky, Aromashevsky flexures and the western side of the Vagay-Ishim depression (Fig. 29.3).

Within the western side of the Vagay-Ishim depression (profile 16, Fig. 29.1), the MT curves ρ_{xy} and ρ_{yx} differ no more than 25%, the additional impedances modulo do not exceed 0.15 of the values of the main impedances. The polar diagrams of the main impedances, as a rule, are close in shape to the circles, and the configuration of the amplitude and phase curves does not significantly depend on the measurement directions. These characteristics gave the basis to the performers of MT works to confine themselves to analyzing the effective MT curves on the profile 16. In the remaining parts of the MT-survey (profiles 17–21), transverse ρ_{xy}, φ_{xy} (H-polarized MT-field, sub-latitude direction, azimuth 70°) and longitudinal ρ_{yx}, φ_{yx} (E-polarized MT field, submeridional direction, azimuth 160°) MT curves were analyzed.

To answer the question of the effect of local near-surface and regional inhomogeneities on the amplitude and phase curves, quasi-3D (thin-sheet) modeling was carried out according to the program (Singer and Fainberg 2003). The scheme of total conductivity of the sedimentary cover in southwestern part of the WSP at a scale of 1:2,500,000, covering the area of the MT studies and surrounding area, was constructed using the data of VES, MT and well logging data as well as the "Maps of the relief of the basement of the WSP" (Kostyuchenko 2004). Within the region under study, the S values continuously increase from southwest to northeast,

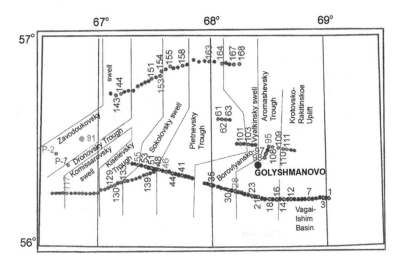

Fig. 29.3 The main geological structures and location of MT sounding points

which makes it difficult to set the S value level outside the MT survey plate. Quasi-3D modeling allowed us to draw two conclusions: (1) in the first stage of the interpretation, we can confine ourselves to a 1D joint inversion of the amplitude and phase curves; (2) to define the parameters of the deep geoelectric section more precisely, a two-dimensional inversion of the amplitude and phase curves is necessary.

At the first stage of the research (Borisova et al. 2013), a joint point-by-point 1D inversion (Barsukov and Fainberg 2010) of the transverse amplitude and phase MT-curves was performed. The results of the inversion are presented here for two profiles: Profile 20 (Fig. 29.4) and Profile 16 (Fig. 29.5). Along the profile 16, the effective curves were inverted (in the absence of MT curves measured along the axes of the MT installation). In the remaining parts of the MT-survey plate (profiles 17–21), the transverse ρ_{xy}, φ_{xy} and longitudinal ρ_{yx}, φ_{yx} MT-curves were analyzed.

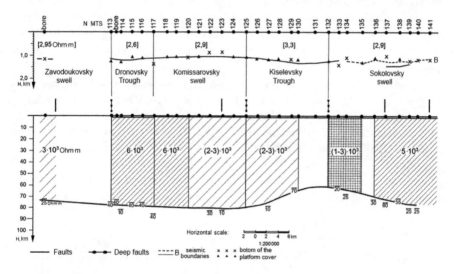

Fig. 29.4 One-dimensional pointwise inversion, profile 20

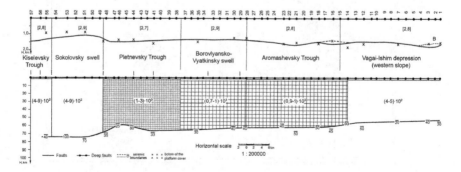

Fig. 29.5 One-dimensional pointwise inversion, profile 16

The upper boundary is in good agreement with the seismic boundary, identified with the platform cover or with the reflecting seismic horizon "B" (volgian upper Jurassic), which is a regionally traceable seismic benchmark and confined to disagreement surfaces (Kunin et al. 1995). The thickness of the sedimentary layer on the profiles 16 and 20 varies in the range of 1.1–1.5 km.

The lower geoelectric boundary, considered by us as the roof of a conducting layer in the mantle lithosphere, on the profile 20 rises from a depth of 70–80 km (Zavodoukovsky swell—Komissarovsky swell) to 60 km, and then descends to the former level (central and eastern regions of the Sokolovky swell). The spatial rise of the roof of the conductive layer corresponds to the zone of the Kiselevsky deep fault separating the Kiselevsky trough and the Sokolovsky swell and is one of the main faults in the territory of the MT research. The values of the specific electrical resistivity of $(1–3)10^3$ Ω-m of the consolidated lithosphere in the region of the maximum rise of the roof of the conducting horizon are somewhat lower in comparison with the background values $(3–6)10^3$ Ω-m.

The north-eastern end of the profile 20, where the depths to the conducting asthenospheric horizon are of H $\sim$ 73–78 km, approaches the northwest section of the profile 16 H $\sim$ 70 km, related to the side zone of the Pletnevsky trough. On the profile 16 in the Pletnevsky trough, the roof of the conducting horizon is first raised to a mark of 58–61 km, and then its immersion to 69 km. This anomalous zone is indicated only by two MT probes (MTS 44, 46) and, if it does exist, it may be a continuation of the anomaly manifested on the profile 20, which cuts the Kiselevsky fault (MTS 132) and extends to the north-east direction at an angle of $\sim 50°$ to the Kiselevsky fault) to the intersection with the profile 16. (Note. It is also possible that this anomaly traces the Kiselevsky fault, which is manifested on the profile 16, not in the contact zone of the Kiselevsky trough with the Sokolovsky swell, as on the profile 20, but in the zone of the Pletnevsky trough. It is noteworthy that the values of the electrical resistivity of the consolidated lithosphere in this region of the rising of the roof of the conducting horizon also amount to $(1–3)10^3$ Ω-m, but in this case the values of the resistivity of adjacent blocks of the lithosphere are lowered).

Profile 16. ***The lower geoelectric boundary*** along the entire length of the profile 16, except for the supposed anomalous zone, spatially confined to the Pletnevsky trough, monotonically rises in the east direction from the level 71–76 km (Kiselevsky trough—Sokolovsky swell) to the marks 53–55 km (west side Vagay-Ishim depression, entering the zone of the Ishim branch of the Triassic rift system of WSP, Fig. 29.3). The rise of the roof of the conducting horizon of the lithospheric mantle is accompanied by a gradual decrease in the electrical resistivity of the consolidated lithosphere from the values $(1–3)10^3$ Ω-m (Pletnevsky block), up to 700–1000 Ω-m (Borovlyansko-Vyatkinsky—Aromashevsky block) and further 400–500 Ω-m—block of the consolidated lithosphere, corresponding to the western side of the Vagay-Ishim depression. At that the resistivity of the lithospheric block corresponding to the Kiselevsky trough region—Kiselevsky deep fault—Sokolovsky swell is characterized by lower values $(4–9)10^2$ Ω-m, which are much smaller than those determined for the roofing area of the conducting horizon in the lithospheric mantle—$(1–3)10^3$ Ω-m on the profile 20.

Discussion

From the analysis of the results of a one-dimensional joint inversion of the amplitude and phase curves MTS curves corresponding to the H-polarized MT field and obtained in the period ranges of 0.25—400–540 s (profiles 17–21) and the results of inversion of the amplitude and phase effective MT curves obtained in the range 0.25–1000 s (profile 16), it follows that in the mantle lithosphere of the Tobol-Ishim interfluve, a conductive horizon with electrical resistivity of tens of Ω-m is traced (Fig. 29.4 and 29.5). Its roof rises from the northern and western sections of the MT-survey area (depths of 70–85 km) towards the Vagay-Ishim depression (depths of 55–65 km). At the same time, the resistivity of the lithosphere decreases from the first thousand Ω-m to the first hundred Ω-m.

The traced rise of the conductive zone in the subcrustal lithosphere is particularly significant in that area of the Ishim branch of the rift system, where, according to geothermal data, anomalously high values of the heat flux (HF), namely, 67–70 mW/m^2 are established (Kostyuchenko 1992). This area extends along the rift zone by $\sim$110 km, and its width, like the width of the Ishim Rift, is of $\sim$35 km. The region of high temperatures on the Mohorovicic surface, contoured by an isoline of 700 °C corresponds to the anomalous region of HF. Inside this high-temperature region, a value of 780 °C is noted. In the eastern direction, the temperature on the Moho surface decreases rapidly: at a distance of only 20 km from the 700 °C isoline an isoline 500 °C passes. To the west of the anomaly, temperature data are absent.

The anomalous HF region is crossed by the marginal suture of the Early Paleozoic Central Turgay deep fault. The temperature on the surface of the Paleozoic basement in the area of the MT survey decreases smoothly from the north-east corner (70 °C) in the south-west direction (up to 40 °C). With the geothermal anomalies, obviously, the features of the deep geological structure of the Ishim rift zone are considered (Kostyuchenko 1992). On the surface of the crystalline basement, a depression of 12 km in depth is observed, consistent in plan with the anomaly of the heat flux and with the temperature anomaly on the Moho surface. The surface of the Mohorovicic is elevated to a depth of 36–38 km.

Thus, the thickness of the crystalline basement complex within the Ishim rift zone is abnormally shortened and amounts to 24–26 km. (For comparison, the geotraverse of Murmansk-Kyzyl in the Central-Western Siberian fold region, pickets 210–310, the minimum thickness of the crystalline part of the crust is pegged within the limits of the Tobol-Nizhniy Irtysh rift zone where it has a thickness of 31 km). There is a powerful folded complex of the lower-middle Paleozoic, about 9 km thick, graben filled with Triassic sediments with a thickness of 1.5 km and a sedimentary cover with a thickness of 1.25 km located on the surface of the crystalline cap.

The anomalous zone in the subcrustal lithosphere of the Ishim branch of the Triassic rift system of WSP is probably associated with granitoid (PZ2-3) and main (PZ2) intrusions confined to the region of its localization (Fault map… 1980).

According to recent studies (Dobretsov 1997), many intrusions common in the vast territories of the WSP and the Siberian Platform and dated to the early Middle Paleozoic can be attributed to the Late Paleozoic- Triassic. The conductive object detected in the spreading zone may be an element of a two-level mantle-crustal fluid paleosystem—"the area of decompression melting in the ascending flow of the mantle substrate (astenolens) and the melting region in the lithosphere under the Moho boundary" (Sharapov et al. 2008). The cited work says that the upper boundary of the melting zone in the continental lithosphere of the Siberian Platform and the WSP is located at a depth of 60–70 km.

Conclusion

To confirm the revealed conductive objects in the depth section of the research area and to refine the parameters of these objects, it is necessary to perform a two-dimensional inversion of the amplitude and phase MT curves. It is highly desirable to organize an array survey of AudioMTS, MTS and deep MTS on the basis of broadband equipment and modern noise-resistant technology for recording electromagnetic fields in order to study the deep geoelectric structure of the Ishim magnetotelluric anomaly. The field MT studies should be aimed at:

(1) determining the configuration and electrical conductivity of an anomalous object (asthenospheric diapir/astenolense/element of a multilevel paleomagmatic system;
(2) investigating the mechanisms of interaction of the asthenosphere and lithospheric mantle in zones of continental spreading of the West Siberian plate.

References

Barykin, S.K. and V.S. Bochkarev (1993). The structure of the basement and the intermediate complex of sediments. Col. Geological and geophysical modeling of oil and gas areas. Moscow. Nedra, pp. 34–41.

Borisova, V. P., T. A. Vasil'eva, S. L. Kostyuchenko, N. V. Narskii, B. K. Sysoev, E. B. Fainberg, A. G. Charushin and A. L. Sheinkman (2013). On Deep Electric Conductivity of the Lithosphere in the Tobol–Ishim Interfluve (West Siberia). Izvestia, Physics of the Solid Earth, V. 49, N3, pp. 363–372.

Barsukov, P.O. and Fainberg E.B. (2010). 1D MT-modeling and inversion (manual). Applied Electromagnetic Research (AEMR), the Netherlands, 36 p.

Dobretsov, N.L. (1997). Perm Trias magmatism and sedimentation in Eurasia as a reflection of superplume. DAN. Geophysics, V. 354. № 2. pp. 220–223.

Fault map of the USSR and neighboring countries territory. Scale 1:2 500 000 (1980). Editor A.V. Sidorenko, M., Ministry of geology of the USSR, VNIIGeofizika.

Kostyuchenko, S. L. (1992). The deep structure of the Earth crust of West Siberian plate by the results of complex geological and geophysical research. PhD Thesis in geological and mineralogical sciences. Moscow, MSU, 247 p.

Kostyuchenko (2004) A relief map of the pre-Mesozoic basement of the West Siberian Plate. Scale 1: 5,000,000 / Ed. Kostyuchenko S.L. M .: GEON. 2004

Kunin, N.Ya., V.S. Safonov and B.N. Lucenko (1995). Fundamentals of the search strategy for oil and gas fields (on the example of Western Siberia). Part I: IPE RAS, Khanty-Mansiysk Committee on Geology and Subsoil Use, «ACCOTTEO», 134 p.

Sharapov, V.N., Yu.V. Perepechko, L.N. Perepechko and I.F. Rakhmenkulova (2008). The nature of the mantle sources of Perm Triassic Traps of the Western Siberian Plate and Siberian Platform. Geology and geophysics, V. 49. №7. pp. 652–665.

Singer, B.Sh. and E.B. Fainberg (2003). User documentation for thin layer modeling code SLPROG and accompanying software, www.aemr.net/slprog/win/slprog.zip.

Part IV
Geological Interpretation

Chapter 30
Study of the Magnetic Properties of Geological Environment in Super Deep Boreholes by the Magnetometry Method

G. V. Igolkina

Abstract **The purpose of the article**: obtain new data of magnetic fields, magnetization, and magnetic susceptibility of rocks at large depths under conditions of their natural location. Solve technological problems associated with the detection of metal in the walls of wells and near borehole space. **Methods**: for magneto metric studies, magnetometers-inclinometers have been developed at the Institute of Geophysics of the Ural Branch of the Russian Academy of Sciences, which allow continuous measurements of the magnetic susceptibility of rocks (χ), the vertical component (Z_a) and whole vector of the horizontal component (H_a) of the geomagnetic field; magnetic azimuth (A_m) and zenith angle (φ) of the well. **Results**: the capabilities of the well magnetometry method for studying the magnetic properties of rocks and the refinement of the litho logical and stratigraphic characteristics of the Kola SG-3, Krivoy Rog SG-8, Ural SG-4, Muruntau SG-10, Saatlinsk SG-1, Timan-Pechora SG-5, Kolvin, Vorotilov, Tyumen SG-6, Novo-Elkhovsk, Tyrnyauzsk super deep wells. **Conclusions**: it was showed the interrelation of magnetic characteristics with depth, age and litho logical composition of rocks was studied. Analysis of well magnetometer and MEP data allowed establishing the boundary of pyrite-pyrrhotine transition in the gold ore zone. It is shown the possibility of borehole magnetometry for solving paleomagnetic problems for determining the magnetization of rocks.

Keywords Borehole magnetometry · Magnetic characteristics · Super deep well Magnetic susceptibility · Geomagnetic field · Magnetization of rocks Method · Dolerite · Pyrrhotite

---

G. V. Igolkina (✉)
Institute of Geophysics UB RAS, Yekaterinburg, Russian Federation
e-mail: galinaigolkina@yandex.ru

Introduction

The superdeep boreholes are the basis for direct study of the composition of rocks at deep horizons of the earth's crust. They allow identify the geological sections to a great depth; to get direct data on state, composition of rocks and to study their variation according to the depth; identify the nature of geophysical boundaries; to study physical properties of rocks in real thermodynamic conditions; and also contribute to the development of new technologies and equipment for in-depth study and use of underground resources etc. (The main results ... 2000; Pelmenev et al. 1991).

The borehole magnetometry, which includes measurement of magnetic susceptibility and magnetic field, is effectively used for investigation of superdeep wells and for the study of deep wells in oil-and-gas regions of Russia (Igolkina 2002). The efficiency of the borehole magnetometry is associated with the development and implementation of magnetometers-inclinometers capable to carry out a simultaneous and continuous investigation with high precision of the magnetic field, the magnetic susceptibility in wells and also of the magnetic azimuth and the well deviation angle. In addition, the developing of new ways and techniques of the interpretation of magnetic fields and of the study of magnetic bodies allowed to solve complex geological problems and to move from a qualitative interpretation of measured magnetic parameters to their quantitative analysis.

The method of boreholes magnetometry was used to study such superdeep wells as: Kola, Krivoy Rog, Ural, Muruntau, Saatly, Timan-Pechora, Kolvin, Vorotilov, Tyumen, Novo-Yelkhov, Tyrnyauz, KTV. This provided valuable data on the magnetic field and magnetic susceptibility features, as well as on the magnetization of the main geoblocks of the earth's crust and its parameters at various horizons in a wide stratigraphic range.

The investigated superdeep boreholes are located in the main geostructures (Fig. 30.1): on the ancient shields (Kola, Vorotilov, Krivoy Rog, Novo-Elkhov), in different-aged infolded facilities (Ural, Muruntau, Tyrnyuz) and in the mantles of ancient (Kolvin, Timan-Pechora) and young platforms (Tyumen SG-6), region of Mesozoic and Cenozoic foldings (Saatly SG-1) (The Main results ... 2000; Pelmenev et al. 1991).

Study Results and Their Discussion

To carry out magnetometry investigations magnetometers-inclinometers were developed at the Institute of Geophysics UB RAS. The instruments allow to carry out continuous measurements of the magnetic susceptibility of rocks (χ), of the vertical (Z_a) and whole vector of the horizontal component (H_a) of the geomagnetic field, the magnetic azimuth (A_m) and the well deviation angle (φ). The developed

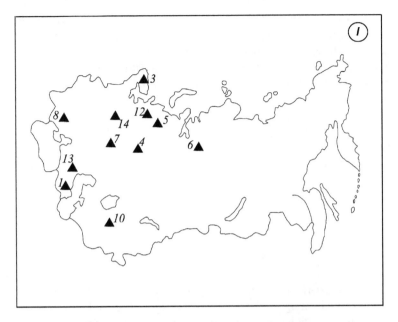
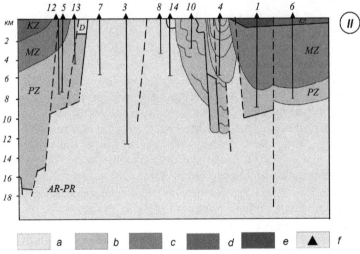

Fig. 30.1 Location map of superdeep and deep boreholes (**I**) and the geostructure diagram (**II**): (a) areas of ancient platforms and shields (3—Kola SG-3, 12261 m, 8—Krivoy Rog SG-8, 3841 m, 14—Vorotilov, 5374 m, 7—Novo-Elkhov, 5100 m), (b) different-aged infolded facilities (4—Ural, 6001 m, 10 - Muruntau SG-10, 4220 m, 13—Tyrnyauz, 4001 m); (c) mantles of ancient platforms (12—Kolvin, 7054 m, 5—Timan-Pechora, 6903.5 m), (d) areas of the young platforms (6—Tyumen SG-6, 7502 m), (e) Mesozoic and Cenozoic foldings 1—Saatlinskaya SG-1, 8324 m), (f) wells

software allows to realize the measurement process with automatic correction input and representation of measurement results on the display. The magnetometer-inclinometer MI-6404 has a thermopressure resistance of 250 °C @ 220 MPa, the whole measurement suite is carried out during two put-out-of-hole operations using a three-core logging cable (Astrakhantsev and Beloglazova 2012).

To improve the reliability of the interpretation together with the results of borehole magnetometry, the results of calipering, inclinometry, of the electrode potential method and other geophysical well logging methods, as well as geological sections along the borehole, the results of paleomagnetic and petromagnetic core studies, other petrophysical information provided by geological services at the wells and by other researchers. But it should be noted that there are some restrictions in the qualitative and quantitative interpretation of logging materials. The effectiveness of the qualitative interpretation and the reliability of the conclusion are based on the following factors: weak dependence of the magnetic field measurements on the parameters of the well and the area adjacent to it; high resolution capacity both in the radial direction and along the well; suitable accuracy of measurements and their stability (Igolkina 2002).

The task of determining the magnetization of rocks under conditions of their natural occurrence along the boreholes of the superdeep wells has the advantage over the study of magnetization on core samples. A complete section of the well is studied instead of individual coring stations. In addition, vector measurements of the geomagnetic field using three components Z, H_x, H_y are bended to the cardinal points, while the core does not have such a binding. Particularly the advantage showed in the cases when the core is not oriented "up-down", and, consequently, the polarity of the remnant magnetization remains unknown (Igolkina 2007; Parker and Daniell 1979; Parker 1974). The features and changes in the magnetization of the rocks discovered by superdeep boreholes are based on the regular relationship of magnetic anomalies with geological factors: lithological type of rocks, the degree of their change, structural and texture features, the type and concentration of magnetic minerals, etc. Their study gives reasons to use these parameters for the geological interpretation of the observed both internal and external magnetic fields.

The author established the features of the magnetization of in situ rocks according to the results of the investigation of the superdeep wells.

For rocks with magnetite and titanomagnetite mineralization, represented by dolerite intrusions, dikes of microdiorites, basaltic andesites, basalts (Ural SG-4, Timan-Pechora SG-5, Kola SG-3, Saatly SG-1, Kolvin parametrical borehole) both normal and inverse magnetization are observed.

The presence of fine-grained monoclinic pyrrhotite in the wells of Ural SG-4 and Muruntau SG-10 changes the rate of the J, J_i, J_n curves (they become rugged and alternating). The inhomogeneity of the magnetization of the rocks is mainly associated with the change in the natural remnant magnetization J_n due to its different stability.

A comparison of the evaluation of the types of magnetic mineralization with petromagnetic, paleomagnetic studies of the core of the superdeep borehole and with data on petrographic and petrochemical descriptions of rocks is necessary to supplement and confirm the interpretation; to assess the possibility of borehole magnetometry for solving paleomagnetic problems (Igolkina 2002).

The use of the borehole magnetometry to correlate geological sections along the boreholes of the Kola and Ural superdeep wells required an individual approach for each well under study and area of work. The successful solution to the problem depends on the specific geological conditions and physical properties of the rocks for each well, measured by core and in natural occurrence. To correlate the geological sections, a set of features of the magnetic correlation was developed, which allows comparing and correlating the magnetic rocks along the boreholes of the superdeep wells with a sufficient degree of reliability (Igolkina 2014).

The magnetic parameters on the basis of which the correlation is made are: the magnitude and degree of ruggedness of the magnetic susceptibility curve; the magnitude, sign and degree of brokening of the internal magnetic field curve; the magnitude and polarity of the full magnetization; the magnitude of the induced magnetization; the magnitude and polarity of the natural remnant magnetization; the magnitude and sign of the Q_Z and Q_H factors; the prevailing rhythms of the frequency of changes in the magnetic properties of rocks along the section; position of the internal field vector in space $\vec{T}_a$; the results of statistical processing of magnetic parameters.

An example of the use of the results of borehole magnetometry for the construction of a volumetric model of the borehole environment can be the correlation of magnetic rocks along three boreholes of the Kola SG-3 superdeep well, along the original and pilot boreholes of the Ural SG-4, along the boreholes of oil-and-gas wells of the Siberian platform and Western Siberia. The deep structural forecast is largely related to the quality of the correlation of rocks in the interwell space and, therefore, the use of borehole magnetometry becomes crucial.

In addition, according to the borehole magnetometry data, in the Ural SG-4 superdeep well, intersections of the magnetic rocks were identified along the original and pilot boreholes in the interval from 300 to 3400 m and their mutual correlation was made (Igolkina 2014).

A study of the effect of artificial isothermal magnetization in areas of sulphide mineralization allows to estimate the complex nature of magnetic mineralization and to give a qualitative evaluation of the distribution of pyrrhotite types in the Muruntau SG-10 (Kalvarskaya et al. 1978). The integration of the borehole magnetometry method and the electrode potential method allowed to separate the sulphide mineralization into pyrite and pyrrhotite, to establish the boundary of the pyrite-pyrrhotite transition, which is especially important for determining the boundaries of gold-ore mineralization in the Muruntau ore field (Fig. 30.2). These transitions coincide with the geological boundaries of the section and metamorphic zoning, indicating their interrelation.

At the Muruntau borehole, the study of artificial magnetization of sulphide mineralization zones became possible due to the technological reasons (frequent use of bottom-hole magnet in drilling operations). The magnetic field of the

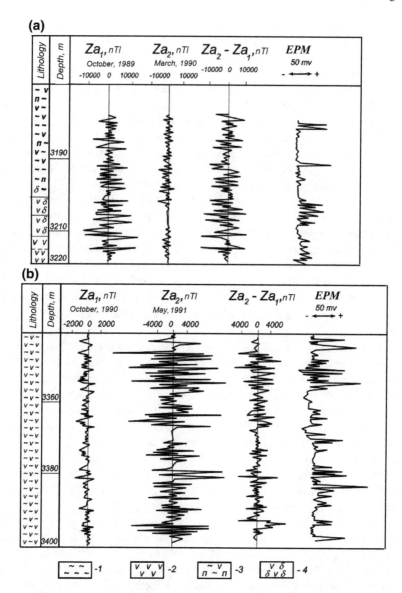

Fig. 30.2 Comparison of the results of the borehole magnetometry in the Muruntau superdeep well SG-10: **a** before (October 1989) and after (March 1990) artificial magnetization of rocks in the interval 3170–3220 m; **b** before (October 1990) and after (May 1991) artificial magnetization of rocks in the interval 3360-3400 m. 1—meta-aleurolites; 2—aleurolites; 3—carbon-mica schists; 4—biotite-plagioclase rocks; χ and Z_a—magnetic susceptibility and anomalous vertical component of the magnetic field; EPM—logging results by the electrode potential method

bottom-hole magnet at a 1 m distance is about 796 A/m. The studies of artificial magnetization of the sulphide mineralization zones were proposed by G. V. Igolkina in investigations of the superdeep wells to solve the geological task of determining the boundaries of pyrite-pyrrhotite transition (Igolkina 1995).

When the magnetization in a magnetic field is up to 796 A/m, the monoclinic modification of pyrrhotite acquires a significant isothermal magnetization. The study of the isothermal magnetization in laboratory and borehole conditions allowed V. P. Kalvarskaya (Kalvarskaya et al. 1978) to propose a new method for the dismembering and mineralogical diagnostics under natural conditions of the types of magnetic mineralization.

The change in the magnetic field and magnetic susceptibility was studied, and the magnetization of the rocks was calculated, and the effect of the bottom-hole magnet on the magnetic characteristics was estimated. The developed device for measuring the magnetic susceptibility was used in the study and allowed to solve the problem of measuring the magnetic susceptibility at depths of up to 12 km at high temperatures (Astrakhantsev and Beloglazova 2012). The advantage of the instrument gives complete independence of the output signal of the magnetic susceptibility sensor from the electrical conductivity of rocks, which is rather important in studying rocks containing iron sulphides (such as pyrite and pyrrhotite).

The magnetization effect $\Delta Z = Z_{a2} - Z_{a1}$ reaches 20,000 nT, and the anomaly of the field is related to the quantitative content of pyrrhotite and depends on its type: monoclinic, hexagonal or intermediate—intergrowth of the first two types.

The comprehensive analysis of the borehole magnetometer and the electrode potential method (EPM) allowed to determine the boundary of the pyrite-pyrrhotite transition at the 480–1200 m depth interval, but according to data (Alekseeva and Kremenetsky 2000) this boundary lies at depths from 0 to 1000 m.

The interrelation of magnetic characteristics with structural and texture features of dolerite intrusions, basalts and gneisses discovered by the Ural, Timan-Pechora, Kolvin, Vorotilov superdeep wells was studied (Igolkina 2002, 2015).

In addition, the use of borehole magnetic prospecting increases the efficiency of geophysical studies of wells in solving technological tasks of in-depth study and utilization of underground resources. The solution of the problem associated with the detection of metal in the hole walls and in the borehole environment allows, in case expansion of the borehole or the change of its direction, to avoid emergency situations.

Many drilling assemblies have been left in the borehole environment of the Kola SG-3 (Igolkina 2002, 2013). Figure 30.3 shows the assumed from the data of the borehole magnetometry position of the drilling devices consisting of two parts. The first part is the tube of the core slicer, the second part is the element of the drilling device of the BHA. The distance to the boundaries of metal objects from the wall to the top of the first part was 13–15 cm, of the second part—23–26 cm (upper end) and 18–21 cm (lower one).

Knowing the magnitude of the magnetic field anomaly, as well as the type of the metal object, it is possible to estimate the distance to it from the borehole of the

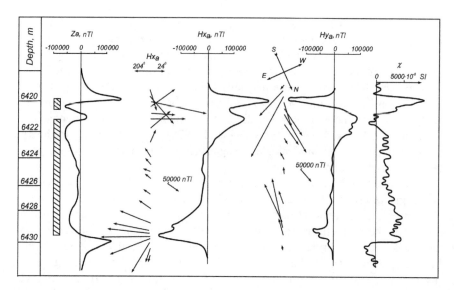

Fig. 30.3 Determination of the location of the drilling assembly left in the wall of the Kola SG-3 in the 6420–6430 m depth interval

investigated wall, which allows to determine the three-dimensional position of the emergency borehole and to solve the practical task of the devices bypass during further well-drilling (Igolkina 2013).

Conclusions

Thus, the borehole magnetometry is one of the most effective geophysical methods for solving many geological tasks and should be included in the standard complex when investigating superdeep and deep boreholes including oil-and-gas ones. The uniqueness of these results, obtained from depths up to 12,261 m, remains and will remain for a very long time, because now the Global Society Of Geologists has no projects of repetition of so deep penetration into the depths of the Earth, but they are necessary.

At present, new geological tasks are given to the borehole magnetometry:

1. Study of modern and paleomagnetic fields: identification of areas of magnetic field inversions along the boreholes section, their analysis and comparison with data of paleomagnetic studies; and also investigation of magnetic field gradients according to the depth.
2. Investigation of the possibility of the borehole magnetometry to solve the paleomagnetic problems and creation of effective methods for integrated interpretation of the materials of borehole magnetometry, of paleomagnetic and petromagnetic studies.

3. Complex analysis of well magnetometer and MEP data allows establishing the boundary of pyrite-pyrrhotine transition in the depth interval 480–1200 of the Muruntau well.
4. The solution of the problem associated with the detection of metal in the walls of the borehole and near wellbore space, allows for the expansion of the wellbore or changing its direction, to avoid emergency situations.

References

Alekseeva A.K., Kremenetsky A.A. (2000). Nature of petrophysical inhomogeneities in sections of gold-ore black-shale strata. International Geophysical Conference. 300 Years of Mining and Geological Service of Russia. Scientific conference abstracts. St. Petersburg. Pp. 212–213.

Astrakhantsev Yu.G., Beloglazova N.A. (2012). Integrated magnetometric instrumentation for the investigation of superdeep and exploratory wells. Yekaterinburg: UB RAS, 2012. 120 p.

Igolkina G.V. (1995) The role of artifical Magnetization of evaluation of sulfide mineralization. Book of abstracts XX1 General Assembly of IUGG, Scientific Program GA 5.19" Magnetic Petrology and Magnetic Signature of Ore Deposits and Ore Environments ". 2–14 July. Boulder; Colorado; USA. GAB51K-13. P. 13.

Igolkina G.V. (2002). Borehole magnetometry in the study of superdeep and deep wells. Yekaterinburg: UB RAS, 2012. 215 p.

Igolkina G.V. (2007) Study of the magnetization of rocks in natural occurrence according to measurements data in superdeep and deep wells. Proceedings of Murmansk State Technical University. V. 10. No. 2. Pp. 244–250.

Igolkina G.V. (2013). Solution of technological tasks in the study of superdeep and oil-and-gas wells by the method of magnetometry. Scientific and technical bulletin "Karotazhnik", No. 230. Pp. 25–40.

Igolkina G.V. (2014) Correlation of magnetic rocks in the interwell space of oil-and-gas and superdeep wells. Oil and Gas Engineering. No. 2. Pp.12 –20.

Igolkina G.V. (2015). The study of the relationship between magnetic characteristics and structural and texture features of gneisses (on the example of the Vorotilov Sputnik-1 well). Geophysics. No. 4. Pp. 53–56.

Kalvarskaya V.P., Filippycheva L.G., Metallova V.V., Petrov I.N. (1978) Logging method with magnetization of rocks in a well. Transactions of Leningrad State University. Geophysics issues. Issue 27. Pp. 77 –85.

Parker P.L.(1974). A new method of modelling marine gravity and magnetic anomalies. J. Of Geophys. Res., vol. 79. No. 14. Pp. 2014–2016.

Parker P.L., Daniell G.J.(1979). Interpretation of Borehole Magnetometer Date.J. Geophys. Res., vol. 84. No 10. Pp. 5467–5479.

Pelmenev M.D., Krivtsov A.I., Khakhaev B.N. (1991). Condition and tasks of deep exploration by deep and superdeep wells. Sovietskaya geologiya. No. 8 Pp. 3–7.

The main results of the deep and superdeep drilling in Russia (2000). St. Petersburg, St. Petersburg cartographic factory of Russian Geological Research Institute. 111 p.

Chapter 31
Underground Water Flows Detection and Mapping by Direct-Prospecting Geoelectric Methods

S. Levashov, N. Yakymchuk, I. Korchagin and D. Bozhezha

Abstract Many years of experience in the practical application of non-classical geoelectric methods of forming a short-pulsed field (FSPEF) and vertical electric-resonance sounding (VERS) for various problems of the near-surface geophysics solving demonstrates their high efficiency in searching and delineating subsurface water flows and aquifers. The results of our study indicate that the zone of rocks moistening, underground water streams of natural and man-caused origin and aquifers are detected and mapped operatively by a real survey with FSPEF method. The depth of lying and thicknesses of water-saturated horizons are determined with a high accuracy by VERS sounding. Field works of such character are often executed very operatively and easily. The results of geophysical studies show the effectiveness of FSPEF survey method and VERS and GPR soundings methods in dealing with the detection and mapping of underground water flows. Practical application of this technology during the geotechnical studies conducting for the construction of large engineering projects can bring significant economic benefits by reducing significantly the duration of exploration work and the drilling activity reduction.

Keywords Geoelectric survey · Electric-resonance sounding · Deposit type anomaly · Zone of moistening · Aquifer · Water flow · Well · Landslide zone

---

S. Levashov (✉) · N. Yakymchuk · D. Bozhezha
Institute of Applied Problems of Ecology, Geophysics and Geochemistry, Kyiv, Ukraine
e-mail: geoprom@ukr.net

I. Korchagin
Institute of Geophysics, NAS, Kyiv, Ukraine
e-mail: korchagin.i.n@gmail.com

Introduction

Mobile geoelectric methods of forming the short-pulsed electromagnetic field (FSPEF) and vertical electric-resonance sounding (VERS) (Levashov et al. 2005) for more than twenty years have been successfully applied to operatively solve a wide class of geological and geophysical problems, the searching for ore and fossil fuels including. During this time, the FSPEF-VERS technology in conjunction with seismic-acoustic and GPR soundings methods were widely used for operative solving the various problems of the near-surface geophysics. In particular, the FSPEF and VERS geoelectric methods have been repeatedly used during (a) exploration and mapping of aquifers, water-saturated horizons and mineral water deposits; (b) identification and mapping of areas of high humidity of soil, naturally occurring and man-made ground water flows, leaks from underground water communications; (c) examination of engineering-geological and hydrogeological conditions and monitoring over their changes on the territory of historical and architectural monuments, buildings and parks location; (d) carrying out geotechnical studies on the sites of construction of bridges, subway lines near surface occurrence, industrial buildings, residential buildings and objects of social and cultural facilities; (e) mapping the oil contaminated areas, etc. (Bokovoy et al. 2003; Levashov et al. 2005, 2010).

Special attention should be paid to the results of using this complex of operative geophysical methods for studying engineering and geological conditions on the sites of construction of new subway lines in the Kyiv, as well as on already constructed and operating sections of it. A large volume of experimental and research work performed and the results obtained have clearly and convincingly shown, on the one hand, the negative (destructive) impact of groundwater on the objects of the transport infrastructure under construction and located near the building and facilities. On the other hand, the results of these studies give good reasons for well-founded conclusions that when carrying out design work for the construction of buildings, industrial facilities and transport infrastructure, it is necessary to take into account the groundwater flows. Underestimation of underground flows leads to significant losses of time and financial resources. On the third side, the detection and mapping of water flows and areas of increased soil moistening can be quickly (operatively) implemented by a complex of geoelectric methods of FSPEF and VERS and georadar sounding. This practically tested complex of methods can also be used to solve specific engineering and geological problems during the construction of housing complexes, new underground lines, other cultural and industrial facilities, as well as for regular monitoring of the engineering and geological condition of the environment in the areas of already built and put into operation objects.

The report describes and analyzes the results of the operative solving of urgent practical problem—the detection and mapping of underground water flows.

General Information

At the initial stage of a residential complex construction in Kyiv on the Krasnopilskaya street unforeseen problems associated with the flooding of the pit dug by groundwater have been arisen (Fig. 31.1). This circumstance led to the need for the operative conducting the geophysical research in the area of construction in order to establish the causes of flooding.

The site is located on a hillside. Some lower along the slope a small lake was formed due to the ground water. This indicates that the migration of ground water takes place partially in the area of the construction site, which led to the trench filling by the water after excavation.

The main objective of the geophysical research was to identify areas of enhanced groundwater filtering, establishing areas of migration and depths intervals of the maximum moistening of the soil. According to geophysical research there were necessary to formulate the recommendations on the organization of drainage of underground water flows away from the construction site.

Geophysical surveys were conducted with using the FSPEF and VERS geoelectric methods and GPR sounding the cross-section with antenna unit AB 250 MHz. The survey by FSPEF method was used for zones of high soil moistening mapping and the routes of water flows migration determining. Methods of VERS and GPR soundings were used to determine the depth and thickness of areas of high soils moistening and the vertical cross-sections of moistened horizons construction (Fig. 31.2).

We must note also that on construction site a full range of engineering and geological surveys was made. Figure 31.3 shows the engineering and geological cross-section along one of the lines, which gives a complete picture of the nature of

Fig. 31.1 The origins of the groundwater at the construction site

Fig. 31.2 Formation of the landslide zone in the area of Western underground water flow

research and detail of the cross-section study on the construction site. We draw attention to the fact that areas of high soils moistening were marked in the cross-section by many wells. They are marked by black on the vertical lines of wells in Fig. 31.3.

Results of Geophysical Work

The Areas of Increased Groundwater Filtration Mapping

According to the survey by the FSPEF method three areas of increased groundwater filtration and soil moistening were identified and mapped on the construction site: there are water flows under the code name "Western", "Central" and "Eastern". These zones are formed by underground water flows that migrate down the hill. Figures 31.3 and 31.4 show the map of moistening zones and contours of underground streams applied on the work plan of construction, and on the satellite image of the area of work.

"Western" stream was traced from Krasnopilskaya Street down the hill to the construction site. The stream passes under part of the house number 2, then between

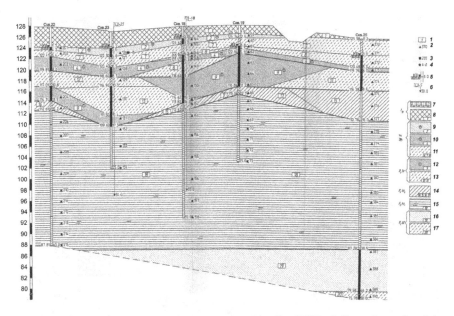

Fig. 31.3 Engineering and geological cross-section along line III-IIIa, built on the results of the survey for the residential complex construction

houses number 2a and 2b and goes in the south-western part of the construction site. At this point in the pit there is ground water sources (Fig. 31.1). Due to the influence of the underground stream a small landslide area on the territory adjacent to the garages site start to form (Fig. 31.2).

At the house 2a the ground subsidence is observed within an underground stream. This drawdown could be formed at the expense of man-made water and leaks from the heating duct. Removal of the soil, in this case, is performed along the zone of the underground flow.

Width of subsurface flow in the upper part, in the Krasnopilskaya street area is 8.0–10.0 m. Downslope flow width is gradually increased, and in the south-western part of the construction area reaches 30.0 m.

Central" stream was traced from the Krasnopilskaya street, where he was recorded near the house number 2. Then the flow passes the house number 2b at a distance of 5–8 m from his foundation and sent to the central part of the construction site (Figs. 31.4 and 31.5). The stream crosses the central part of the site. In the southern part of the pit within this flow the groundwater sources are observed.

The width of the soil moistening zone along an underground stream in the Krasnopilskaya street area is 5–8 m. Towards the construction site the stream width increases to 25–27 m.

"Eastern" stream is defined in the south-eastern part of the construction site. It follows from the forest slope zone and only partially crosses a small area of the construction site. The stream width in the area of the construction site is 15-20 meters.

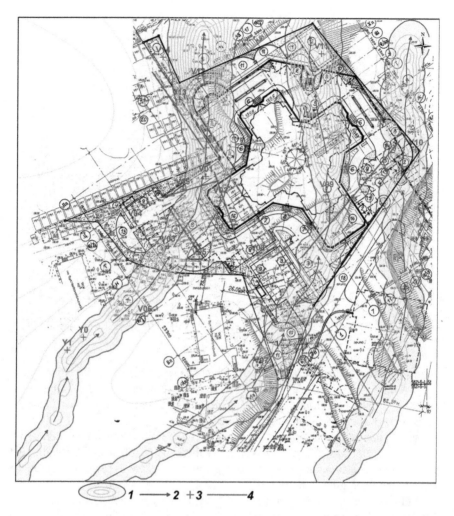

Fig. 31.4 The map of soil moistening areas and migration routes of groundwater flow in the area of the housing complex construction (according to the survey by FSPEF). 1—zone of soil moistening; 2—direction of groundwater migration; 3—VERS points; 4—profiles of GPR sounding

Determination of the Depth Interval of Increased Filtration Zones

The depth and thickness of the soil moistening zones were defined according to VERS and GPR soundings. Results of VERS sounding are given in Table 31.1. According to the GPR sounding three vertical cross-section of soil moistening zones were constructed, one of which is shown in Fig. 31.6.

Fig. 31.5 Map of soil moistening areas and migration routes of underground water flows at the site of a residential complex construction on the satellite image of area. 1—zone of soil moistening; 2—direction of groundwater migration; 3—VERS points; 4—profiles of GPR sounding

Two main soil moistening intervals were set at the construction site by sounding: number 1 and number 2 (Fig. 31.6; Table 31.1). Third moistening interval (number 3) is defined in part at some sounding points (Table 31.1). The "surface" water is moistening interval number 1. In the area of the western and southern parts of site the depth of the wetting zone roof within an interval number 1 is about 3.5 m. The

Table 31.1 Depth intervals and thickness of aquifers according to VERS sounding

VERS points	Aquifer # 1	Aquifer # 2	Aquifer # 3
V1	3.2–4.7	6.4–10.7	12.0–13.1
V2	3.1–3.9	7.7–10.6	–
V3	**3.5–4.8**	6.4–10.7	12.8–13.8
V4	3.4–4.0	7.6–10.2	–
V5	3.2–3.8	7.8–9.2	–
V6	3.6–5.0	6.0–10.7	10.7–12.5
V7	3.2–4.5	5.6–8.8	–
V8	4.6–5.4	8.5–11.0	–
V9	3.4–4.7	6.2–9.0	11.9–13.0
V10	**2.1–4.0**	6.4–9.5	–
V11	**3.1–4.8**	8.0–11.2	13.1–14.0
V12	**2.9–5.0**	6.2–10.6	12.7–13.5
V13	**3.3–5.1**	6.6–10.1	12.3–13.3
V14	3.4–4.0	7.0–9.2	12.3–13.5
V15	3.5–4.8	6.6–9.6	11.9–12.9

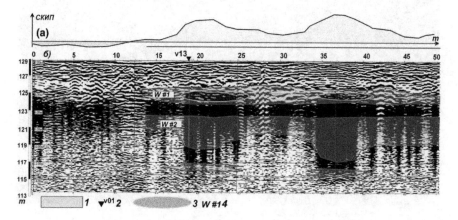

Fig. 31.6 Graf of FSPEF field (a) and GPR cross-section of soil moistening zone. 1—zone of soil moistening; 2—VERS points; 3—zone of high filtration of groundwater; 4—number of aquifer

thickness of filtration area is 1.3–1.5 m. In the southern part of the site the thickness of the soil moistening interval number 1 reaches 2.1 meters. The water filtration is carried out on the horizon of fine sand set according to drilling. In the area of excavation pit the soil is removed before the water flow roof in the interval number 1. The source of the water in the pit is carried out from a given interval.

The moistening horizon number 2 lies in the depth range of 6.0–10.0 m. The average thickness of the interval is 3.5 m. Water filtration is carried out on the sand horizon set by drilling.

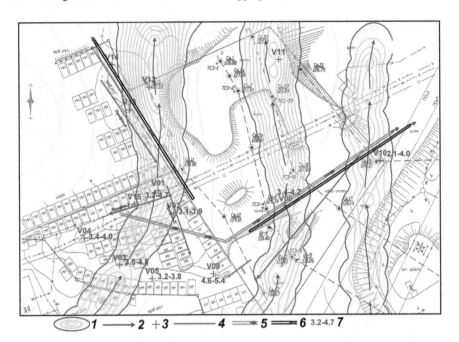

Fig. 31.7 Possible variants of drainage construction for the groundwater draining from the construction site. 1—soil moisture zone; 2—direction of groundwater migration; 3—VERS points; 4—profiles of GPR sensing; 5—variant of the drainage system number 1; 6—variant of the drainage system number 2; 7—depths for the top and base of the first aquifer

The third horizon of soil moistening is traced only at certain points in the depths interval of 12.0–13.0 m. The horizon thickness is 1.0 m. The moistening soils of this interval are located above the clay.

Results and Recommendation

In general, conducted geophysical studies revealed the causes of flooding of the pit at the construction site. The zones of increased filtration of groundwater and ways of their migration have been discovered and mapped. It is found that the water filtration is performed along the local zones, passing through the construction site. The depths of the main aquifers were determined.

The results of the research indicate the need for the drainage arrangement on the western and southern part of the site to divert the water from the construction site. To eliminate the water in the pit by theirs drainage it is necessary to intercept the first aquifer, the thickness of which is of 1.5 m in average.

Figure 4.10 shows two proposed options of drainage system laying to drain groundwater from the construction site.

Conclusion

The research materials of surveying character demonstrate again quite convincingly and clearly the devastating effects of underground water flows on various objects (including those under construction) of a modern city. They also point on the objective necessity of the destructive underground streams identifying and mapping at the stage of geological engineering survey for the construction of engineering structures, buildings and facilities of various purposes. This need ignoring leads, in most cases, to enormous time and material (financial) cost.

On the other hand, the results of geophysical studies show the effectiveness of FSPEF survey method and VERS and GPR soundings methods in dealing with the detection and mapping of underground water flows. Practical application of this technology during the geotechnical studies conducting for the construction of large engineering projects can bring significant economic benefits by reducing significantly the duration of exploration work and the drilling activity reduction.

References

Bokovoy V.P., Levashov S.P., Yakymchuk M.A. Korchagin I.N., Yakymchuk Ju.M. (2003) Mudslide area and moistening zones mapping with geophysical methods on the slope of the Dniper river in Kyiv. 65nd EAGE Conference & Exhibition. Stavanger, Norway, 2003. Extended Abstracts, P208, 4 p.

Levashov S.P., Yakymchuk M.A. Korchagin I.N., Pyschaniy Ju.M. (2005) Express-technology of geoelectric and seismic-acoustic investigations in ecology, geophysics and civil engineering. Near Surface 2005 - 11th European Meeting of Environmental and Engineering Geophysics, Palermo, Italy, 2005. Extended Abstracts P046, 4 p.

Levashov S.P., Yakymchuk N.A., Korchagin I.N., Pischaniy Ju.M., Bozhezha D.N. (2010) Application of mobile geophysical methods for the examination of areas of landslide processes formation and development. Near Surface 2010 – 16th European Meeting of Environmental and Engineering Geophysics, Zurich, Switzerland, 6–8 September 2010. Extended Abstracts P70. 5 p.

Chapter 32
Areas of Negative Excess Density of the Earth's Crust as Sources of Energy for Ore Formation

M. B. Shtokalenko, S. G. Alekseev, N. P. Senchina and S. Yu. Shatkevich

Abstract It has been calculated that the potential energy of one cubic kilometer of rocks with an excess density of -0.02 g/cm^3 is sufficient to form an ore deposit with medium reserves.

Keywords Energy · Potential energy · Gravity field · Density
Ore deposits · Negative gravity anomalies

Safronov (1966), Ivanov et al. (1978) proposed to consider ore formation as a transition of chemical elements in the Earth's crust from the diffused state to a concentrated and schematically describe this transition as the compression of an ideal gas. So N. I. Safronov proposed to estimate the energy of ore formation using the formula of work (A) on the compression of an ideal gas:

$$A = \frac{m}{\mu} RT \ln \frac{V_1}{V_2}, \qquad (32.1)$$

where m is the reserve of metal in the deposit, kg; μ is kg-mol of the metal, kg/kg-mol; R—universal gas constant, $R = 8.3144598$ J/(kg-mol°K); T—temperature, °K; V_1—volume of kg-mol of the metal in the diffused state, in rocks, m^3; V_2—volume of kg-mol of the metal in the concentrated state, in ore, m^3;

$$V_1 = \frac{\mu}{C_1 \rho_1}, \; V_2 = \frac{\mu}{C_2 \rho_2}, \qquad (32.2)$$

M. B. Shtokalenko
Geological Survey of Estonia, Tallinn, Estonia
e-mail: mihkel@egk.ee

S. G. Alekseev · N. P. Senchina (✉)
Saint Petersburg Mining University, Saint Petersburg, Russia
e-mail: n_senchina@inbox.ru

S. Yu. Shatkevich
AS "Geologorazvedka", Saint Petersburg, Russia

C_1—content of the metal in rocks, part per one; ρ_1—density of rocks, kg/m³; C_2—content of the metal in ore, part per one; ρ_2—density of the ore, kg/m³. Substituting (32.2) into (32.1), we obtain the calculation formula:

$$A = \frac{m}{\mu} RT \ln \frac{C_2 \rho_2}{C_1 \rho_1}. \qquad (32.3)$$

Table 32.1 shows the results of calculations using formula (32.3) for the medium reserves of deposits of several ore elements. The value of the reserve (m) was chosen at the upper limit of the established medium reserves.

The temperature T was assumed equal to the critical temperature of water 374.15 °C = 647.3°K with rounding to 700°K. The medium reserves of metals in deposits are adopted in accordance with Order No. 50 of the Ministry of Natural Resources of the Russian Federation of March 31, 1997 (Order 1997). The values of ρ_2 and C_2 are taken from Betekhtin's book (1951), with densities taken on the upper limit. Thus, we mentally enrich the deposits to massive ores, increasing the estimated energy of ore formation. Values of C_1—from the Quick Reference Book of the Geochemist (1977) (average contents of elements in the Earth's crust—according to Vinogradov 1962). The rock density (ρ_1) is assumed to be 2.67 g/cm³.

The calculations showed that the energy expenditure on the formation of the medium ore reserves does not exceed 10^{12} J for all the roundings to the greater side. Assuming that the efficiency of natural systems is a value of the order of 1%, we obtain a total energy expenditure of 10^{14} J. For comparison: 1 kiloton of TNT, i.e. railroad train, produces at an explosion of 4.184×10^{12} J of energy.

Where does the energy of ore formation come from? At present, this question remains debatable (Sendek and Chernyshev 2015). The association of ore deposits of gold, uranium, copper, lead and zinc in Russia to negative gravity anomalies (zones of

Table 32.1 Results of calculations of energy of ore formation

Element	Mineral	Formula	m		C_2		C_1	A
			th. t	kg	%	g/cm³	ppm	J
Co	Cobaltine	CoAsS	15	58.94	35.4	6.5	18	1.60×10^{10}
Cu	Chalcocite	Cu$_2$S	1000	63.57	79.8	5.8	47	9.63×10^{11}
Cu	Chalcopyrite	CuFeS$_2$	1000	63.57	34.57	4.3	47	8.59×10^{11}
Zn	Sphalerite	ZnS	1000	65.38	67.1	4.3	83	8.43×10^{11}
Mo	Molybdenite	MoS$_2$	50	95.95	60	5	1.1	4.20×10^{10}
Ag	Argentite	AgS	3	107.88	87.1	7.4	0.07	2.81×10^9
Sn	Cassiterite	SnO$_2$	50	118.7	78.8	7	2.5	3.34×10^{10}
Au	Gold	Au	0.05	196.97	90	18.3	0.083	3.11×10^7
Hg	Cinnabar	HgS	15	200.61	86.2	8.2	0.0043	7.52×10^9
Pb	Galena	PbS	1000	207.21	86.6	7.6	16	3.36×10^{11}
U	Uraninite	UO$_2$	20	238.03	88	10.6	2.5	6.92×10^9

negative excess density of rocks) is shown by the authors (Alekseev et al. 2010). Similar results were obtained for the territories of the USA, Canada, and Australia. As an example, Fig. 32.1 shows the map of the regional component of the US gravity (filter size 164 × 164 km) with the location of ore deposits.

The regional component of the field is created not only by deep sources, for example in the east of the Russian plate the regional component of gravity, obtained by averaging over a radius of 200 km, has a statistically significant positive correlation with the elevation of the crystalline basement.

An area of negative excess density has a positive potential energy W

$$W = Mgh, \qquad (32.4)$$

because the negative excess mass M is multiplied by the negative elevation h; g is the acceleration of gravity. A visual model of the negative excess mass is a ball immersed in water. When we release the ball, it jumps out of the water, converting potential energy into kinetic energy.

It would seem that the potential energy is greater if the negative excess mass is deeper, but excess mass is a relative concept. For example, the intrusion of diorites with a density of 2.8 g/cm^3 in the lower crust will have an excess density of 2.8–3.0 = −0.2 g/cm^3, in the middle crust it will have zero excess density: 2.8–2.8 = 0. and in the upper crust acquires a positive excess density: 2.8–2.7 = +0.1 g/cm^3. In view of this circumstance, the elevation in formula (32.3) is measured from the level at which excess mass becomes zero and is divided by 2 under the assumption that the excess mass changes linearly with elevation.

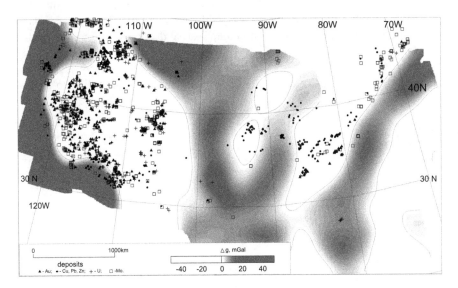

Fig. 32.1 Placing of large ore deposits of the USA on the background of the regional component of gravity. The field matrix (us_bouguer.grd file) was obtained from the US Geological Survey website (https://mrdata.usgs.gov/gravity/bouguer)

To estimate the elevation h in the formula (32.4), we can propose the following rough approximation. The density of the Earth's crust varies from 2.67 to 3.27 g/cm^3 per 45 km of thickness, i.e. with an average gradient of 0.0133 g/cm^3/km. Consequently, for the excess density of -0.01 g/cm^3, the elevation in formula (32.4) will be approximately $-0.01/0.0133/2 = -0.375$ km. For -0.02 g/cm^3, respectively, -0.75 km.

It is known that 1 cubic kilometer of rocks contains as many metals as an industrial ore deposit. Let us verify this situation for the chemical elements considered here (Table 32.2).

Let us now find out what excess density should have 1 km^3 of rocks lying from the surface, so that the potential energy of the mass considered is equal to the energy of ore formation, taking into account the efficiency of 1%. To do this, in formula (32.4) we represent the excess mass M as the product of the excess density σ by volume $V = 1$ km^3: $M = \sigma V$, then we assume $W = 10^{14}$ J, $h = -500$ m (depth of the center of mass) and find $\sigma = -0.02$ g/cm^3. It proves that not much is needed to get the required energy.

It is easier to imagine the considered volume of rocks at some depth, for example, at a depth of the center of 1500 m, the volume has an excess mass M, at a depth of 500 m the excess mass would be zero, since the density of the considered volume would become equal to the density of the enclosing rocks. The elevation difference of 1000 m should be divided in half to take into account the reduction in excess mass.

The anomaly from the cube 1 km^3 with an excess density of -0.02 g/cm^3, lying from the surface, will be -0.629 mGal above the center of the cube. For verification: according to the formula of gravity of an intermediate layer with a thickness of 1000 m with the indicated density, we obtain $0.0419 \cdot (-0.02) \cdot 1000 = -0.838$ mGal; from the excess mass, concentrated at a depth of 500 m, we obtain -0.544 mGal.

The potential energy of mass distribution in the Earth's layers is the most important energy source of the planet. If on the continents, according to modern

Table 32.2 The contents of metals in 1 cubic kilometer of rocks

Element	Content th. t	Size of deposit
Co	48	Large
Cu	125	Medium
Zn	222	Medium
Mo	3	Small
Ag	0.2	Small
Sn	7	Medium
Au	0.01	Medium
Hg	0.2	Small
Pb	43	Small
U	7	Medium

concepts, about 80% of the natural heat flow is created by radioactive decay, then under the oceans, where there is no granite layer (the upper crust in which the radioactive elements are concentrated), the potential energy of masses released during tectonic movements becomes the leading source of heat. Since the ore is heavier than the surrounding rocks, ore formation reduces the negative excess mass, filling them with positive ones.

The use of negative gravity anomalies for prospecting of ore deposits is physically justified by comparing the energy of ore formation and the potential energy of negative excess masses of the Earth's crust.

Negative gravity anomalies, to which ore deposits are connected, are created in particular by fluid-permeable tectonic zones. A fluid is a superheated liquid that becomes a super-solvent (Course of Mineralogy 1936). For this reason, the calculated temperature of ore formation was assumed to be somewhat higher than the critical temperature of water.

The potential energy of the observed gravity anomaly can be estimated by multiplying the excess mass of the anomaly source by the depth of the center of mass with a minus sign. From the depth it is necessary to subtract the depth for which the density of the source of the anomaly is average, and divide by 2. The excess mass of the source is determined by the integration of an anomaly previously isolated from the field (Andreev and Klushin 1965), the depth of the center of mass is estimated by known methods of interpretation, a tomographic section at full densities is required to correct the depth or the above rough estimate.

Conclusions

One cubic kilometer of rocks with the excess density of -0.02 g/cm^3 contains not only metals in an amount equal to the reserves of an industrial deposit, but also the potential energy sufficient to form an ore deposit with medium reserves.

Negative gravity anomalies may be used in prospecting for ore deposits because the negative anomalies indicate the presence of favorable conditions for the formation of the deposit.

References

Alekseev S.G. et al. (2010). Indications of hydrocarbon and ore systems of various ranges in gravity and magnetic fields. In: Theoretical and practical aspects of geological interpretation of gravity, magnetic and electric fields: Proceedings of the 37th session of Uspensky International Geophysical Seminar. Moscow: Institute of Physics of Earth of RAS, pp. 15–20. (in Russian).

Andreev B.A., Klushin I.G. (1965). Geological interpretation of gravity anomalies. Leningrad: Nedra. Leningrad Branch, 495 p. (in Russian).

Betekhtin A.G. (1951). The course of mineralogy. Moscow: State Publishing House of Geological Literature, 542 p. (in Russian).

Course of Mineralogy. (1936). The collective of authors (Betekhtin A.G., Boldyrev A.K., Godlevsky M.N., Grigoriev D.P., Kiselev A.I., Levitsky O.D., Razumovsky N.K., Smirnov A. A., Smirnov S.S., Sobolev V.S., Soloviev S.P., Uspensky N.M., Chernykh V.V., Shatalov E.T. and Shafranovsky I.I.) Moscow: ONTI, 1051 p. (in Russian)

Ivanov N.P., Meshcheryakov S.S., Safronov N.I. (1978). The energy of ore formation and the prospecting for minerals. Leningrad: Nedra, 215 p. (in Russian).

Order of the Ministry of Natural Resources of the Russian Federation of March 31, 1997, № 50 On instructions on the payment of remuneration for the identification of mineral deposits (in the edition of the Order of the Ministry of Natural Resources of the Russian Federation No. 112 of April 24, 2000). (in Russian).

Quick reference book on geochemistry. (1977). Ed. 2./ Voitkevich G.V., Miroshnikov A.E., Povarennykh A.S., Prokhorov V.G., Moscow: Nedra, 180 p. (in Russian).

Safronov N.I. (1966). Basic thermodynamic regularities in the study of the energy of ore formation. Proceedings of the Mining Institute (Journal of Mining Institute), vol. 50, № 2, pp. 17 – 31. (in Russian).

Sendek S.V., Chernyshev K.E. (2015). Theoretical justification for analyzing distribution of gold concentrations in ores of hydrothermal gold deposits. Proceedings of the Mining Institute (Journal of Mining Institute), V.212, pp. 30–40. (in Russian).

Vinogradov A.P. (1962). Average contents of chemical elements in the principal types of igneous rocks of the Earth's crust. Geochemistry № 7, pp. 641–664. (in Russian).

Chapter 33
Distribution of Sources of Magnetic Field in the Earth's Core Obtained by Solving Inverse Magnetometry Problem

V. Kochnev

Abstract The purpose of this work is to create a model of sources of Earth's magnetic field. Main magnetic field in IGRF-2005 model is used as initial condition for solving inverse magnetometry problem using adaptive method of solving systems of equations. Distributions of effective magnetization, magnetic moments and volumetric currents in the Earth's core, consistent with observed magnetic field, are obtained.

Keywords Magnetic field of earth · Effective magnetization · Inverse magnetometry problem volumetric currents · Geodynamo

Introduction

The sources of Earth's magnetic field are usually approximated by dipoles and current loops. (Peddie 1979). In this paper they are represented by magnetized prisms, filling the planet's core. This allows to transition from magnetization to magnetic moment and volumetric current. This transition is valid due to the equivalence between magnetized objects and currents, established in (Alpin et al. 1985). Distribution of effective magnetization within prisms is obtained while solving the inverse magnetometry problem with Z-components of the main magnetic field of the Earth (as per IGRF-2005 model) in geocentric coordinate system as initial condition. Resulting system of 2450 equations with 354 unknowns is resolved using adaptive method (Kochnev 1988, 1993, 1997; Kochnev and Khvostenko 1996).

V. Kochnev (✉)
Institute of Computational Modeling SB RAS, Krasnoyarsk, Russia
e-mail: kochnev@icm.krasn.ru

Initial Conditions

Magnetic field lines are presented in a geocentric coordinate system, with center of the Earth as the origin, Z axis is aligned with Earth's rotational axis and X and Y axis lying within equatorial plane, crossing equator at longitudes of 90° and 180° correspondingly. Figure 33.1 a,b shows contour lines of Earth's magnetic field as seen from the north pole according to IGRF-2005 model, calculated on an even grid with spacing of 400 km at height of 1 km above the surface. For reference, figures contain projections of the geographic pole (polus) and certain cities: Tokyo (TOK), Krasnoyarsk (KRS), Yekaterinburg (EKT), London (LON), Ottawa(OTW) in Northern hemisphere and Punta Arenas (PAR) and Canberra (CNB)—in southern.

For the northern hemisphere, contour line 56,270 nT (absolute value), corresponding with maximum of magnetic field highlights two well-defined areas of higher magnetic field strength. In equatorial area magnetic field strength varies between 29,500 to 34,800 nT (absolute value). In southern hemisphere maximum lies between southern pole and Canberra, while a well-defined minimum lies opposite of it in the equatorial area.

Z-component of the magnetic field in geocentric coordinate system (Fig. 33.1 c,d) has its maximum in an elliptic shape around the pole in the north hemisphere. For southern hemisphere, similarly to the overall field maximum lies in an area between pole and Canberra.

Magnetization and Magnetic Field

Magnetic induction potential, created by the magnetized body is defined as (Aleksidze 1987; Bulakh and Shuman 1998).

$$\Delta U(\xi,\zeta,\eta) = \frac{\mu_0}{4\pi} \int\int\int_G (\vec{J}(x,y,z), grad\frac{1}{r}) dG, \qquad (33.1)$$

where $\vec{J}(x,y,z)$ is magnetization (A/m) of a unit of volume dG, with coordinates x, y, z, r is distance between the unit of volume and the point of observation, and the integral is taken over the entire magnetized volume G.

For numerical integration the volume is divided into rectangular prisms for which magnetization is assumed to be constant. Components of magnetic field within these prisms are calculated via differentiation of Eq. (33.1)

Resulting system of equations generally is viewed as an ill-posed problem. This was circumvented by using the adaptive method (Kochnev 1983, 1988, 1993).

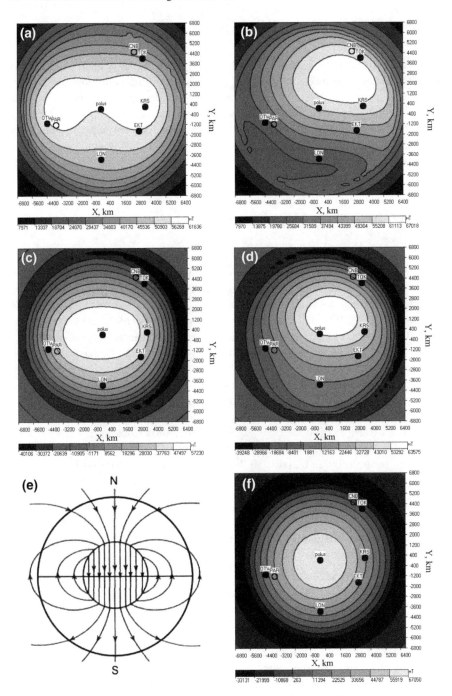

Fig. 33.1 Contour lines of Earth's magnetic field in geocentric coordinate system: **a** northern hemisphere, **b** southern hemisphere. Z-component of Earth's magnetic field: **c** northern hemisphere, **d** southern hemisphere (viewed from the northern pole), **e** field lines of Earth's magnetic field with evenly magnetized core, **f** contour lines of Z-component of magnetic field with evenly magnetized core

Adaptive Method for Solving Systems of Linear Equations

Key quality of adaptive method is its ability to deal with the systems of equations by refining the unknown values based both on the initial conditions and newly found values for each iteration. This refinement is performed successively for all the equations and is based on the discrepancy between the initial and projected values.

Method is realised in the ADM-3D application for solving magnetometry problems (Kochnev 2006), that is being used for solving various scientific and engineering problems in geophysics.

Solving the Inverse Magnetometry Problem

The purpose of this work is to come up with a model of the sources for Earth's magnetic field that results in and accurate representation of actual Earth's magnetic field. Assuming these sources, as mentioned above, to be uniformly magnetized rectangular prisms, based on the equivalence of currents and magnetized objects (Alpin et al. 1985), the calculation uses two layers of rectangular prisms (Fig. 33.2e) with 177 prisms in northern and southern hemispheres each. Prisms dimensions in equatorial plane are 400 by 400 km. Effective magnetizations of the prisms are considered the unknown values.

Using even distribution of magnetization for initial condition results in an uneven distribution very similar to the one obtained with no initial magnetization. Root mean squared discrepancies showed significant reductions as iterations went on. This indicates stability and convergence of obtained solution.

Figure 33.2 a,b depicts effective magnetizations in northern and southern hemispheres, varying between 0 and 1580 A/m for northern and −605 and 1315 A/m for southern. Obtained distributions differ significantly from the initial values.

Estimation of Magnetic Moments and Currents

Having obtained effective magnetization i for each prism and knowing their volume V, we can calculate the magnetic moment (Alpin et al. 1985):

$$M = iV. \tag{33.2}$$

Figure 33.2 c,d depicts distributions of magnetic moment for prisms in northern and southern hemispheres correspondingly. Obtained values vary between -222×10^{18} and 512×10^{18} A m². Total magnetic moment for northern hemisphere is 4.87×10^{22} A m², for southern 3.54×10^{22} A m². Resulting total magnetic moment of Earth $M_E = 8.41 \times 10^{22}$ A m². Negative magnetic moment is

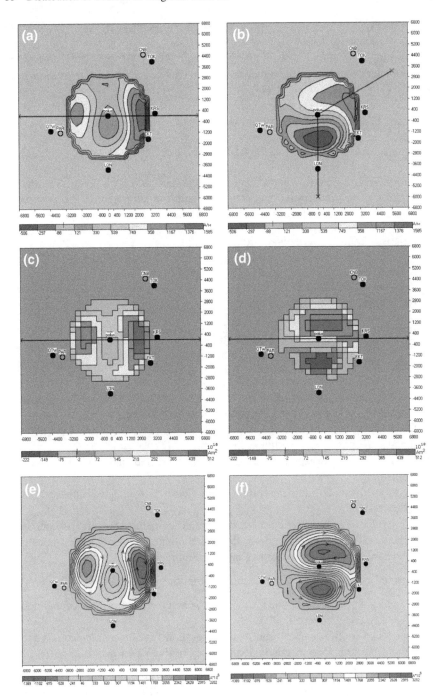

Fig. 33.2 Magnetization of **a** northern and **b** southern hemispheres of the Earth's core, obtained by solving inverse magnetometry problem; magnetic moments for **c** northern and **d** southern hemispheres; Volumetric currents for **e** northern and **f** southern hemispheres

only observed in a small segment of southern hemisphere and amounts to -3.63×10^{21} A m^2 total.

Resulting estimation for the entire Earth's core 8.65×10^{22} A m^2 is close to the value obtained from evenly magnetized model: 8.45×10^{22} A m^2 and estimation of 10^{23} A m^2 by (Kaufman 1997).

Having obtained the values for magnetic moments we can estimate currents necessary to create them. For rectangular prism $M = IS$, therefore:

$$I = M/S \text{ or } I = ih, \tag{33.3}$$

where I is total current flowing around the lateral surface of the prism, S is area of the prism's base, i is effective magnetization and h is the height of the prism.

Figure 33.2 e,f depict currents in northern and southern hemispheres correspondingly. The highest current value reaches 3.2×10^9 A, while volumetric currents lie within the 0.004–0.03 A/m^2 margins.

Conclusion

Following new scientific results were obtained:

1. Using Earth's global magnetic field on the surface, inverse magnetometry problem was solved to estimate sources of Earth's magnetic field in its core.
2. Z component of the global magnetic field model IGRF-2005 in the geocentric coordinate system was used for solving inverse problem.
3. The Earth's core was modeled with two layers of rectangular prisms, with even effective magnetization within each prism.
4. An ADM-3D-earth application package was created and used to solve the problem.
5. Model of magnetic moments in the Earth's core was obtained, consistent with earlier results.
6. The model of volumetric core currents is and constructed, which is consistent with observed Earth's magnetic field, including its main global anomalies: Canadian, Asian-Siberian, Australian and South Atlantic.

Obtained results allow the following assumptions to be made:

1. The volumetric current in the Earth's core is assumed to represent of the motion of weakly charged core fluid.
2. The clockwise (if observed from north pole) motion of fluid in the core is due to the and the influence of gravitational forces of the Moon and the Sun.

In (Kochnev 2013, 2017) tidal forces are calculated and linear relationship between tidal forces and a magnetic field is established for the planets of the solar system. (The correlation coefficient is 0.997).

Acknowledgements The author expresses his gratitude to L. Tabarovsky, A. S Dolgal, Y.I. Bloch for advice and constructive criticism, as well as I.V. Goz and A. E. Koroleva for help with editing and formalization.

References

Aleksidze M.A. (1987) Approximation methods for solving direct and inverse problems of gravimetry. Moscow, Nauka, 1987, 336 p.

Alpin L.M., Daev D.S., Karinsky A.D. (1985) Theory of fields used in exploration geophysics. Textbook for Universities. Moscow, Nedra, 1985, 407 p.

Bulakh E.G., Shuman V.N. (1998) Fundamentals of vector analysis and field theory. Kiev, Naukova Dumka, 1998, 359 p.

Kaufman A.A. (1997) Introduction to the theory of geophysical methods. Gravitational, electric and magnetic fields. Moscow, Nedra, 1997, 520 p.

Kochnev V.A. (1983) Adaptive tracking of reflected waves and an estimation of their parameters according to the data of multiple observation systems.// Geologiya i Geophysika, 1983, no. 2, p. 95–103.

Kochnev V.A. (1988) Adaptive methods of interpretation of seismic data.// Novosibirsk, Science. Sib., 1988, 152 p.

Kochnev V.A. (1993) Adaptive methods for solving inverse geophysical problems.// Krasnoyarsk, ICM SB RAS, 1993, 130 p.

Kochnev V.A., Khvostenko V.I. (1996) Adaptive method for solving inverse problems of gravimetry// Geology and Geophysics, 1996, No. 7, p. 120–129.

Kochnev V.A. (1997) Adaptive method for solving systems of equations in inverse geophysical problems// Proceedings of the Siberian Conference on Applied and Industrial Mathematics, dedicated to the memory of LV Kantorovich. Novosibirsk, 1997, p. 129 – 137.

Kochnev V.A., Goz I.V. (2006) Unsolved possibilities of magnetometry// Geophysics, 2006, no. 6, p. 51–55.

Kochnev V.A. (2013) Kinematic-gravitational model of geodynamo// Geophysical Journal, 2013, v. 35, No. 4, p. 3–15.

Kochnev V.A. (2017) Dynamo models created on the planets under the influence of the tidal forces of the satellite and the Sun// 17th International Multidisciplinary Scientific GeoConference SGEM 2017, 2017, vol. 17, issue 62, Section Space Technologies and Planetary Science, p. 899–906.

Peddie N.W. (1979) Current loop models of the Earth's magnetic field// Journal of Geophysical Research, 1979, vol. 84, No. B9, p. 4517.

Chapter 34
Efficiency of High-Precision Gravity Prospecting at Discovery of Oil Fields at Late Stage of Development

Z. Slepak

Abstract This paper contains information about technologies and results of solving a number of oil geology problems based on data of high-precision gravity measurements in various structural levels of sedimentary complex in the Urals and Volga region. The examples of predicting oil-and-gas bearing structures, undiscovered formations in areas of previously discovered oil fields, areas of increased rock porosity in carbonate complexes, to which non-structural fields may be confined, as well as accumulations of high-viscosity oil (natural bitumen) are provided here. This paper shows that high-precision gravity prospecting and well survey results at the late stage of oil field development allow effectively discovering the features of geological structure of predicted objects and deposits, inclosing them.

Keywords High-precision gravity prospecting · Well survey · Oil fields New prediction technologies

Introduction

The majority of oil fields in the Urals and Volga region are confined to local structures, discovered in the second half of the last century in Devon and Carbon deposits. At present, the late stage of these fields development is performed. The methodology of their prediction and hydrocarbon extraction has changed. Much attention is paid to the development of new drilling technologies, using horizontal, inclined and other types of wells (Muslimov 2011).

Successful application of high-precision gravity measurements in prediction of oil fields requires the creation of geophysical technologies characterized by reliable physical and geological justification. In order to solve each specific problem of oil geology, it is necessary to use well-known technologies or create new ones to

---

Z. Slepak (✉)
Institute of Geology and Petroleum Technologies, Kazan Federal University, Kazan, Russia
e-mail: Zakhar.Slepak@kpfu.ru

interpret anomalous gravitational field. The author developed an effective method of geological and geophysical modeling (GGM). Unlike many methods of qualitative interpretation of gravitational anomalies, the method of geological and geophysical modeling allows creating density models of predicted objects and deposits, inclosing them. This paper presents examples of discovering high-precision gravity prospecting of oil-and-gas bearing structures, areas of increased rock porosity in carbonate deposits, to which non-structural fields and natural bitumen accumulations can be related.

Methods of Geological and Geophysical Interpretation and Research Results

Prediction of Oil-and-Gas Bearing Structures

The solution of oil geology problems requires studying physical and geological features of predicted objects. The lateral heterogeneity of composition and physical properties of sedimentary complex rocks is characteristic for the Urals and Volga region (Andreev 1957; Proceedings of All-Russian Geological Research and Development Oil Institute 1974; Slepak 1989, 2015). The density measurements were performed for 6000 core samples from well intervals with continuous sampling and according to data on oil-and-gas bearing complex and gas bearing complex for 300 wells. This allowed establishing regular deconsolidation of rocks above oil reservoirs in areas of oil-and-gas bearing structures discovered in individual sedimentary complex horizons. The variability of average values of total porosity factors K_p and densities σ determined from well survey data on crests and flanks of 23 structures of the South Tatar crest within sulfate-carbonate complex was studied. The thickness of intervals studied for oil-and-gas bearing complex and gas bearing complex was 200–600 m or more. It has been established that in crests of 17 out of 23 structures the decrease in average values of densities compared with flanks is −0.01 to −0.07 g/cm^3 (74%). The increase in the parameter by 0.03–0.06 g/cm^3 (26%) observed for 6 structures is related to the features of secondary processes, occurring within them. This allowed stating (with high reliability) the presence of regular deconsolidation of rocks within local structures above oil fields (equal to the hundredths of g/cm^3), the gravitational influence of which is the main prospecting indicator of their prediction. (Slepak 2005; 2014; 2015).

The presence of regular deconsolidation of rocks is discovered in carbonate deposits in sections of many oil-and-gas bearing structures of the Urals and Volga region. It is significantly associated with the formation of Karstic forms. When drilling wells, there are often "cavings" of drilling tool and the absorption of drilling mud and it is not always possible to eliminate its withdrawal. Karstic areas with thickness of tens of meters may match with the structural plan. Their influence is associated with increase in the intensity of local minima above structures created by rock deconsolidation. The deconsolidation of rocks is also observed in separate horizons of local highs not only in carbonate, but also in terrigenous deposits of other regions (Proceedings of All-Russian Geological Research and Development Oil Institute 1974; Slepak 1989, 2015).

The author developed an effective method of geological and geophysical modeling (GGM), which consists in solving the inverse linear problem (Strakhov 1995) of gravity prospecting with simultaneous creating density models of geological media and predicted objects directly from Bouguer anomalies without dividing the field into components. Unlike many methods of qualitative interpretation of gravitational anomalies, the geological and geophysical modeling method allows extracting information about features of the structure of oil fields and deposits, inclosing them (Slepak 1989, 2005, 2015).

The method is successfully applied to solve two oil geology problems at quantitative level: prediction of oil-and-gas bearing structures in sedimentary complex and study of block structure of crystalline basement and fractures, dividing them, which affect the formation of structures in deposits, overlapping them. Both problems are being solved on quantitative level. The technologies of the geological and geophysical modeling method allow creating physical and geological 2D and 3D models of oil-and-gas bearing structures of different morphological and genetic types and identifying areas of rock deconsolidation above oil reservoirs. The results of inverse modeling depending on physical and geological features of the structure of objects under study are presented in digital form with their prismatic approximation in isolines of density variability in vertical plane and graphic representation of density variability depending on depth. At the same time, the reliability of solutions based on priori data is estimated (Slepak 2005, 2015).

Stepnovskaya structure (Saratov region) is a high of tectonic type, which is a brachianticlinal box fold with wide flat crest located on the southern side of the Marxian draw-down. The crystalline basement lies here at the depth of about 3.0 km according to geophysical data. The amplitude of the high on the surface of Devon deposits is 0.35 km and it is characterized by large slope of flanks. Up the section the amplitude of the high decreases up to 0.050 km. The structure is industrial oil-and-gas bearing deposit concentrated mainly in Devon and Carbon deposits. The total thickness of oil-and-gas bearing formations exceeds 0.10 km (Slepak 2015).

Based on the interpretation results presented in digital form along the profile, which crosses the structure across strike (Fig. 34.1), the complete correspondence of the observed and matched fields is stated. The local gravity minimum between the 7th and 11th km can be observed above the structure crest against the background of local field decrease. The lateral change in densities of model layers characterized by different composition of deposits with prismatic approximation of geological section presented in digital image (Fig. 34.2) allows estimating deconsolidation of rocks in upper layer by value of -0.07 g/cm^3, on average layer -0.03 g/cm^3, in lower layer -0.01 g/cm^3.

The features of lateral variability of rock densities in structure model layers (Fig. 34.2) represented in graphic image indicate that the deconsolidation of rocks

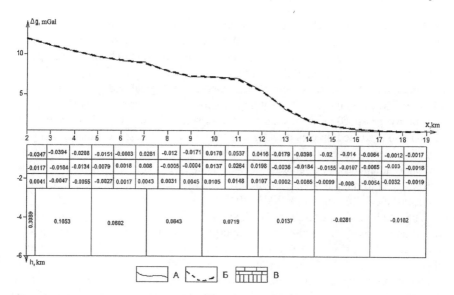

Fig. 34.1 Physical and geological model of Stepnovskaya structure: A is Bouguer anomaly, B is theoretical (matched) anomaly, C is structure density model

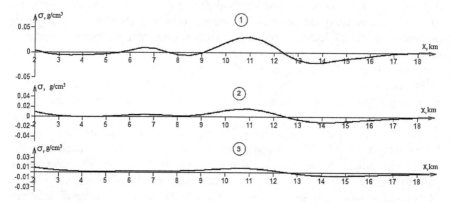

Fig. 34.2 Graphs of density change in layers of Stepnovskaya structure model. 1—changes in densities in upper layer; 2—changes in densities in middle layer; 3—changes in densities in lower layer

in the upper layer is the closest to the actual value and gradual decrease of this value in the second and third layers can be explained by their distant location from the ground surface. Judging by the model the average value of lateral deconsolidation of densities on the structure crest in sedimentary complex represented by three layers, differing in the rock composition, is -0.04 g/cm^3. The influence of the field created by the lower layer varies within the thousandths of g/cm^3 and can be attributed to its background component.

The geological section in the field area of one of the Solikamsk draw-down structures of the Pre-Ural for deep also has three-layered structure. The upper layer with thickness of about 0.25 km (P2) is represented mainly by terrigenous formations. The salt deposits (P1 ir) underlay under them with thickness of 0.30 km. Permian and Tournaisian carbonate deposits are located below in the section (the third layer) with total thickness of about 1.5 km (P1 ar - C1t), overlapping the Upper Devonian oil-bearing reef with dimensions of 5 × 5 km and amplitude of more than 0.1 km. The values of densities in chosen layers are determined approximately and taken correspondingly equal to: 2.45–2.50, 2.05–2.15 and 2.64–2.67 g/cm^3 (Bychkov 2010). The inverse modeling results obtained by the geological and geophysical modeling method are presented in the form of graphs of density variability along the profile in the model layers (Fig. 34.3).

The example of 3D model creation by the geological and geophysical modeling method is the Aktanyshskaya structure located in the northeast of the South Tatar crest, which is represented by local intensity minimum of about 1.0 mGal (Fig. 34.4). The subsequent drilling allowed determining the structure contours.

Some features of lateral variability of densities are observed in the model. The rock deconsolidation section is observed in its central part. It is limited by rock thickening along the edges and their local changes, characterizing the geological structure components, which are of undoubted interest for determining the drilling location. Similar 3D model is constructed according to the data of high-precision gravity measurements for the Yamashinskaya structure section, the reliability of which is confirmed by well survey results (Slepak 2005, 2015).

The creation of 3D models by the geological and geophysical modeling method based on profile-areal high-precision gravity measurements gives evidence of high geological information content of applied technologies for interpretation of

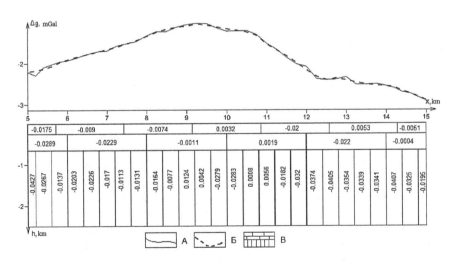

Fig. 34.3 Physical and geological model of structure A is Bouguer anomaly, B is theoretical (matched) anomaly, C is structure density model

Fig. 34.4 Density 3D model of sedimentary complex in area of predicted Aktanysh-skaya structure (isolines of densities in g/cm^3)

anomalous gravitational field. They allow identifying physical and geological heterogeneities of geological media under study. Also simultaneously with 2D models they can serve as justification for using the geological and geophysical modeling method for studying the structure and predicting oil fields.

Discovery of Areas of Increased Porosity of Deposits in Carbonate Complexes

The comparison of physical and geological models, created as a result of inverse modeling by the geological and geophysical modeling method for high-precision gravity measurements, with well survey results is especially effective in studying the geological structure and predicting "missing" oil formations at the late stage of their development. There is the possibility of gaining information on the spatial variability of porosity and density according to the data on oil-and-gas bearing complex in deposits under study. The areas of increased porosity of rocks within sulfate-carbonate complex, traced by graphs of oil-and-gas bearing complex along regional geological and geophysical profiles are of particular interest.

The regional profile of about 80 km in length extends from the southwest to the northeast through the western slope of the South Tatar crest (Fig. 34.5) as presented as an example. The total porosity factor Kp is on average 5–7% along the profile within the Tournaisian-Devon deposits. The average weighted capacity in

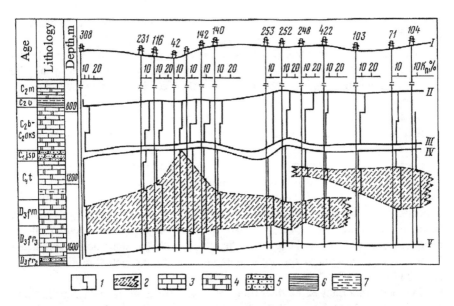

Fig. 34.5 Geological and geophysical profile along line of wells No. 308–104: surfaces: I—ground, II—Bashkirian stage C2, III—Yasnaya Polyana superhorizon C1, IV—Tournaisian stage C1, V—Middle Frankish substage D1; 1—diagram; 2—areas of increased porosity; rocks: 3—limestones, 4—dolomites, 5—sandstones, 6—clays, 7—siltstones

individual wells varies from 5 (well No. 103) to 12% (wells No. 71, 116) and reaches 14–15% in individual intervals of the section (wells No. 116, 252). (Slepak 2005, 2015).

Abnormal increase in Kp by 3–4% is observed throughout the carbonate thickness in regions of local highs discovered along the profile (wells No. 308, 116–142, 253–248). The most significant increase is observed in the area of the Yamashinskaya structure (well No. 231–140), where the area of increased porosity of lenticular rocks with value of Kp of 10–12% is clearly observed. It is possible that the area surface can be a contact—"cap rock" of sharp reduction in Kp by 7–8% for accumulation of oil reservoirs of non-structural type. They can be discovered with more detailed processing and interpretation of well survey data for drilled wells. It is reasonable to perform high-precision gravity measurements on certain sections of the profile and determine drilling location. Areas of increased porosity (similar to that shown in Fig. 34.5) discovered in other areas, should also be considered.

Discovery of Ultra-Viscous Oils (Natural Bitumen)

Ultra-viscous oils (natural bitumen) are discovered on the western and south-eastern slopes of the South Tatar crest and associated with carbonate and sandy-siltstone formations with thickness up to 40 m. Ultra-viscous oils of the majority of discovered reservoirs (80–85%) are liquid, semi-liquid, travelling (Khisamov, Fayzullin 2011).

Since the density of bitumens is close to the groundwater density, it is practically impossible to determine their gravitational influence. The sand lenses in the Ufa deposits, which are bitumen reservoirs, are characterized by high factor of total rock porosity (22–46%) and low density (1.5–2.2 g/cm^3), which differ significantly from parameters of deposits, inclosing them. Therefore, the gravitational anomalies created by the influence of sand lenses are characterized by local field decreases, which are prospecting indicators of possible accumulation of natural bitumen reservoirs. The resultative example of discovering the sand lenses with high-precision gravity prospecting is the Studeno-Klinskoe field of natural bitumen (Slepak 2005, 2015). Since the field, sand lenses differ significantly in density from deposits, inclosing them, the method of recalculating the measured gravitational field into the lower half-space was used to discover them. The method allows analyzing the change in the gravitational field with the approach to predicted objects, using maps and graphs of gravitational field constructed in vertical plane (Fig. 34.6).

The most important is the thickening of field isolines in the center of the sand lens base (Fig. 34.6) in so-called "singular point" at depth of 117 m, completely matching with the drilling data, which undoubtedly indicates high reliability of the prediction results.

The sand lenses with accumulation of natural bitumens are often located in areas of rock deconsolidation above oil reservoirs in underlying deposits of sedimentary complex and reach ground surface. Such deconsolidation areas are typically

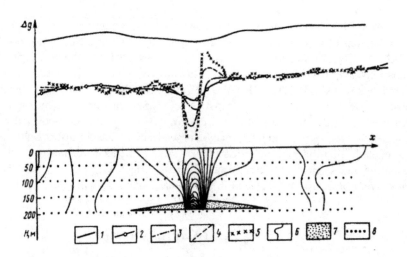

Fig. 34.6 Results of recalculating the measured gravitational field to the depth along the profile above the Studeno-Klinskoe field of natural bitumens: 1—measured gravitational field; 2, 3, 4, 5—graphs of measured field anomalies recalculated at depths of 50, 100, 150, 200 m correspondingly; 6—isolines of field in vertical plane

associated with crests of structures and inevitably appear in gravitational field in the form of local gravity minima, which are the prospecting indicator of their prediction.

The influence of the Ufa sand lenses may be added to the gravitational influence of the deconsolidation created by the areas, which can enhance the intensity of their representation in the gravitational field. When performing high-precision gravity measurements in monitoring mode, this necessarily should increase the extraction of bitumen and allow correcting the drilling process at the late stage of development.

Conclusion

The geological and geophysical modeling (GGM) method developed by the author is characterized by high efficiency in discovering oil-and-gas bearing structures and recommended for prediction of oil fields in different regions.

The above examples of solving the oil geology problems allow stating that high-precision gravity prospecting and well survey results should be used to specify geological structure features of predicted objects and deposits, inclosing them.

During the development of previously discovered oil fields the physical and geological features of sedimentary complexes significantly change under the influence of groundwater dynamics, secondary rock reorganizations, lithological-facial reorganizations of rocks and neo-tectonic movements. As a result, the conditions are created for the migration of oil reservoirs to other layers. Therefore, the performance of high-precision gravity measurements in monitoring mode in oil-and-gas bearing areas during the late stage of field development should undoubtedly contribute to discovering new hydrocarbon accumulations.

References

Andreev B.A. (1957) Layer-by-layer zonality of physical properties of sedimentary rocks and its relationship with platform area structures / Andreev B.A. - Soviet geology - 1957 - No. 61 - pp. 112-120.

Bychkov S.G. (2010) Methods of processing and interpretation of gravity measure-ments in solving problems of oil-and-gas geology. - Ekaterinburg: Ural Division of Academy of Sciences of Russia, 2010. - p. 188.

Lateral variability of composition and physical properties of sedimentary thick-ness within structures and its reflection in zoning of geophysical fields/Proceedings of All-Russian Geological Research and Development Oil Institute, 1974. - Issue 160 – p. 417.

Muslimov R.Kh. (2011) Oil production: past, present, future. - Kazan: FEN publishing house, Academy of Sciences of the Republic of Tatarstan, 2011, Russia. - p. 663.

Slepak Z.M. (1989) Use of gravity prospecting in search for oil-bearing structures. Moscow: Nedra, Russia. 1989. - p. 200.

Slepak Z.M. (2005) Gravity prospecting in oil geology. Kazan, Kazan University Pub-lishing House.Russia. - 2005. p. 222.

Slepak Z.M. (2015) Gravity prospecting. New technologies for prediction of oil fields. - Kazan: Publishing house of the Kazan University, Russia. 2015. - p. 168.

Strakhov V.N. (1995) Major trends in the development of theory and methodology of geophysical data interpretation at boundary of XXI century. Part I // Geophys-ics, No. 3, 1995, pp. 9–18.

Khisamov R.S., Fayzullin. (2011) Geological and geophysical study of oil fields at late stage of development. Kazan: Academy of Sciences of the Republic of Tatarstan, FEN publishing house, Russia. 2011, p. 228.

Z.M. Slepak. (2014) New Opportunities of High-Resolution Gravimetry for the Studies of Subsurface Geology and Prediction of Oil Fields. - 14th International Multidisciplinary Scientific GeoConference on Science and Technologies in Geolocy Exploration and Mining. Volume 1. Bulgaria, 2014, pp. 743 – 750.

Chapter 35
Geophysics in Archeology

Z. Slepak and B. Platov

Abstract This paper contains information about possibilities of exploration geophysics in solving archeological problems. High effectiveness of electromagnetic sounding at revealing ancient buildings remnants in urban conditions, dating the layers of anthropogenic formations and tracing their bottoms using Impulse-auto M-1/0-20 electromagnetic surveying system is demonstrated herein. The measurement procedure and developed computer technologies for processing geophysical information, which allow solving particular archeological problems and evaluating their veracity, are stated in this paper. The results of conducted researches on the examples of archeological objects of the Kazan Kremlin, Bogoroditsky Monastery, Bilyar and Bulgar settlements of the Republic of Tatarstan are considered here.

Keywords Geophysics · Electromagnetic sounding · Cultural layer Archeology

Introduction

Geophysical research methods have no negative impacts on anthropogenic layer. They are widely used in solving geological problems, in engineering geology and archeology. Because of limited space and numerous interferences in modern cities (presence of buildings, ground and underground pipelines, moving vehicles, etc.) the results of geophysical operations are characterized by significant errors and require using specific measurement procedures and interpretation of the data obtained. The study of anthropogenic layer structure and revealing archeological objects is a difficult task, as weak changes in physical fields created thereby are practically equal to errors of used geophysical instruments.

Z. Slepak (✉) · B. Platov
Institute of Geology and Petroleum Technologies, Kazan Federal University,
Kazan, Russia
e-mail: Zakhar.Slepak@kpfu.ru

High effectiveness of the procedure for solving archeological problems developed by the author was demonstrated in the territory of historical center of Kazan (and in areas of ancient settlements of Volga Bulgaria). The main object of research was the Architectural ensemble of the Kazan Kremlin located on a high hill. Its peak reaches 30 m relative to the surrounding streets from the north, east and west, which created certain difficulties for geophysical researches. Despite this the results of predicting remnants of ancient buildings in anthropogenic layer using geophysical methods were confirmed by subsequent archeological excavations.

Measurement Procedure and Research Results

The geophysical technologies used in the territory of the Kremlin for the first time allowed determining the location of lost ancient buildings in conditions of urban area, study the features of anthropogenic layer structure practically from the earth surface and solve other problems (Slepak 1997, 1999, 2007, 2010).

The testing of various geophysical methods for solving archeological problems in the territory of the Kremlin allowed selecting the most effective geophysical method, i.e. the method of transient electromagnetic sounding using Impulse-auto M-1/0-20 electromagnetic surveying system developed and constructed at the Siberian Research Institute of Geology, Geophysics and Mineral Resources (SNIIGGiMS) of the Russian Academy of Sciences, which allows researching near-surface section.

Simultaneously with the observation of current source behavior in the process of measurements the secondary field is registered at super-short times, which allows electromagnetic sounding of uppermost layers of geological section. The electromagnetic surveying system includes current generator, transmitter and receiver antennas, measuring unit and software. After interruption of the current pulse created by the generator, the measuring unit with the receiver antenna registers the electromotive intensity of transient processes induced in earth at field changes (signal fall). The signal is displayed on the monitor in real time beginning from tens of nanoseconds, which allows sounding the upper part of geological section practically from the earth surface.

Optimal distance between electromagnetic sounding points was selected according to the current problems. The use of 4×4 m square transmitter antenna and 1.3×1.3 m receiver antenna (loop in loop) allowed sounding the anthropogenic layer all the way down to its bottom. The electromagnetic surveying system was used for discrete and continuous measurements (in the process of motion). When studying lateral in-homogeneities of the anthropogenic layer and revealing archeological objects, the pitch of soundings along the profiles was 1.0–2.5 m.

The sounding results were presented in the form of vertical sections of total electrical conductivity S(H) along profiles, horizontal sections of total electrical conductivity S(H) at different depths, maps of constant S(H) values in isolines of absolute depths (and in the axonometric projection) and maps of variability in the thickness of anthropogenic layer. The comparative analysis of these data allowed controlling the materials processing reliability, and the comparison of the sounding results with the drilling data and the results of subsequent archeological excavations allowed assessing the geological interpretation reliability.

A number of archeological problems confirmed by archeological excavations were successfully solved in some sections of the Kazan Kremlin based on the electromagnetic sounding results. The wall remnants were revealed (Fig. 35.1), the map of anthropogenic layer bottom was plotted and archeological dating of extracted layers in anthropogenic deposits in the area of the former Junker Academy (Fig. 35.2) was performed. Also some other tasks were solved (Slepak 2007).

The developed technologies of geophysical researches for prediction of archeological objects of the Kazan Kremlin were applied during the survey of the territory of the former Kazan Bogoroditsky Monastery, which is practically located at the eastern slope of the Kremlin Hill. The geophysical prediction here was possible only along separate profiles between the buildings and backyard buildings, as well as in some open areas and driveways covered with asphalt. The research objectives were to reveal the location of the Cathedral in the name of Icon of Our Lady of Kazan (the Cold Church), churches and towers of the monastery destroyed in the 30s of the twentieth century.

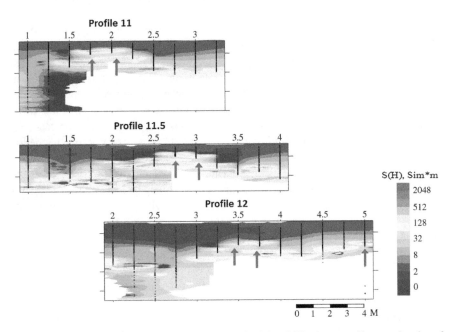

Fig. 35.1 Vertical sections of total electrical conductivity S(H) along profiles on the site of Suyumbeki Tower and the Governor's Palace

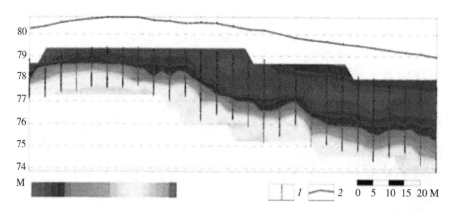

Fig. 35.2 Vertical sections of total electrical conductivity S(H) along profiles on the site of former Junker College

The only area where it was possible to perform areal geophysical measurements was the central part of the Kazan tobacco factory, which was covered with asphalt. According to archival data, the location of previously destroyed Cathedral in the name of Icon of Our Lady of Kazan was supposed to be there. The electromagnetic sounding was performed over 2.5 × 2.5 m grid. Vertical sections of total electrical conductivity identified with geological and geophysical profiles were constructed along longitudinal and transverse profiles, which allowed studying the features of anthropogenic layer structure. The results of measurements along the profiles are presented in the form of vertical sections of total electric conductivity S(H) (Fig. 35.3). Vertical lines in sounding points show the electromagnetic signal penetration at various depths. Minimum signal depths characterize the predicted buildings remnants closest to the earth surface. In observation points 6.25, 6.50 and 6.75 of profiles 2.25 and 2.75, as well as in observation points 6.25, 6.50 and 6.75 of profile 2.50 the sharp decrease in penetration depths of signal is observed (indicated by arrows). These data allowed assuming the possible presence of remnants of arc-shaped stone building belonging to the eastern part of the Cathedral under the earth surface.

During subsequent archeological excavations the location of the eastern apse of the Cathedral was revealed, which confirmed the geophysical prediction results (Fig. 35.4). The use of obtained data and archival drawings allowed identifying the location of the entire building, which is being reconstructed at the present time. The remnants of the churches and the southeast tower of the monastery were revealed along separate profiles (Slepak 2007, 2010, 2016). The recommendations are given to perform archeological excavations in the areas of minimum penetration depth of electromagnetic signal, which will allow limiting the amount of excavation work upon their renewal.

The geophysical measurements in open areas significantly increase the possibility of studying the bottom of anthropogenic layer and revealing details of its structure. The results of tracing the bottom of anthropogenic layer in the territory of

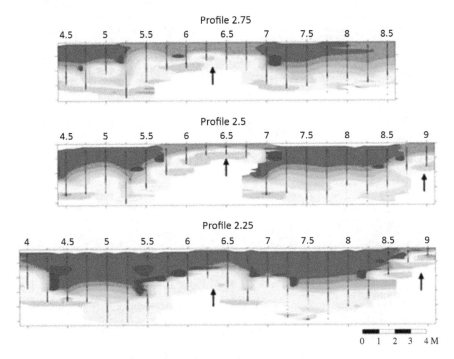

Fig. 35.3 Vertical sections of total electrical conductivity S(H) along profiles on the site of former tobacco factory in the Kazan Bogoroditsky Monastery

Fig. 35.4 Archeological excavation with the remains on the geophysical data of the surviving remnants of apse of the Cathedral in the name of Kazan Icon of Our Lady

the Bilyar settlement, the ancient capital of the Volga Bulgaria (X—beginning of the XIII century), can be the prime example. The anthropogenic layer here is characterized by small thickness, and its bottom is extracted fairly confidently (Fig. 35.5).

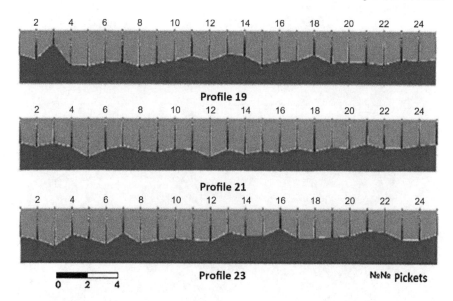

Fig. 35.5 Vertical sections of total electrical conductivity S(H) along profiles 19, 21, 23 on the site of the Bilyar settlement, displaying the bottom of anthropogenic layer

According to the data of electromagnetic soundings the traces of the ancient city streets were displayed in the area of the ancient settlement (Fig. 35.6). Along the profiles 20, 22, 27, 29 the "chains" of depressions, which are separated by local elevations of the relief on the extracted surface and show traces of streets or roads directed towards the city center, are especially clearly observed. According to the archeological researches, the depressions may display traces of pantry pits, barn-cages, cellars, etc. located near former buildings. Their dimensions are practically the same as those determined during archeological excavations (Khuzin 2001; Khuzin and Sitdikov 1996).

Significant possibilities of the applied electromagnetic sounding method were demonstrated in one of the areas of Balynguz settlement necropolis (Fig. 35.7) located south of Bilyar settlement, where the features of anthropogenic layer bottom and the place of ancient burials are traced.

The results of the archeological dating within the cultural layer in the form of extraction of individual strata, which differ in color from the areas of the Bulgarian settlement (Fig. 35.8), are of particular importance, as during excavations the archeological dating can be performed only fragmentarily according to archeologists' data.

Because the layers age determination of separate periods is carried out only fragmentary by archaeological data (Khusin 2001, Khusin and Sitdikov 1996), results of age determination using geophysics on the territory of Bolgarian Historical and Archaeological museum-preserve are of particular importance. Here is traced the earliest Bulgarian layer (X–XI centuries), premongolian layer

35 Geophysics in Archeology 309

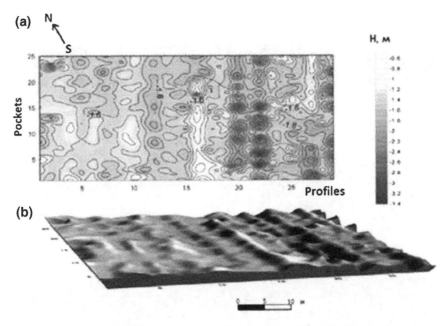

Fig. 35.6 Surface of the bottom of anthropogenic layer on the site of the Bilyar settlement, **a** in isogipses, **b** in isometric image

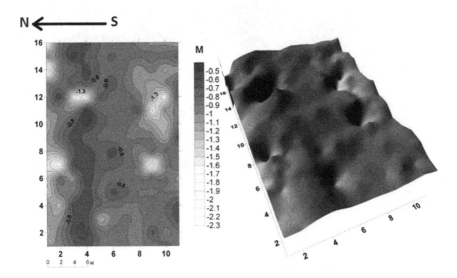

Fig. 35.7 Forecasting results of surface of the bottom of anthropogenic layer in one of the areas of the Balynguz settlement necropolis, where ancient burials are traced

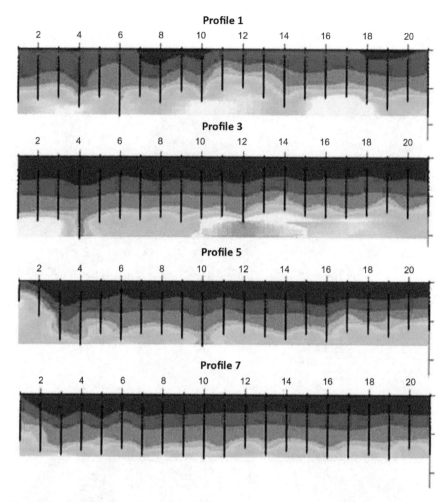

Fig. 35.8 Vertical sections of total electrical conductivity S(H) along profiles 1–7 on the site of the Bulgarian settlement

(XII–XIII centuries), Bulgarian layer (XIII–XIV centuries), layer of desolation period (XV–XVI centuries) and two Russian layers (XVI–XIX centuries and XX century) simultaneously with the selection of the anthropogenic layer. The three most ancient layers are clearly traced (Fig. 35.8) using the electromagnetic sounding data (Slepak 2010).

Summary

The confirmation of geophysical prediction results by archeological excavations is extremely important. Joint geophysical and archeological research allows promptly adjusting the performance of archeological excavations, reducing negative impacts on geological environment and significantly reducing the amount of excavation and material costs.

Thus, the results of solving archeological problems presented in the article are based on the application of new geophysical technologies developed by the author. They give evidence of the geophysics possibilities in solving specific archeological problems in modern cities between buildings and allow studying the features of the structure of the ancient settlements in details.

In the practice of solving archeological problems several different geophysical methods are often applied simultaneously without their proper physical-and-geological justification. The author is deeply convicted that to solve the problems of archeology it is extremely important to apply these methods to known objects and choose the ones that can effectively solve particular task or several tasks. The electromagnetic sounding method appeared to be such method. Some of its application results are considered in this article.

Acknowledgements This study was funded by the subsidy allocated to Kazan Federal University as part of the state program for increasing its competitiveness among the world's leading centers of science and education.

References

Khuzin F.Sh. (2001) The Bulgarian city. – Kazan: Master Line, 2001.
Khuzin F.Sh., Sitdikov A.G. (1996) Ancient Kazan through the eyes of contemporaries and histo-rians. – Kazan: Fest, 1996.
Slepak Z.M. (2007) Urban geophysics, – M. EAGO, GERS Publishing house, 2007, p. 239.
Slepak Z.M. (2010) Exploration geophysics in archeology, Publishing house of Kazan University, 2010, p. 222.
Slepak Z.M. (2016) Geophysical researches to identify the remnants of the Cathedral in the name of Icon of Our Lady of Kazan and churches in the territory of the Kazan Bogoroditsky Monastery. Collection of selected articles based on the results of the International Practical Conference: The Miraculous Kazan Image of the Virgin in the fates of Russia and World Civilization. Kazan, Center for Innovative Technologies, 2016. pp. 139–144.
Slepak Z. (1997) Complex geophysical investigations for Studying the Cultural Layer and Re-mains of Ancient Buildings in the Territory of Kazan Kremlin // Archeological prospection. John Wiley & Sons, Ltd. 1997. Vol. 4. P. 207–218.
Slepak Z. (1999) Electromagnetic Sounding and Highprecision Gravimeter Survey Define Ancient Stone Building Remains in the Territory of Kazan Kremlin // Archeological prospection. John Wiley & Sons, Ltd. 1999. Vol. 6. P. 147–160.

Chapter 36
Using of Probabilistic-Statistical Characteristics in the Interpretation of Electrical Survey Monitoring Observations

L. A. Khristenko, Ju. I. Stepanov, A. V. Kichigin,
E. I. Parshakov, A. A. Tainickiy and K. N. Shiryaev

Abstract It is within the Verkhnekamsk salt deposit, on the potentially dangerous sections of the mine fields the geologic-geophysical monitoring, which includes the electroprospecting researches by the methods NF and SEP on three spans of the power line AB is carried out regularly. The analysis of statistical characteristics of values of potentials of the natural field and apparent resistance by means of the theory of estimates allows to increase significantly the volume of useful information and more accurately to trace the features of a geological structure which are implicitly expressed in the observed fields. For a more distinct allocation of the hidden regularities of change of amplitude of the apparent resistance (AR) field, a fast wavelet-transformation (FWT) of discrete values of AR by means of the HAAR_2 program was executed previously. The statistical characteristics of SEP values and potential of NF were calculated by various methods realized in the COSCAD 2D software package (in the sliding window, in one-dimensional and two-dimensional dynamic windows), and with different sizes of windows. The statistical characteristics of values of NF potential were combined in turn with the statistics of AR obtained at AB 100, 200 and 400 m, i.e. three multi-attribute spaces were formed. Their structure was analyzed by means of various methods of non-standard classification. The using of procedures of non-standard classification allowed to break the analyzed sets on homogeneous, by formal mathematical criteria, the classes spatially answering to sites of possible engineering-geological complications, that it is extremely difficult by results of only the qualitative analysis of field observations.

L. A. Khristenko (✉) · Ju. I. Stepanov · A. V. Kichigin · E. I. Parshakov
A. A. Tainickiy · K. N. Shiryaev
Laboratory of Surface and Underground Electrometry, Mining Institute
of the Ural Branch Russian Academy of Sciences, Perm, Russia
e-mail: liudmila.hristenko@yandex.ru

Keywords Verkhnekamsk salt deposit · Monitoring observations
Electroprospecting researches · Wavelet-transformation · Statistical
characteristics · Methods of non-standard classification

The results of electrometric observations by the methods of the natural field (NF) and those of the resistances in the modification of the symmetric electroprofiling (SEP) allow to allocate and delineate sites of engineering-geological complications that promotes their wide use in the conditions of the urbanized territories. It is within the Verkhnekamsk salt deposit, on the potentially dangerous sections of the mine fields the geologic-geophysical monitoring, which includes the electroprospecting researches by the methods NF and SEP on three spans of the power line AB is carried out regularly. The main objective of the carried-out works is the control of the condition of the waterproof strata and the identification of the negative changes in the mountain massif. However, the interpretation of the field measurements data by the methods of electroprofile is most often based either on the simplest methods of estimating the depths and sizes of solid or on qualitative analysis. Such a method entails ambiguous of the conclusions about the presence and nature of the phenomena which caused changes of observed parameters. The data obtained as a result of the geologic-geophysical researches, owing to the objective reasons, can be considered as a selection of one or several random variables. It allows to analyzing their statistical characteristics by means of the theory of estimates. Such an analysis significantly increases the volume of useful information contained in observations, and also allows to underline features of change in geofields and to evaluate the patterns of distribution of the studied parameter. Thus, there is an opportunity to increase the efficiency of the geological interpretation process and the quality of the final results (Petrov and Soloha 2006).

As practice showed, the analysis of statistical characteristics of values of potentials of the natural field and apparent resistance by means of the theory of estimates allows to increase significantly the volume of useful information and more accurately to trace the features of a geological structure which are implicitly expressed in the observed fields (Khristenko and Stepanov 2014, 2015; Khristenko 2015; Khristenko et al. 2017).

In 2015–2016 years on the mine field Solikamsk-2 monthly observations on 4 profiles (Fig. 36.1) were being executed from May till October.

The spans of the power line AB at measurements by the SEP method were 100, 200 and 400 m. The length of the receiving line MN and a step along a profile for the SEP and NF were 20 m.

For a more distinct allocation of the hidden regularities of change of amplitude of the apparent resistance (AR) field, a fast wavelet-transformation (FWT) of discrete values of AR by means of the HAAR_2 program was executed previously (Dolgal 2004). The program implements the compression of the signal, presented in the form of amplitude values, using decomposition in the Haar basis.

The short impulses–wavelets constructed on the basis of orthogonal basis functions are applied to execute so-called discrete direct and inverse wavelet

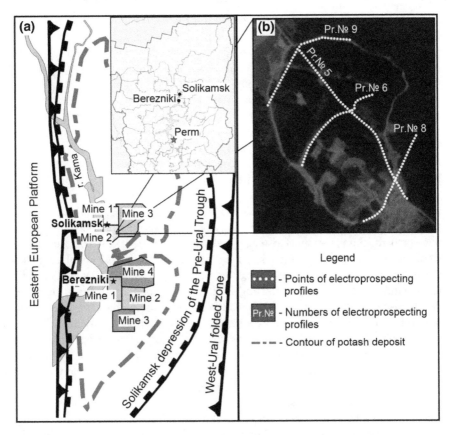

Fig. 36.1 Scheme of the research site: **a** scheme of the Verkhnekamsk salt deposit; **b** plan of electroprospecting profiles

transformation with the use of FWT. Haar's functions are the elementary an example of orthogonal wavelets. In the practical analysis of signals, they (and their more difficult variants) are called broadband and narrowband filters, respectively, because they filter out signal components at big and small scales (Dremin et al. 2001).

Now, algorithms of the wavelet-transformation (WT) are presented in the widespread systems of computer mathematics (SCM), such as Mathcad, MATLAB, Mathematika. International standards JPEG-2000, MPEG-4, graphic software Corel DRAW and many others widely use WT for processing of images. In spite of the fact that in Russia the first works on application of WT appeared almost with a ten-year delay and originally had overview character, now practical use of wavelets is so extensive, and results so impressive that it prompted on the assumption of the possibility of obtaining more informative results of complex interpretation of electrical survey data. There is a fairly successful experience of use of the wavelet-analysis in problems of interpretation of geopotential fields (Dolgal 2004; Dolgal and Pugin 2006; Dolgal and Simanov 2008).

Processing of a signal by means of FWT allows to significantly compress an information content, to discard its small details and allocate the most significant features. The authors investigated efficiency of application of wavelet-transformation as a procedure of preliminary preparation of the electrometric observations data for further estimation and calculation of their statistical characteristics and for using of classification procedures for the solution of problems of borders mapping on the sites of engineering-geological complications within the areas of complex geologic-geophysical monitoring observations potentially dangerous on karst and technogenic manifestations. The analysis of the obtained results showed that using of wavelets allows to gain an impression of schedules of apparent resistance, convenient for the analysis, and, as a result, more exact space position of the class contours with the subsequent using of procedures of the non-standard classification (Hristenko and Stepanov 2014; 2015). Calculations were carried out on each profile for 3, 4 and 5 levels of decomposition of a signal at the set error of approximation of observed values 1 and 2 Ωm. Calculations at 4 levels with an error of approximating 2 Ω made it possible to obtain the signal decomposition, which is preferable for the solution of the problem (to reject many relatively small features of a signal when saving its main characteristics).

To allocate the zones of possible engineering-geological complications by uniform criteria, it is reasonable to carry out classification of the general character space for all 4 research profiles. Therefore values of potential of NF and the values of AR reconstructed after applying FWT procedure were interpolated into the nodes of regular network. The statistical characteristics of SEP values and potential of NF were calculated by various methods realized in the COSCAD 2D software package (in the sliding window, in one-dimensional and two-dimensional dynamic windows), and with different sizes of windows. (Petrov and Soloha 2006). The algorithms of adaptive linear filtration realized in the complex give the opportunity of correctly processing geofields, which are non-stationary on spectral-correlation characteristics.

The statistical characteristics calculated in the sliding window of 3 × 3 points of network, in our opinion, in the best way reflect features of change of the analyzed parameters. They were selected for multi-attribute space.

The statistical characteristics of values of NF potential were combined in turn with the statistics of AR obtained at AB 100, 200 and 400 m, i.e. three multi-attribute spaces were formed. Their structure was analyzed by means of various methods of non-standard classification (total distance, dynamic condensations and the method of Petrov).

The using of procedures of non-standard classification allowed to break the analyzed sets on homogeneous, by formal mathematical criteria, the classes spatially answering to sites of possible engineering-geological complications, that it is extremely difficult by results of only the qualitative analysis of field observations.

The method of the general distance allowed to receive the most informative classes. The classification algorithm by method of the general distance realized in the KOSKAD program complex is a typical example of a heuristic algorithm. The heuristic methods of classification are based on splitting range of values of each sign into the set number of gradations and, in the majority, come down to calculation of complex parameter, which is the corresponding number of gradation interval linear combination on set of the analyzed signs in each point of the observations.

The essential lack of the heuristic methods is that circumstance that they are under construction in the assumption of independence of separate signs among themselves. However the existence of separate shortcomings of classification algorithms does not reduce their importance in the processing of geologic-geophysical observations.

In the applied problems of automatic classification (in the absence of the reference objects), the heuristic algorithms began to be applied by one of the first and still keep great value thanks to visualization of the received results and simplicity of implementation. The main idea of the general distance method is that the set of the objects which are at identical distance from each of k-standards forms compact group (Petrov et al. 2010).

Figure 36.2 shows results of performing of the classification procedure with data of NF and SEP received in August, September and October of 2015 and 2016 years with the span of power electrodes AB = 100 m. It is visible that in 2016, in general, the space provision of contours 2–4 of the classes allocated in 2015 remains. This can indicate the presence of time-stable zones of the water vertical migration. Some increase in the area of class contours in comparison with 2015, which is possibly caused by the intensification of migration processes is observed. The results of implementation of the classification procedure with data of NF and SEP at AB = 200 and 400 m showed that the depth of vertical overflows is limited by the terrigenous-carbonate stratum. I.e. the class contours obtained at various power line spans reflect the dynamics of disturbance areas distribution in the rock massif.

The space position of the contours of classes was compared with the areas of the wavefield complications revealed by the results of the seismic surveys observations. Their coincidence is noted on Pr. 5 between pickets 70–300; on Pr. 6 between pickets 400–600; Pr. 8 between pickets 240–520 and on Pr. 9 in the interval between pickets 480–620 (Fig. 36.3). It can be regarded as indirect evidence about sufficiently high degree of the reliability of the executed transformations.

For forecasting of the direction of possible cracking zones development on the site gas-geochemical profile observations were carried out. By results of observations the anomalies reflecting zones of demultiplexing of upper-salt stratum on which intake of migratory gases in the near-surface part of the section are marked out. Geological structures and also different structure complications of the upper-salt part of the section (the zones of natural and technogenic cracking) revealed by the results of seismic observations are followed by gas-geochemical anomalies (Bachurin and Borisov 2013).

Fig. 36.2 The results of classifications by the method of the general distance on statisticians of potentials NF and AR with a span of AB = 100 m in the periods from August to October in 2015 and 2016 years

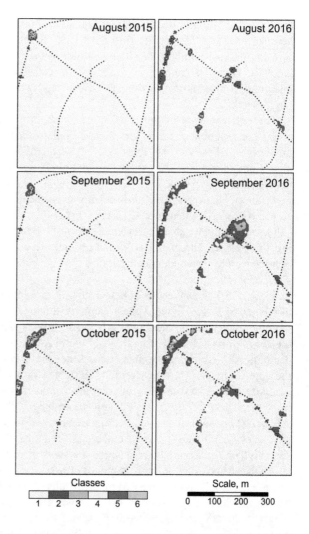

By comparison of seismic and electrometric data it is necessary to consider that the high porosity, leading to considerable reduction of seismic waves speed, can have little effect on unit resistance changing therefore depths of wave guides and conductor layers do not always coincide. The structure of the pore space has a special effect on the conductivity in different layers. It matters whether the pores are joined together or not. Only through the canals give a contribution to conductivity. The deadlock branches and the isolated pores absorb most of the solutions and, increasing porosity, do not affect the electrical conductivity.

The characteristics of classes within the Verkhnekamsk salt deposit for one more site—SKRU-2 are as a result received that will significantly add already available data on other sites (Khristenko and Stepanov 2014; Khristenko et al. 2017) for their analysis and definition of a possibility of using classifications results at the solution

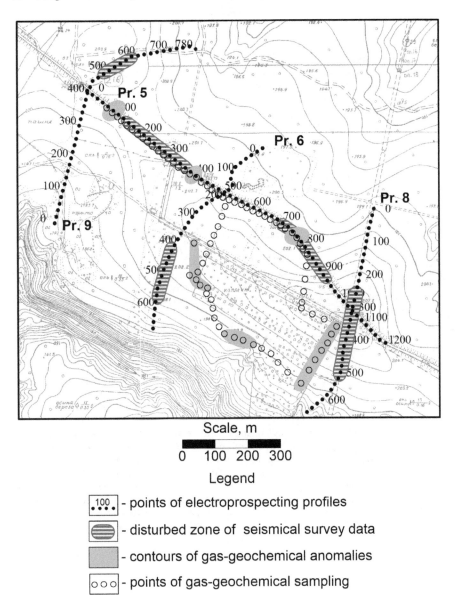

Fig. 36.3 The results of the complex interpretation of seismic and gas-geochemical data

of the tasks set for monitoring researches. The effective set of procedures and the sequence of their application during the processing and interpretation of electrometric observations allowing to carry out more authentically mapping of borders of engineering-geological complications sites are defined.

Acknowledgements This work was supported by a grant RFBR № 16-45-590046.

References

Bachurin B.A., Borisov A.A. (2013) Modern gas-geochemical technologies for controlling technogenesis processes in the development of the resources of the Verkhnekamsk region. Gornyj zhurnal, 2013, no. 6, pp. 78–82.

Dolgal A.S. (2004) The use of the fast wavelet transform in the solution of the direct problem of gravity prospecting. Report of the Academy of Sciences, 2004, vol. 399, no. 8, pp. 1177–1179.

Dolgal A.S., Pugin A.V. (2006) Construction of analytical approximations of geopotential fields taking into account their fractal structure. Report of the Academy of Sciences. 2006, vol. 410, pp. 1152–1155.

Dolgal A.S, Simanov A.A. (2008) The application of multi-scale wavelet analysis for analytic approximations of geopotential fields. Report of the Academy of Sciences, 2008, vol. 418, no. 2, pp. 256–261.

Dremin I.M., Ivanov O.V., Nechitaylo V.A. (2001) Wavelets and their use. Physics-Uspekhi, 2001, vol. 44, no. 5, pp. 447–478.

Petrov A.V., Yudin D.B., Syueli Khou. (2010) Processing and interpretation of geophysical data by methods of probabilistic and statistical approach with use of the computer technology «KOSKAD 3D». Journal KRAUNTs. Science of Earth, 2010, no. 2, issue. 16, pp. 126–132.

Petrov A.V., Soloha E.V. (2006) Technology for analyzing geofields in sliding windows. The theory and practice of geological interpretation of gravity, magnetic and electric fields: 33-rd SES. Intern. workshop them. D. G. Uspensky. Ekaterinburg, Russian Acad., Sci., Inst. of Geophysics Ural Branch, 2006, pp. 272–275.

Khristenko L.A., Stepanov Ju.I. (2014). Electrometric observations when assessing the influence of a goaf of the subsoil at the base of the railway embankment. Natural and technical Sciences, 2014, no. 7, pp. 58–62.

Khristenko L.A., Stepanov Ju.I. (2015). A fast wavelet transform with Haar basis functions in interpreting electrical profiling data. International conference "Eighth scientific reading of Yu. Bulasevich" Deep structure, geodynamics, thermal field of the Earth, interpretation of geophysical fields. Ekaterinburg, Russian Acad., Sci., Inst. of Geophysics Ural Branch, 2015, pp. 354–356.

Khristenko L.A. (2015) Use of a fast wavelet transform in the complex interpretation of electrical prospecting data. Strategy and processes of development of geo-resources: Collected papers. Russian Acad., Sci., Mining Inst. of the Urals Branch, Perm, 2015, issue 13, pp. 220–222.

Khristenko L.A., Kichigin A.V., Parshakov E.I., Stepanov Y.I., Tainickiy A.A., Shiryaev K.N. (2017) Improvement of interpretation of the monitoring data electrical investigation by means of the theory of estimates. 13th International Scientific and Practical Conference and Exhibition "Engineering Geophysics 2017" on 24–28 April. Kislovodsk, EAGE. https://doi.org/10.3997/2214-4609.201700419. Available at: http://www.earthdoc.org/publication/publicationdetails/?publication=88140 (Accessed 25 April 2017).

Chapter 37
Multi-Electrode Electrical Profiling Results in the Northern Ladoga Area

V. E. Kolesnikov, M. Yu. Nilov and A. A. Zhamaletdinov

Abstract DC electrical profiling with the use of Method of External Sliding Dipole (MESD) has been made along the profile Sevastyanovo-Hijtola-Sujstamo of 200 km length. The profile crosses the Ladoga anomaly of high conductivity. Interpretation of MESD results is performed in two stages. At the first stage, a one-dimensional inverse problem is solved using the Zohdy procedure (Zohdy 1989). The starting model is set in the form of a multilayered section with the number of layers equal to the number of deletions from the feeding line plus one (first) layer for the results of measurements inside the line. The depths to the base of the layers and their resistivity are taken equal to the corresponding values of the effective distances and to the apparent resistivity at corresponding distance. Further, an iterative selection of the theoretical model of the medium occurs before obtaining agreement with the experimental data within the error of observations. At the second stage, a quasi-2D section is constructed by symmetrically shifting the deep sounding centers with relation to the supply line AB to the southwest and northeast as the receiving dipoles MN move in opposite directions. The results of profiling, as well as the results of the study of samples from outcrops and boreholes, indicate the connection of anomalies with the presence of electron-conducting sulphide-carbonaceous rocks. Obviously, they explain the nature of the upper part of the regional Ladoga conductivity anomaly.

Keywords Multi-electrode electromagnetic sounding · DC resistivity method Modeling

---

V. E. Kolesnikov (✉) · A. A. Zhamaletdinov
Geological Institute of the Kola Science Centre of RAS, Apatity, Russia
e-mail: vk51@list.ru

A. A. Zhamaletdinov
e-mail: abd.zham@mail.ru

M. Yu. Nilov
Institute of Geology of the Karelian Centre of RAS, Petrozavodsk, Russia

A. A. Zhamaletdinov
St. Petersburg Filial of IZMIRAN, St. Petersburg, Russia

© Springer Nature Switzerland AG 2019
D. Nurgaliev and N. Khairullina (eds.), *Practical and Theoretical Aspects of Geological Interpretation of Gravitational, Magnetic and Electric Fields*, Springer Proceedings in Earth and Environmental Sciences, https://doi.org/10.1007/978-3-319-97670-9_37

Object, Aims and Tasks of the Research

An electrical anomaly in the North Ladoga area has been revealed by magnetotelluric (MT) investigations (Lazareva 1967) and magnetovaritional profiling (MVP) (Rokityansky et al. 1979) (Fig. 37.1). At the next years the territory was further on studied with the use of MT and AMT methods by groups of researchers in Finland (Adam et al. 1982) and Russia (Vasin 1988; Kovtun 1989; Kovtun et al. 2004; Sokolova et al. 2016). Investigations made by different research groups provided contradictory results.

That required frequent revision of parameters of the Ladoga anomaly. The situation happens due to absence of the data about electrical conductivity of rocks near to the day time surface. To obtain this information our team has made the DC geoelectric investigations with the use of multi-electrode installations. Two techniques have been used—Method of External Sliding Dipole (MESD) and the Method of the Inner Sliding Contact (MISC). At the first stage there have been performed MESD measurements along the Sevastyanovo-Hiitola-Suistamo profile of 180 km length with the step of 500 m. That profile crosses the entire width of the Ladoga anomaly. The profile location is shown on the Fig. 37.1. At the second stage detailed measurements with the step of 50 m have been performed with the use of MISC method over the most contrast anomalies detected by MESD method.

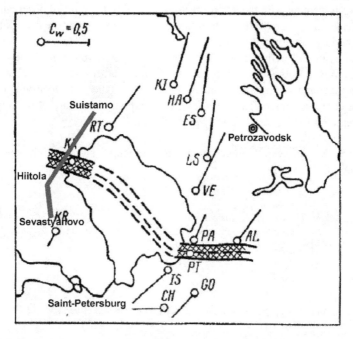

Fig. 37.1 Location of the Ladoga anomaly after (Rokiryansky et al. 1981) and the DC profile Sevastyanovo-Hijtola-Suistamo

Technique of Measurements and Data Processing

Method of External Sliding Dipole (MESD)

The method of electrical profiling according to the MESD scheme was developed especially for observations along the winding roads of the Northern Ladoga area. The MESD installation is shown in the Fig. 37.2.

Arrows show measuring lines, displaced along and across he road. A thin dotted lines show position of the diagonal measuring lines N1–N2.—the angle between the line AB and the lines OMi connecting the centers of the receiving lines MNi with the center AB. A light dashed line is a road.

The principle of the MESD installation is based on dipole-dipole sounding and profiling with the measurement of the totall vector of the horizontal electric field intensity (the mode of vector observations). Feeding line AB of 500 m length is located in the central part of the installation. Current up to 1–2 A is supplied from an EMAK generator with a power of 1 kW at a frequency of 4.88 Hz.

Measurements of electric field strength are performed with the ANCH-3 equipment. The measuring lines moved in two opposite directions in steps of 500 m at distances up to 2 km from the nearest supply grounding (Fig. 37.2). MESD technique relies on the use of the satellite positioning system (GPS) and complete computerization, starting from the digital input of the coordinates and results of observations into the computer's memory and finishing with data processing and graphical display of the results straight in the field conditions.

When processing the MESD results a following equation was used to calculate the absolute value of the total vector of electric field strength (Zhamaletdinov 2012).

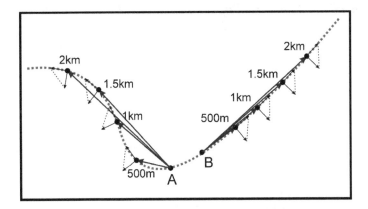

Fig. 37.2 Scheme of work with the installation of MESD (method of external sliding dipole) Legend: AB—current line of 500 m long. Black points—position of groundings of receiving lines and AB

$$|\vec{E}_{tot}^{theor}| = \frac{\rho \cdot I}{2\pi} \cdot \left[r_A^{-4} + r_B^{-4} - \frac{r_A^2 + r_B^2 - l_{AB}^2}{(r_A \cdot r_B)^3} \right]^{1/2}, \qquad (37.1)$$

where l_{AB}—the length of transmitting line AB, r_A and r_B—distance from the central (common for two receiving lines) electrode to groundings A and B, respectively.

Calculation of apparent resistivity ρ_k^{tot} values was carried out by the formula

$$\rho_k^{tot} = K_{tot} \cdot E_{tot}^{\exp}/I, \text{ where } E_{tot}^{\exp} = \sqrt{E_{M1N1}^2 + E_{M1N2}^2},$$

I —the strength of current in AB line, geometric factor

$$K_{tot} = 1/E_{tot}^{theor}, \quad E_{MN} = \Delta U_{MN}/l_{MN}.$$

Interpretation of MESD results is performed in two stages. At the first stage, a one-dimensional inverse problem is solved using the Zohdy procedure (Zohdy 1989). The starting model is set in the form of a multilayered section with the number of layers equal to the number of deletions from the feeding line plus one (first) layer for the results of measurements inside the line. The depths to the base of the layers and their resistivity are taken equal to the corresponding values of the effective distances and ρ_{ki}. Further an iterative selection of the theoretical model of the medium occurs before obtaining agreement with the experimental data within the error of observations. At the second stage, a quasi-2D section is constructed by symmetrically shifting the deep sounding centers with relation to the supply line AB to the southwest and northeast as the receiving dipoles MN move in opposite directions.

Method of the Inner Sliding Contact (MISC)

The method of internal sliding contact (MISC) is a multi-electrode profiling scheme realized by a special installation with variable spacing AM, BM and MN (Zhamaletdinov et al. 1995).

The MISC installation (Fig. 37.3) consists of a fixed length feeding line and several receiving lines with a common electrode N, assembled in a "scythe". On each pair "electrode Mi-electrode N" a signal fed through the supply line is measured. The ANCH-3 equipment is used for current generation and signal measurement. Changing the distance between the supply electrode A and the receiving electrode M_i ("internal sliding contact" in the receiving line) gives information about the change in the properties of the medium with depth (sounding).

Moving the entire MISC installation with fixed spacings between the electrodes along the profile with regular step allows to prform the conductivity studies along

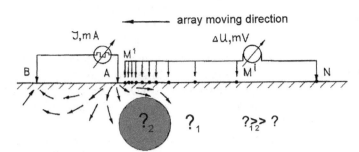

Fig. 37.3 Scheme of the installation of MESD. The explanations are given in the text

the earth-air surface (profiling). Thus, two-dimensional investigation (scanning) of the upper part of the geoelec-trical section is provided.

The location of the receiving electrode M_i provides a change in the configuration of the installation from the potential to the three-electrode and then to the quasi dipole-dipole. During the field work the following parameters were used: $AB = 50$ m, $AN = 100$ m, $AM_i = 5, 7, 10, 14, 20, 30, 50, 70$ m.

The initial processing of the field data of the MISC consists in calculating the apparent resistivity values, plotting the graphs in the profiling format (depending on the distance along the profile), and sounding curves in the VES format (depending on the effective spacing). The apparent resistivity is calculated using expression

$$\rho_k = K \cdot \Delta U/I = \frac{2\pi \cdot AM \cdot AN \cdot BM \cdot BN}{MN \cdot (BM \cdot BN - AM \cdot AN)}. \quad (37.3)$$

Effective spacing (effective depth of sounding) is determined with the use of empirical formula

$$r_{\ni\phi\phi} = AM + 0.5MN \cdot \left(\frac{AM}{AN}\right)^2 + 0.5AB \cdot \left(\frac{AM}{BM}\right)^2 \quad (37.4)$$

The inverse problem of the MISC is solved by two-dimensional numerical modeling. The modeling was performed using ZondRes2D software for direct and inverse problems of the DC resistivity method (Kaminsky 2010). The simulation is performed by constructing a polygonal two-dimensional model of the resistance section, in which each polygon is assigned the value of the resistivity. For the constructed model and the electrode configuration used in the experimental works, the solution of the direct problem (calculation of an array of apparent resistivity values) is performed with the subsequent calculation of the misfit between the model and experimental data. Taking into account the distribution of the misfit with respect to the configuration of the installation and the experimental data, the operator corrects the model and subsequently repeats the calculation cycle to achieve the required misfit values (as a rule, the discrepancy should not exceed 5%).

Results

The first stage of the study was the implementation of work on the 180 km profile by the MESD method. The total examination of graphs 2 and 4 on the loiwer part of the Fig. 37.4 shows that geoelectric section on the profile is characterized on average by high values of rock resistivity (range from the first hundreds to the first tens of thousands of Ω m).

Two anomalies of lowest resistivity (up to 1 Ω m and less) were fixed on the profile. One anomaly (about 7 km wigth) is situated to the south of Ihala village, between sites 60–67 km on the Fig. 37.4. It has been conventionally called as Grand anomaly. The second anomaly of 0.2 km width is situated in the extreme north-eastern part of the profile, near to Suistamo village at the site 175 km on the

Results of MISC studies over the Suistamo anomaly (Fig. 37.5) indicate the complex shape of anomalous body with resistivity falling till tenths of Ω m.

2D modeling of geoelectrical MISC results created the model in the shape of two plates in-clined in opposite directions to each other. Such structure of the anomaly can be caused by the geological structure of the site located at the junction of the massives of Archean (from the north-east) and Proterozoic (south-west) rocks and the proximity of the Janisjärvi fault zone.

Over the Grand anomaly (Fig. 37.4) collation of MESD results with geological data shows that it is located in the field of distribution of mica gneisses, shales, sandstones and siltstones of the Naatselka suite (Sviridenko et al. 2017) At 15 km

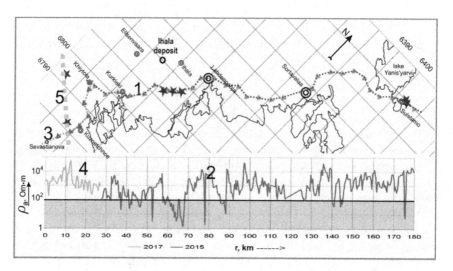

Fig. 37.4 MESD profiling results on the Sevastyanovo-Hiitola-Suistamo profile Legend: 1—MESD profile and 2—apparent resistivity graphs for 2015 on the Hiitola-Suistamo profile, 3—MESD profile and 4—apparent resistivity graphs for 2017 on the Sevastyanovo-Hiitola profile, 5—position of a hypothetical conductivity anomaly in the Priozersk fault zone. Asterisks indicate the AMT–MT sites

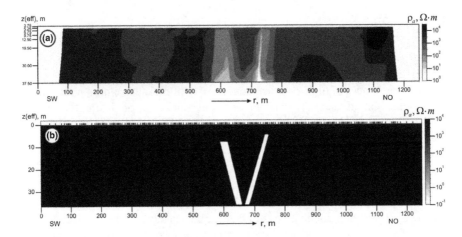

Fig. 37.5 Presentation of the MISC results on Janisjärvi anomaly: **a** apparent resistivity pseudosection, **b** resistivity section by results of 2D digital modeling

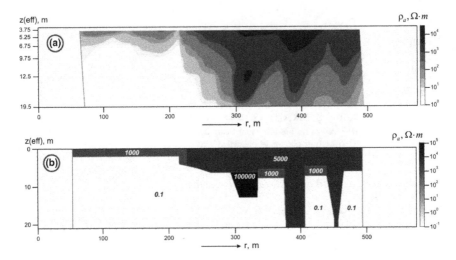

Fig. 37.6 MISC results on Ihala site: **a** apparent resistivity pseudo-section, **b** resistivity section by results of 2D digital modeling

to the north from the Grand anomaly there is a graphite mineralization site in the Ihala settlement area. The MISC profiling on Ihala graphite mineralization site (Fig. 37.6) showed a complex structure of the section, different thickness of the moraine deposits overlapping the conductive rocks and the presence of interbedded structural elements of high (up to 5 thousand of Ω m) and low (tenths of Ω m) resistivity.

Conclusion

Electro profiling with multi-electrode installation of MESD was performed along the 180-km profile crossing the Ladoga electrical conductivity anomaly. Two, the most contrast zones of high electrical conductivity of rocks have been noted, one of which—the Grand anomaly has an apparent thickness of 7 km, the second, Janisjärvi anomaly has an apparent thickness of 200 m. The high resistivity has been registered over the Priozersk fault zone that was supposed as conductive structure. The detailed MISC profiling was carried out on Janisjärvi, Ikhala and Grand anomalies. On all three profiles, the resistivity values drop to units of ohmmeter and less. The results of profiling, as well as the results of the study of samples from outcrops and boreholes, indicate the connection of anomalies with the presence of electron-conducting sulphide-carbonaceous rocks. Obviously, they explain the nature of the upper part of the regional Ladoga conductivity anomaly.

Acknowledgements This work was supported by RFBR grant No 18-05-00528. Authors are thankful to A.A. Sko-rokhodov, A.N. Shevtsov and P.A. Ryazantsev for the help in the field work.

References

Adam A., Kaikkonen P., Hjelt S.E., Pajunpaa K., Szarka L.,Vero J., Wallner A. (1982) Magnetotelluric and audiomagnetotellurics measurements in Finland. Tectonophysics 90, 77–90.

Kaminsky A.E. (2010). ZondRes2D. Software for two-dimensional interpretation of DC re-sistivity and IP data. Zond Geophysical Software, Saint-Petersburg (Russia). 139 p.

Kolesnikov V.E. (2016). Multi-electrode electrical exploration using the internal sliding contact technique - the experience of numerical 2D modeling. Seismic equipment 52 (3), 27–34.

Kovtun A.A. (1989). The structure of the crust and upper mantle in the northwest of the East European Platform based on magnetotelluric sounding data. Leningrad University, Leningrad (USSR). 284 p.

Kovtun A.A., Vardaniants I.L., Legenkova N.P., Smirnov M.Yu., Uspensky N.I. (2004). Features of the structure of the Karelian region according to geoelectric studies. Deep struc-ture and seismicity of the Karelo-Kola region and its surroundings. (edited by Sharov, N.V.).Karelian Science Center of RAS, Petrozavodsk (Russia). pp. 102–130.

Lazareva N.V. (1967). Some features of natural electromagnetic field behaviour on the southern slope of the Baltic Shield. Issues of exploration geophysics 6. Nedra, Leningrad (USSR).

Rokityansky I. I., Vasin N. D., Golod M. I., Novitsky G. P., Rokityanskaya D. A., Soko-lov S.Ya. (1979). Anomalies of electrical conductivity in the south of Karelia. Geophysical collection 89. Naukova dumka, Kiev (USSR). pp. 36–39.

Sokolova E. Yu., Golubtsova N. S., Kovtun A. A. et al. (2016). Results of synchronous magnetotelluric and magnetovariational soundings in the area of the Ladoga electrical con-ductivity anomaly. Geophysics 1. pp. 48–61.

Sviridenko L. P., Isanina E. V., Sharov N. V. (2017). The deep structure, volcanoplutonism and tectonics of the Ladoga area. Proceedings of the Karelian Research Center of RAS 2. Pe-trozavodsk (Russia). pp. 73–85.

Vasin N.D. (1988). Geoelectric characteristics of the section of southwestern Karelia. Notes of Mining Institute. 113. pp. 57–63.

Zhamaletdinov A.A., Ronning J.S., Vinogradov Yu.A. (1995). Electrical profiling by the MISC and Slingram methods in the Pechenga-Pasvik area. Norges Geologiske Undersokelse, Special publication 7. pp. 333–338.

Zhamaletdinov A.A. (2012). Theory and methodology of deep electromagnetic soundings with powerful controlled sources (the experience of critical analysis). SPbSU, Saint-Petersburg (Russia). 163 p.

Zohdy A.A.R. (1989). A new method for the automatic interpretation of Schlumberger and Wenner sounding curves. Geophysics 54 (2), 245–253.

Chapter 38
The Indication in the Potential Fields of Structures Controlling Diamondiferous Magmatism

S. G. Alekseev, P. A. Bochkov, N. P. Senchina and M. B. Shtokalenko

Abstract The general features of spatial distribution of excess density and magnetization along profiles crossing major diamond deposits of Russia, Canada, Australia and the USA are revealed. Geophysical and geological explanation of these regularities is given. Possibility of their application is shown at the prognosis of the various structures controlling kimberlite and lamproite magmatism.

Keywords Gravity and magnetic survey · Tomography · Structure Kimberlite · Diamond deposits

Formulation of the Problem

The recommendations of the workshop (Meeting 2017) note the need for the development of prognosis and prospecting models of mineralogenous taxa of various rank. The development of these models is based on the integrated use of the results of different-scale geological and geophysical work. Naturally, the main task of prognostic geophysical studies is to distinguish structures that control diamondiferous magmatism. Among geophysical methods, an important role is played by deep seismic exploration, MTS, and gravity and magneto-prospecting.

In recent years, various ways of interpretation-tomographic techniques on potential fields have appeared (Babayants et al. 2004, Dolgal et al. 2012). The authors developed a variant of tomographic method, which makes it possible to obtain layerwise sections of spatial distributions of excess density or magnetization of rocks (Shtokalenko and Alekseev 2013). Comparison of the results with seismic sounding on a number of regional profiles showed their satisfactory correspondence (Kozlov et al. 2009).

S. G. Alekseev · P. A. Bochkov · N. P. Senchina (✉)
Saint Petersburg Mining University, Saint Petersburg, Russia
e-mail: n_senchina@inbox.ru

M. B. Shtokalenko
Geological Survey of Estonia, Tallinn, Estonia

Tomographic models of deposits of a number of ore elements (gold, uranium, lead, zinc, copper) in Russia have shown their common features, namely the confinement to extensive regional structures with lower values of density and magnetization of rocks (Alekseev et al. 2010a, b). At the same time, the sizes of the structures controlling the unique deposits and ordinary deposits differed among themselves. Similar data were obtained for a number of ore deposits in the USA, Canada, and Australia. The aim of this paper is to consider tomographic models of diamond-bearing areas constructed on gravity and magnetic data.

Progress of Work

For modeling, the areas of the largest diamond deposits in Australia, Canada, the USA and Russia are taken. To construct models, we used matrices of gravity (in the Bouguer reduction) and magnetic fields with a step of 4 km (USA), 2 km (Canada and Russia) and 1 km (Australia). The data processing technique included a number of sequential operations (Shtokalenko and Alekseev 2013; Alekseev et al. 2016, 2017). Calculations are carried out in 3D geometry. In this variant, the calculated depth depends on the geometry of the field source and may differ from the true value (Alekseev et al. 2011). However, unlike 2D geometry, this variant of calculations allows one to obtain for each point of the study area the distributions over the depth of excess density and magnetization, which do not depend on the direction of the profile. The final results of the processing are maps of the spatial distribution of excess density (for the gravity) and magnetization (for the magnetic field) of rocks in horizontal sections (over the entire area of the matrix) and sections (along the lines of given profiles). Diamond deposits are chosen as the anchor points for the profiles. This made it possible to exclude the diversity of the boundaries of mineralogical diamondiferous taxa described in the literature.

The Results

There are shown the distributions of excess density and magnetization along 3 deep cross-sections with approximately W-E direction for 3 regions of Russia in the Fig. 38.1. The lines of the sections are chosen to cross kimberlite pipes of the known deposits (Vaganov 2000). Dashed lines in the figure denote Moho boundary on the depth of about 40 km and asthenosphere surface on the depth of about 200 km.

The cross-sections show common features of the regions such as funnel-shaped structures of various sizes. For Yakutia (Fig. 38.1a–d) the size of the structures of low density is 500–600 km along the profile. These structures are surrounded with the rocks of greater density. Supposed contours of the structures are shown with short-dashed lines in the Fig. 38.1. The structures are characterized of low values of magnetization of rocks. Within the structures, which can be assigned to the rank of

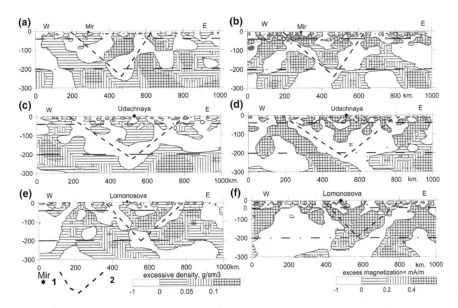

Fig. 38.1 Spatial distributions of excess density (**a, c, e**) and magnetization (**b, d, f**) along the cross-sections with about meridian direction over the regions of kimberlite magmatism in Russia. The section lines cross the pipes Mir (**a, b**) and Udachnaya (**c, d**) in Yakutia and Lomonosov (**e, f**) in Arkhangels region 1—location of kimberlite pipes with their names on the section line; 2—contours of supposed funnel-shaped structures; long-dashed lines denote Moho and asthenosphere surfaces; vertical scale—in calculated kilometers

regional ones, separate blocks of dense rocks are observed. The size of the blocks in the width and in the depth of about 100–150 km. There are two dense blocks in Fig. 38.1a. To the western one the Mir pipe is confined. Within these densified blocks, the location of the pipes is connected with inclined stem-shaped zones of increased magnetization, tracing through the entire Earth's crust, practically up to the day surface directly to the pipes (Fig. 38.1b, d). The pipes themselves are associated with small outcropping blocks of rocks of low density.

The Lomonosov deposit in the Arkhangelsk region is also associated with the regional funnel-shaped structure formed by the rocks of reduced density (Fig. 38.1e). The size of the structure along the profile line is 400 km. The structure includes a funnel-shaped block of dense rocks. The size of this block along the profile line is 120 km, the Lomonosov deposit itself is confined to its western flank (Fig. 38.1e). The structure is marked by increased magnetization of the rocks. From a depth of about 40 km to the Lomonosov deposit is traced a zone of rocks of increased magnetization (Fig. 38.1f).

Funnel-shaped structures of decreased density and magnetization with the size of 400–600 km are shown in the Fig. 38.1 according to data of gravity and magnetic surveys of the scale of 1:1,000,000. These structures are surrounded with the rocks of relatively increased density and magnetization. Analogical structures are characterized for the diamond deposits of Australia (Argyle), Canada (Victor, Ekati) and the USA (Kelsey Lake). Vertical projections of the structures onto the day surface can be attributed to the rank of the diamondiferous subprovince.

The diamond deposits themselves are observed within the funnel-shaped structures above the blocks of rocks of higher density of size of 100–150 km, lying in the Earth's crust and upper mantle. The projections of these blocks onto the day surface can be attributed to the rank of diamondiferous kimberlite control zones. Inside the blocks, linear channels of increased magnetization of rocks are noted, tracing through the entire Earth's crust almost to the day surface. In some cases, fields of reduced density and magnetization are directly under the deposits. To distinguish such fields, it is necessary to involve data from larger-scale surveys.

As an example in Fig. 38.2, the sections of the distribution of excess density and magnetization calculated by the method of interpretation tomography along a profile across the Lomonosov and Grib pipes are given (Arkhangelsk region). The sections are calculated using data of surveys of a scale of 1:200,000. In this figure the upper part of the section is more clear than in Fig. 38.1e. In the block of dense rocks, the Lomonosov and Grib pipes are located above the channels of reduced density, traced from the southwest to the northeast from a depth of 20 km to the day surface. The diameter of these channels is 10–15 km (Fig. 38.2a). The upper part of this channel and can be attributed to the rank of the diamondiferous kimberlite field. In this case, two such channels are observed, extending directly to the pipes of Lomonosov and Grib.

From the north-east, from the depth of 20 km to the day surface stretch linear zones of increased magnetization (Fig. 38.1b). These zones coincide with increased values of density of rocks.

In general, the results of gravity interpretation can be presented in two variants. These variants are conventionally divided into "ore" and "structural" ones. In the first case, the gravity field is interpreted as the presence of masses in a

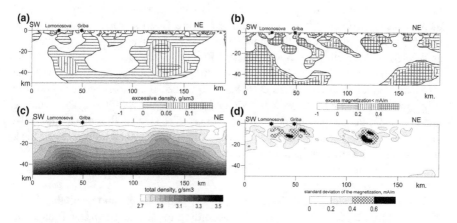

Fig. 38.2 Spatial distributions of excess density (**a**), magnetization (**b**), total density (**c**) and standard deviation of magnetization with the filter size 5–5 km (**d**) in the section along the profile across Lomonosov and Grib pipes according to data of surveys of the scale of 1:200,000. The legend is in Fig. 38.1

homogeneous medium. This version of the representation allows us to visualize extremely insignificant excess density anomalies, whose range of variation is hundredths and tenths of g/cc. Such a representation allows confidently fixing ore bodies, tectonic faults, fluid-conducting zones, and the like. It is in this variant the data in Figs. 38.1 and 38.2a.

The second variant assumes that the entire anomalous gravity field is created by the plicative structures located in a medium with the density depending on depth. Usually, the density of rocks on the surface is taken as 2.67, and 3.3 g/cc at the Moho boundary. Such a structural transformation of the section from Fig. 38.2a is shown in Fig. 38.2c. Within the diamondiferous zones under the Lomonosov and Grib pipes the raise of the upper mantle and according thinning of Earth's crust is observed. The lateral size of the raise is about 30 km.

Above, at depths of 20–30 km, the amplitude of the raise increases. And higher, at depths of 5–10 km, this single raise is divided into two anticlinal structures, which are obliquely traced to the Lomonosov and Grib deposits. Linear zones of increased magnetization of rocks are confined to these structures (Fig. 38.2b).

The parts of the Earth's crust confined to the kimberlite fields are sharply distinguished by increased values of the dispersion of the magnetization of the rocks at depths of 5–15 km (Fig. 38.2d).

By analogy with known deposits, we can assume potential prospects for the discovery of a new kimberlite field with a length of about 15 km in the area at the 120 km of the profile. Within this new field, channels of reduced density, anticlinal structures of total density and channels of increased magnetization and an increase in its dispersion are also observed.

Discussion of the Results

One of the first questions arising when considering the results of gravity and magneto-prospecting is the depth of the interpretation. According to the Gauss theorem, the entire gravity and magnetic field of deep sources can be explained by the distribution of the density and magnetization of the rocks on any surface above the source of the field. However, the same type of sections obtained for diamond deposits in various regions of the world testify the commonness of regional structures that determine kimberlite magmatism and the possibility of their mapping by means of interpretation tomography. With such mapping, small deviations of the true depths from the calculated depths due to the application of 3D geometry are possible.

The second question is the depth of magneto-prospecting, which is limited by the Curie isotherm. Our data on the change of the magnetization of rocks with depth show that their magnetic properties begin to decrease sharply in terms of ancient platforms with a depth of 40–60 km, but do not disappear at all. The question of changing of the values of the Curie point with depth is still controversial. Some experimental data indicate an increase in the Curie point as pressure increases in metal-hydrogen systems (Poniatowski et al. 1982), which may exist in the Earth's mantle.

When processing gravity and magnetic prospecting data at a scale of 1:1,000,000, the funnel-shaped structures are distinguished by the method of interpretation tomography, characterized by lower values of density embedded in the structures of increased density. Within the structures of reduced density, similar regions of rocks of reduced or increased magnetization are observed. The projection of such structures on the day surface may be attributed to the rank of the diamondiferous subprovince.

Within the subprovince, isolated blocks of dense rocks are distinguished, most likely with the basic or ultrabasic composition, above which the deposits themselves are observed. The projections of such blocks on the day surface are related to the rank of diamondiferous kimberlite control zones. We can assume the participation of these dense rocks in the direct formation of diamonds (catalytic effect) or their conservation (in the process of slowing the migration of diamond-bearing intrusions from depth to day surface).

A more detailed analysis of the structure of the blocks of dense rocks based on the results of surveys of a scale of 1:200,000 shows the presence within their boundaries of channels with size of 10–15 km, characterized by both increased and decreased density, traced directly to the deposits. In these channels, there is a direct correlation between the values of excess density and magnetization of the rocks. Channels with rocks of increased magnetization and increased dispersion of the magnetization are noted in the sections. The deposits themselves in the upper part of the section are confined to relatively small zones of decreased density. The area of the projection of these channels on the day surface is about 100–200 km^2 and corresponds to the rank of the diamondiferous kimberlite field.

Within the kimberlite-controlling zones on the surface of Moho there is a rise of the upper mantle and, correspondingly, a decrease of the thickness of the Earth's crust. Above, at the level of the middle crust, the amplitude of the rise increases. In some cases, at the level of the upper crust, single rise are divided into separate anticline structures traced to the deposits.

A site at the 120 km of the profile in the Arkhangelsk region was distinguished because its structural plan, according to geophysical data, is similar to the criteria of the kimberlite field considered above.

In the work carried out, the results of small-scale gravity and magneto-prospecting are used. For a detailed study of the upper part of the section, it is necessary to involve the results of surveys of a scale of 1:50,000 and larger.

Conclusions

1. Methods of interpretative tomography in processing of small-scale data gravimetric and magnetic surveys (scale 1:1,000,000) allow to allocate funnel-shape mantle structure relating to the rank of diamond subprovince and characterized by reduced values of rock density. These structures are also distinguished by the distribution of the magnetization of the rocks. It's quite difficult to identify such

structures in spatial distribution of density and magnetization in the Earth's crust and for their mapping is necessary to involve the measurement of gravity and magnetic fields in the area of at least many hundreds of thousands of sq km. It should be noted that the described structures correspond to the hypothesis of S. E. Haggerty on the origin of diamonds, published in 1986.
2. Within the subprovinces (funnel-shaped structures), isolated blocks of dense rocks, probably of the basic or ultrabasic composition, are distinguished. The considered diamond deposits are observed only in the zones of the projection of these blocks on the day surface, which can be attributed to the rank of diamondiferous kimberlite control zones. It is of interest to evaluate the role of the rocks of these blocks to the migration of kimberlite magmas, to possible synthesis or conservation of diamonds.
3. When processing data of a larger scale (1:200,000) channels are observed within the blocks of dense rocks characterized by both reduced and increased values of excess rock density. At the same time, channels of increased density are spatially associated with areas of increased values, both the magnetization of rocks and its dispersion. At the very top of the section these channels reach the small areas of low density that can be attributed to the rank of the kimberlite field.
4. The performed works show the effectiveness of the application of interpretation tomography of gravity and magnetic fields for the detecting of structures controlling diamond-bearing magmatism, and also prove the similarity of such structures in different regions of the world. During the work, new potentially prospective sites were distinguished in the rank of kimberlite fields. In the course of further research, it is necessary to compare the results obtained with the data of other geophysical (seismic and electrical prospecting MTS) and geological methods (Pospeeva 2017).

References

In Russian

Alekseev S.G., Voroshilov N.A., Margovich E.G., Kozlov S.A., Shtokalenko M.B. (2010a) Indications of heterogeneous hydrocarbon and ore systems in gravity and magnetic fields. In: Questions of the theory and practice of geological interpretation of gravity, magnetic and electric fields: Proceedings of the 37th session of the International Conference. sci. Seminar named after D.G. Uspensky. Moscow, IPE RAS, pp. 15–20.

Alekseev S.G., Kozlov S.A., Shtokalenko M.B. (2010b) Peculiarities of the geological interpretation of the results of gravity prospecting and magnetic prospecting. In: Proceedings of the 37th session of the Intern. sci. Seminar named after DG Uspensky. Moscow, IPE RAS, pp. 20–25.

Alekseev S.G., Kozlov S.A., Smirnov V.E., Shtokalenko M.B. (2011) Features of 2D and 3D interpretation of anomalies of potential fields. In: Notes of the Mining Institute, vol. 194, pp. 128–131.

Alekseev S.G., Senchina N.P., Shatkevich S.Yu., Shtokalenko M.B. (2016) Advantages and disadvantages of tomography of potential fields. In: Materials of the 43rd session of the Intern. sci. Seminar named after D.G. Uspensky. Voronezh. "Nauchaya kniga", pp. 10–13.

Alekseev S.G., Senchina N.P., Shatkevich S.Yu., Shtokalenko M.B., Movchan I.B. (2017) Rationale of the adequacy of tomography of potential fields. In: Materials of the 44th Session of the Intern. Sci. Seminar named after D.G. Uspensky. Moscow, IPE RAS, pp. 21–26.

Babayants PS, Blokh Yu.I., Trusov AA (2004) Interpretation tomography according to gravimetric and magnetic prospecting data in the "SIGMA-3D" package. In: Materials of 41st session of the Intern. sci. Seminar named after D.G. Uspensky, IPE RAS, Moscow, pp. 6–7.

Dolgal A.S., Bychkov S.G., Kostitsyn V.I., Novikova P.N., Pugin A.V., Rashidov V.A., Sharkhimullin A.F. (2012) On the theory and practice of tomographic interpretation of geopotential fields. In: Geophysics, No 5, pp. 8–17.

Kozlov S.A., Alekseev S.G., Lebedkin P.A., Savitsky A.P., Shtokalenko M.B. (2009) Comparison of seismic sections by regional profiles with distributions of singular sources of potential fields, effective excess density and effective magnetization of rocks. In: Proceedings of the 36th Session of the Intern. sci. Seminar named after D.G. Uspensky, Kazan, Kazan State University, pp. 164–166.

Meeting 2017 "Scientific and methodological and technological problems of prognosis and prospecting for low contrast kimberlite pipes in the East European and East Siberian diamondiferous provinces" June 8–9, 2017 St. Petersburg, FSUE "VSEGEI". http://vsegei.kiji.men/ru/conf/summary/index.php?ELEMENT_ID=98213.

Poniatowski EG, Antonov VE, Belash IT. (1982) Properties of high-pressure phases in metal-hydrogen systems. IN: Uspekhi Fizicheskikh Nauk, vol. 137, No. 4, p. 684.

Pospeeva E.V., Conducting lithospheric heterogeneities as a criterion of predictive assessment for promising diamond areas (on the example of Siberian kimberlite province), Journal of Mining Institute. 2017. Vol. 224. p. 170–177.

Shtokalenko MB, Alekseev SG (2013) New technology of density and magnetic tomography from gravimetric and magnetic surveys. Algorithms and application. In: The 83rd session of the Scientific and Methodological Council on Geological and Geophysical Technologies for Prospecting and Exploration of Mineral Deposits (NMS GGT) pp. 164–167.

Vaganov V.I. (2000) Diamond deposits of Russia and the world. Moscow, ZAO Geoinformmark, 371 p.

In English

Haggerty S.E. (1986) Diamond genesis in a multiplay-contrained model. In: Nature, vol. 320, pp. 34–38.

Chapter 39
Horizontal Shear Zones and Their Reflection in Gravitational Field

V. Philatov, L. Bolotnova and K. Vandysheva

Abstract Deposits and ore-deposits of various minerals are genetically and spatially related to geological structures. Therefore, prospecting and exploration are impossible without their mapping, the study of the internal structure and determine the mechanism of education. Horizontal shears are great importance among the large variety of geological structures and their certain parts are connected ore mineralization. The study of the structural paragenesis of shear zones and their density characteristics showed gravimetry is that an effective method of studying these areas. The gravimetry allows to define the sign of shear (right, left), width of a shear zone, position of its active and passive wings and rheological conditions of a shear formation. Thus, this method allows for an unambiguous mapping of the shear zone in the gravitation field and studying their internal structure, especially in confined areas. In the article petro density basing for the use of gravimetry for the study of shear zones and examples of their mapping for different regions of the Urals are suggested.

Keywords Gravitation field · Shear zones · Modeling · Structural paragenesis Mapping · Tominsky ore zone · Durinsky Trough

Horizontal shear zones (as well as zones of ruptures of other genetic types) are three-dimensional geological objects with a complex internal structure. When forming horizontal shear in a zone of his dynamic influence as a result of redistribution of mechanical stress conditions for formation of various secondary fault and folded structures are created. A large number of theoretical and experimental

---

V. Philatov
Vladimir State University, Vladimir, Russia
e-mail: filatov47@bk.ru

L. Bolotnova (✉) · K. Vandysheva
Ural State Mining University, Yekaterinburg, Russia
e-mail: lb63@bk.ru

K. Vandysheva
e-mail: vandysheva_ksenya@mail.ru

works is devoted to questions of studying stresses, processes of deformation and secondary structurization in shear zones (Gzovsky 1975; Stoyanov 1977; Gintov and Isay 1988; Tapkin 1986; Sherman 1977; Shear tectonic … 1988).

Having analysed the results of numerous experiences of physical modeling, Stoyanov (1977) has established the minimum set of structural elements or a structural paragenesis which can be formed in a shear zone. He has referred two conjugated chips R and R'; cleavage cracks T which are oriented at elastic deformation at an angle 45° to the direction of moving; the echeloned folds (Fd) which are formed at deformation of layered environments (medium); the backslashes chips (P) and longitudinal chips (L) characteristic of a final stage of process of a shear formation to elements of a structural paragenesis.

Results of studying of an internal structure of shear zones in the earth's crust and the data of physical modeling demonstrate that in both cases not all elements of a structural paragenesis are formed. At physical modeling the lack of some elements of a paragenesis is explained by the properties of material of the model and inconstancy of speed of deformation by a number of random factors. The earth's crust the condition of shear formation are even more complicated and difficult estimation. Therefore the answer to the question of why some elements of shear tectonics in field conditions weren't shown, in some cases is not answered. An example of incomplete development of a shear paragenesis is the zone of the shear which was becoming more active in one of regions of the Spitak earthquake in 1988 (Fig. 39.1). When mapping this zone only separation cracks (T), the echeloned folds (Fd) and cleavage cracks (R) were established.

Studying the shear process on models of bentonite clay, Stoyanov, S. has established that the interfaced chips R and R' arise and develop at the same time, repeating along the axial line of shear; R' cracks are shorter and meet more often then R cracks. In the process of development of the crack shear R; the displacement like slays, is the result crack R crosses R' cracks which lose the independence and do not gain development.

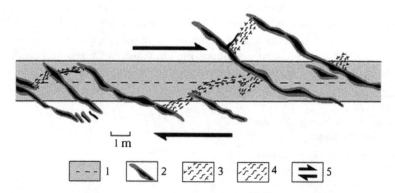

Fig. 39.1 A structural paragenesis of a zone of shear by the results of field observations around the earthquake in Spitak of 1988: 1—axial line of a fault; 2—the separation crack (T); 3—the echeloned folds (Fd); 4—cracks without the shear (R cleavage cracks); 5—direction of shering

According to Sherman et al. (1983) the research of a shear process was made on the models of clay paste and two stages of deformation were revealed: plicate and disjunctive-plicate. In the second stage, there is a formation of two systems of cracks: cracks along the strike, which make a corner of about 85° with the axial line displacement and diagonal, forming a corner of 15°–30° with the same axial line. Both systems consist of the set of smaller cracks (primary) on which there is a displacement.

In results of comparison of physical modeling show that the systems of cracks of R and R' and the cross and diagonal are identical. These are the systems of the cleavage cracks. Higher detail of the researches described in Sherman et al. (1983), allowed to trace dynamics of the development of a shear zone as a process of coalescence of primary cracks in large cracks.

Planes of incidence of the integrated cross cracks in an axial part of the shear are vertical, and on the periphery, there is a decrease in dip and they acquire the "propeller" shaped form. The same plane of incidence, as the cross, characterizes diagonal cracks, arising after the cross and gaining intensive development.

According to the researches made on models of clay, damp sand and a petrolatum with stearine and some other materials formation along the line of future shear of system echelon located "S" shaped separation cracks which in the shear formation process of unite in the wavy rupture representing alternation of zones of stretching and compression has been established (Gzovsky 1975). These results are confirm by researches on models from pyrophyllite (Sobolev 1980).

Researches of process of shear formation with use of a luminescent method have shown that after emergence the echelon crack at further increase of the load from mouth of cracks the cracks parallel to the direction of shearing begin to develop. These are cleavage cracks (P and L). Merging, they form one or two rectilinear rupture parallel to the axial line of shearing.

According to Spencer (1981) the shear zone consists from anticlinal and the sinklinal folds which are on one or both sides of the axial line of displacement, which was broke by normal fault. These faults are form in structures of stretching. Therefore, they were consider as the transformed separation cracks.

The analysis of a results of studying of the shear zones on models and in operating conditions, has allowed drawing a number of conclusions on regularities of their development and an internal structure:

1. The development is gained generally by cracks of a separation and a cleavage cracks from elements of a structural paragenesis, which form the ordered echelon sequences of the same structures of relatives in a form and orientation and remote from each other at identical distance. This structural-morphological sign «is the direct instruction on the development of a shear formation» (Voronin 1988).
2. Separation cracks and a cleavage cracks consist of big set of smaller primary cracks, which promote improvement of permeability of the geological medium and growth of her dilatation on the layer thickness involved in the shear process (Stakhovskaya 1988). On these permeable zone of destruction as on the channels there is a transportation of crustal and mantle fluids and are formed of a chain chains of explosion tubes (Voronov 1988), small Intrusive bodies of a

granitoids (Pavlinov 1988; Sintsov 1988), developing endogenous ore genesis (Morakhovsky 1988) and other geological phenomena.
3. Permeable sections of destruction are offset from the axial line of the displacement on the periphery of a shear zone towards the active, offset wing. Process of a crack formation in a fixed wing occurred with significantly smaller intensity (Pavlinov 1988; Lobatskaya 1987; Seminsky 1988).
4. The elements of a structural paragenesis are shown in development the plicate of structures in a sedimentary cover, and in increase in porosity and permeability of cover rock over zones of intensive crushing of the base rocks in the closed areas (Yure 1988).
5. The structure of the shear zones is same at all large-scale levels both for deep shears and for shears of higher orders arising in areas of dynamic influence of deep shears. Results of studying of the shear cracks on samples of rocks and in the regions of Gobi-Altai and San Francisco earthquakes, in particular, demonstrate to it (Stakhovskaya 1988).

Big role of the shear zones as structural-ore-controls a factor was predetermine considerable interest in their studying. Mapping of these zones and studying of their internal structure presents considerable difficulties even in open areas. The major role when studying shear belongs to geophysical methods among which basic is the gravimetry in the closed areas.

The crack formation, is widely shown in the shear process (as well as in general at deformation of the geological environment at structurization) and leads to considerable change of density of the geological medium that is a physical basis of application of a gravimetry for studying of shears. Density of rock in a sample and in the massif significantly differ. This distinction well comes to light by comparison of data the petro density of the measurements with results of the determination of density by integrated methods (registration of intensity of space radiation in excavations, results of gravitational logging, etc.).

Results of the detailed geological and petrophysical researches executed in zones of dynamic influence of Pervomaisky, Talkovsky Faults of Ukraine, Trans-Baikal and other regions (Belichenko and Isay 1987), demonstrate to what the main mechanism of plastic and fragile deformations of the rocks is cataclasis, expressed in crushing of the minerals grains and development of micro fracturing. Thanks to this process density of rocks decreases by $(0.01–0.07)$ g/cm^3. Communication between density of rocks and volume of open cracks is well described by the linear equation of regression (Filatov and Bolotnova 2015). In the structural relation zones of development of cleavage cracks are expanded.

When studying physical properties of the tektonite developed in shear zones to which many ore fields of Tajikistan are dated it is established that deformation is practically not resulted by change of mineral density of tektonite. Considerable reduction of density of tektonite (up to 0.40 g/cm^3) is caused only by their fracturing. At the same time sites of development of tektonite are genetically connected with mineralization. Namely, the industrial ore mineralization is in a shear zone in places of distribution the tektoni rudaceous of big thickness. She is almost not in finely dispersed of tektonit (Kuddusov 1988).

Thus, structures of a separation and a cleavage crack represent objects, abnormal on density. These objects located along strike of a shear zone have the close sizes and the elliptic form in the plan. Big half shafts of ellipses are parallel and form an acute angle with a shear axis. Distance between ellipses approximately identical. In the anomaly field of gravity to such distribution of density heterogeneity there corresponds systems an echelon of the local negative anomalies close on intensity, a form, the sizes and orientation. Existence of systems of such anomalies (systems, but not separate anomalies) is reliable morphological feature of a shear zone. Density heterogeneity (anomalies) have a steep drop and their lower edges are at a considerable depth, which reaches the layer thikness (involved in the shear process) to several kilometers even tens of kilometers (Filatov and Bolotnova 2015), if the shear process captures the earth's crust, for all its layer thickness.

Expressive examples of shear zones are the Tominsky ore zone and the Durinsky Trough on a roof of salts on the Verkhnekamsky field of potassium-magnesium salts (Filatov and Bolotnova 2016a, b). Importance of both examples is that they give an idea not only of common features of morphology of the field of gravity of shear zones, but also about their distinction, the caused different rheological situation in which the Tominsky ore zone and Durinskoy Trough was formed.

The Tominsky ore zone was created in volcanogenic and sedimentary thickness; she has north northwest pro-deleting; its width is about of 3–5 km, along the strike about of 25 km. A characteristic complex of local anomalies in gravitational and magnetic fields maps this zone. The faults, which limiting structure are fixed by systems the cleavage of the linear magnetic anomalies close on intensity and the sizes. Long axes of these anomalies are parallel each other and have the northeast direction. Anomalies are caused by the systems of a separation cracks in which the streaky disseminated magnetite mineralization is developed. In the field of gravity a part of faults is mapped by anomalies like "gravitational step" (Fig. 39.2a).

The gravitational field of internal parts of a zone is characterized by the system cleavage of the located local negative anomalies of relatives on intensity. The form of anomalies is close to elliptic; the sizes in the plan change from 1.0×0.5 km to 2.0×1.0 km; long axes of ellipses are parallel and focused in the northeast direction under corners of $(40°–45°)$ along the strike of the axial line of a zone—to the direction of displacement; almost all anomalies are displaced to northeast border of a zone (Fig. 39.2b).

These interpretations of gravity anomalies and drilling have shown that sources of anomalies are small intrusive bodies of diorites, quartz diorites and quartz of porfirit which lie in more dense volcanogenic rocks of sedimentary and neutral structure of Silurian age. Central parts of bodies stock shaped; intrusions are followed the apofizam complicating their form; horizontal sections of intrusions are close to elliptic to the sizes of axes from 2.0×1.0 km to 2.0×0.5 km and to the same orientation, as at local anomalies of gravity. The lower edges of intrusions with which porphyry-copper mineralization is connected are at a depth in the first kilometers. Rocks of intrusive bodies are strongly crushed.

Brought given and about a character of the gravity field and geological data demonstrate that the Tominsky zone represents left-side shear.

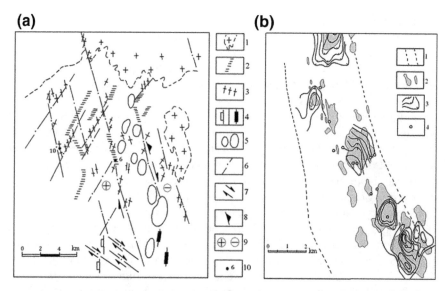

Fig. 39.2 Tominsky ore zone; scheme of local anomalies of gravitational and magnetic fields (**a**): 1—granitoid massifs and their borders; 2—provision of gradient zones of the gravitational field; 3—axes of local gravitation anomalies; 4—axes of local anomalies of magnetic field: (a) negative; (b) positive; 5—the local decreases in the gravitation field caused by small intrusive bodies and zones of a decompaction; 6—the ruptures allocated according to geologic-geophysical data; 7—the directions of shearing on ruptures; 8—direction of falling of a surface of ruptures; 9—signs of vertical shears on ruptures: "+"— a raising; "−"—lowering; 10—position of wells of drilling; scheme of an internal structure (**b**): 1—zone borders; 2—contours of small intrusive bodies; 3—isolines of local gravitation anomalies; 4—a copper ore-deposits with the maintenance of 0.3% and above

The Durinskoy Trough on a roof of salts of the Verkhnekamsky field of potassium-magnesium salts was created in other rheological situation, than the Tominsky ore zone. The represents the imposed structure in a zone of dynamic influence of the width deep fault of the before Palaeozoic age crossing a platform part of Cisural area and a part of structures of the folded Urals.

The gravitational field of a deflection is raised in comparison with the neighboring territories. On this background the chain of the located positive local anomalies "S" shaped form is mapped. Interpretation of anomalies has shown that their sources are immersions in a roof of salts which are filled with terrigenous material, more dense, than salt. Drilling has confirmed this Trough conclusion: immersions have also "S"—shaped form in the plan; long axes and immersions and anomalies are parallel and extended in the northeast direction under corners of 10°–20° along the strike and trough, and a deep fault (Fig. 39.3a, b).

Proceeding from morphology of the gravitation field follows that the eshelon system of local anomalies and the system of local immersions were formed as a result of left-side shear which has happened on the Durinsky Deep Fault. This conclusion was confirmed with tool measurements on leveling lines of Tyulkino-Chusova and Serov-Kushva. These lines cross a fault and in places of their crossing anomalies of shear from 25 to 40 km wide are noted.

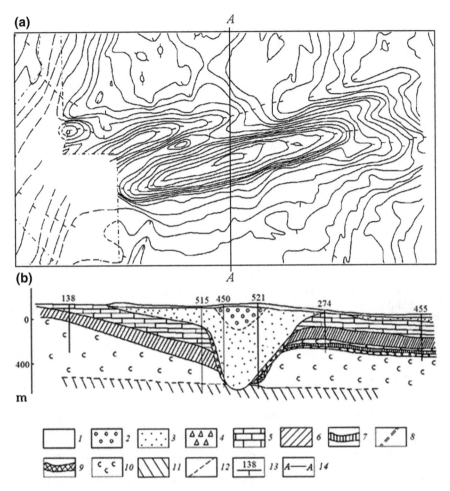

Fig. 39.3 Scheme of gravitational field (**a**) and geological section of the Durinsky Trough (**b**): 1—quaternary deposits; 2—deposits of the Belebey suite; 3—deposits of Sheshminsky suite; 4—tectonic breccia; 5—deposits of Verkhnesolikamsky subsuite; 6—deposits of Nizhnesolikamsky subsuite; 7—sylvinite-karnallite thickness of Verkhnepopovsky subsuite; 8—sliding mirrors; 9—silvinite thickness of Verkhnepopovsky subsuite; 10—top and lower galite thicknesses of Verkhnepopovsky subsuite; 11—Nizhnepopovsky subsuite, halogen and terrigenous and carbonate thickness; 12—lines of tectonic displacement; 13—position of the well and its number; 14—line of a section

The most important question of establishment of a trough genesis is the question of formation of local immersions. These immersions were formed as a result of leaching of potash salts by subsalt waters and brines with the subsequent filling of emptiness with overlying dense terrigenous deposits. The movement of subsalt waters and brines happened on zones of fracturing and loosening of salt rocks as a result of a shear formation. Therefore spatially process of leaching happened in a trough not everywhere, and is selective in compliance with situation in a shear zone of elements of a structural paragenesis, first of all, of cleavage cracks and separation cracks.

The essential difference in orientations of big axes of local gravitation anomalies in the Tominsky zone and in the Durinsky Trough is caused by distinction of rheological situations in which these structures were formed. At formation of the Tominsky zone of rocks at a shear formation experienced significantly elasto-plastic deformations. In such situation, the main axes of deformation form the corner close to $40°$–$45°$ with the direction of displacement. In the Durinsky Trough the major role was played by plastic deformation of rocks. In this case the ellipsoid of deformation becomes strongly extended with a deviation by length of an axis of rather elastic component of deformation on a corner of $20°$–$25°$. Therefore in a trough all systems of cracks are focused under smaller corners to the direction of displacement.

Generalizing qualitative results of the tektonofizic analysis of the gravitation field of two shear structures, it is possible to draw the following conclusions: the gravimetry is the effective method of mapping of shear zones allowing to define unambiguously the sign of shear (right, left), width of a shear zone, position of her active and passive wings, an internal structure and rheological conditions of a shear formation.

References

Belichenko P., Isay V. (1987). Tektonofizic studying the dilatation of effects in the Central part of the Ukrainian board//Experimental tectonics in the solution of problems of theoretical and practical geology: II symposium in USSR. Kiev (USSR), pp. 112–113.

Filatov V., Bolotnova L. (2015). Gravimetry. Method of the tektonofizic analysis of gravitational field. – Yekaterinburg (RF). 284 p.

Filatov V., Bolotnova L. (2016a) Nature and dynamics of the Durinsky Trough. Yekaterinburg (RF). pp. 111–119.

Filatov V., Bolotnova L. (2016b) Genesis of the Tominsky ore zone according to geologic-geophysical data. Yekaterinburg (RF). pp 111–119.

Gzovsky M. (1975). Tektonofizic bases. Moscow (USSR). 536 p.

Gintov O., Isay V. (1988). Tektonofizic of a research of faults of the consolidated crust.– Kiev (USSR). 225 p.

Kuddusov H, (1988). Tectonics shear violations of ore fields and their role in an ore deposits (on the example of Tajikistan)//Shear tectonic violations and their role in formation of mineral deposits: I meetings on shear tectonics. Issue 3. – Leningrad (USSR), pp 85–88.

Lobatskaya R. (1987). Structural zonality of faults. Moscow (USSR). 128 p.

Morakhovsky V. (1988). Shear deformations and shears in crust//Shear tectonic violations and their role in formation of mineral deposits. I meetings on shear tectonics in USSR. Issue I. General questions of shear tectonics, results of laboratory modeling. Leningrad (USSR), pp. 41–43.

Pavlinov V. (1988). A role of deep shears in an arrangement of granite bodies//Shear tectonic violations and their role in formation of mineral deposits: I meetings on shear tectonics in USSR. Issue I. General questions of shear tectonics, results of laboratory modeling. – Leningrad (USSR), pp. 45–47.

Stoyanov S. (1977). Mechanism of formation of explosive zones. Moscow (USSR). 144 p.

Sherman S. (1977). Physical regularities of development of faults of crust. – Novosibirsk (USSR). 102 p.

Shear tectonic violations and their role in formation of mineral deposits (1988): I meetings on shear tectonics in USSR. Leningrad (USSR), Issue 1. - 95 p. Issue 2.-108 p. Issue 3. -144 p.

Spencer E. (1981). Introduction to structural geology. Leningrad (USSR). 308 p.

Sherman S., Bornyakov S., Buddo V. (1983). Areas of dynamic influence of faults. – Novosibirsk (USSR). 112 p.

Sobolev G. (1980). Studying of education and harbingers of a rupture of shear type in vitro// Physical processes in the centers of earthquakes Moscow (USSR), pp. 86–99.

Stakhovsky I. (1988). Crack formation and superficial deformations in a zone of the formed shear trough in a sample of rocks. No. 5. Khabarovsk (USSR), pp. 88–94.

Sintsov A. (1988). A structural paragenesis of shears of Patom Highland//Shear tectonic violations and their role in formation of mineral deposits: I meetings on shear tectonics in USSR. Issue 2. Planetary, regional and local manifestations of shear tectonics in a lithosphere of Earth and planets. Leningrad (USSR), pp. 72–75.

Seminsky K. (1988). Modeling of large shear zones and specifics of development of their wings// Shear tectonic violations and their role in formation of mineral deposits. I meetings on shear tectonics in USSR. Issue I. General questions of shear tectonics, results of laboratory modeling. Leningrad (USSR), pp. 74–77.

Tyapkin K. (1986). Studying faulted and folded structures of the Precambrian by geologic-geophysical methods. – Kiev (USSR). 168 p.

Voronin P. (1988). The principles of shear tectonics//Shear tectonic violations and their role in formation of mineral deposits: I meetings on shear tectonics in USSR. Issue I. General questions of shear tectonics, results of laboratory modeling. Leningrad (USSR), pp. 8–22.

Yurel G. (1988). Tectonic interpretation of results of experimental geomechanics//Shear tectonic violations and their role in formation of mineral deposits: I meetings on shear tectonics in USSR. Issue I. General questions of shear tectonics, results of laboratory modeling. Leningrad (USSR), pp. 68–71.

Chapter 40
Intermediate Conducting Layers in the Continental Earth's Crust—Myths and Reality

A. A. Zhamaletdinov

Abstract The nature and scale of intermediate conducting layers propagation in the continental earth's crust are discussed in this paper. The myth on the existence of intermediate conducting layers in the Earth's crust at the depth of 10–20 km firstly appeared owing to results of the super-deep dipole-dipole sounding performed in the Gulf of Finland in 1946. Since then for the many decades a large number of anomalies of electrical conductivity in the earth's crust have been detected. The authors of these studies interpreted the anomalies as the existence of intermediate (sub horizontal) conducting layers of fluidal (or temperature) origin at the depths of the first tens of kilometers, same as in the Gulf of Finland. But analysis of experimental data presented in the article allows to conclude that in most cases the anomalies of electrical conductivity in the Earth's crust appears due to presence of a steeply dipping electronically conductive sulfide and carbon (graphite) bearing rocks of organic nature («SC-Layer» of Semenov). Fluids exist only in the uppermost part of the continental Earth's crust in the depth interval from 2–3 to 7–10 km. They are detected as intermediate conductive area associated with the influence of dilatancy-diffusive processes and named as "DD-layer". The nature of electrical conductivity anomalies in the Earth's crust represents fundamental problem in the interpretation of the deep sounding data. Solution of the problem determines the role of crustal conductors in the study of geological structure and composition of the Earth's interior.

Keywords Sulfide and carbon bearing rocks · SC-layer · Fluids
Earth crust · Dilatancy-diffusive layer · DD-layer

A. A. Zhamaletdinov (✉)
St. Petersburg Branch of Pushkov Institute of Terrestrial Magnetism,
Ionosphere and Radio Wave Propagation of the Russian Academy of Sciences,
St. Petersburg, Russia
e-mail: abd.zham@mail.ru

A. A. Zhamaletdinov
Geological Institute of the Kola Science Center of RAS, 184209, Apatity, Russia

Introduction

The myth on the existence of intermediate conducting layers in the Earth's crust firstly appeared owing to results of the super-deep dipole-dipole sounding performed by the team of Physical Department of the Leningrad University in the Gulf of Finland in 1946 (Fig. 40.1) (Kraev et al. 1947).

A feeding line of 1000 m length was located along the southern coast of the Kronstadt island. The current strength reached to 1000 A. A battery of accumulators assembled from submarines of the Baltic Navy served as a source of the current. The soundings were conducted along the marine and land traces at distances of up to 70 km (Fig. 40.1a). The apparent resistivity curve at distances up to 40 km had an ascending appearance, and then, at more distances (more then 40 km) underwent a sharp decline (Fig. 40.1b). On the marine trace the first decline happened at distance 9 km (Fig. 40.1a). But at more distances the apparent resistivity curve again has asquired an ascending shape.

Professor A. P. Kraev, scientific supervisor of this experiment, being a theoretical physicist, interpreted the decrease in apparent resistivity at spacings of more then 40 km by existence of the conducting layer in the earth's crust at the depths of 10–20 km. He explained the nature of this "layer" by increase of temperature with depth. However the second participant of the experiment (professor A. S. Semenov) have had another opinion. He believed that the lowering of apparent resistivity at distance 40 km (same as at 9 km on the Fig. 40.1b) is due to the lateral influence of a steeply dipping conductive objects, a zones of sulfide or graphite mineralization.

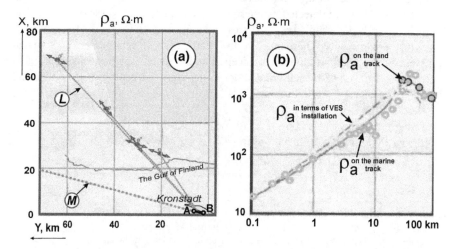

Fig. 40.1 Results of the first in the world super-deep sounding that have fixed the seeming conductive layer at the depth of 10–20 km Kraev et al. 1947). **a** Location of transmitting line AB and receiving tracks—sea track (M) and land track (L). The arrows indicate electric field vectors on the land route. **b** Diagram of apparent resistivity curves (Ωm) versus spacing between transmitter AB and receiving sites on the sea and land tracks

That idea of A. S. Semenov has been published only many years later (Semenov and Zhamaletdinov 1981) and for a long time did not find supporters.

Since then for the many decades a large number of experimental studies have been made on the deep electrical sounding with controlled and natural sources. They indicated existence of anomalies of electrical conductivity in the earth's crust. The authors of these studies, using a formal, one-dimensional model, interpreted the nature of anomalies by existence of intermediate conducting layers at the depths of the first tens of kilometers. A summary diagram of geoelectrical sections of the continental lithosphere from results of soundings with natural and controlled sources is given in Fig. 40.2.

This property of the crustal conducting anomalies is of planetary scale of spreading. The depths of anomalously conductive objects vary from units to tens of kilometers. Their influence substantially limits the possibilities for studying of electrical conductivity in deeper horizons of the upper mantle of the Earth. At the same, time crustal conductors are of special interest for fundamental and applied geology. They are indicators of physical state and geodynamic development of corresponding blocks of lithosphere. The nature of electrical conductivity anomalies in the Earth's crust represents fundamental problem in the interpretation of the deep sounding data (Joedicke 1992). Solution of the problem determines their role

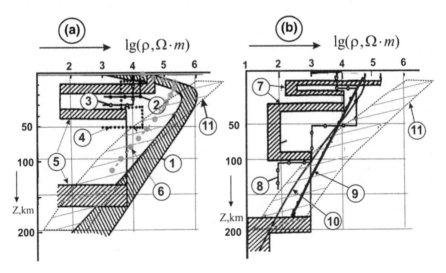

Fig. 40.2 Summary diagram of geoelectrical sections of the continental lithosphere from results of the deep soundings with controlled (**a**) and natural (**b**) sources. **a** The soundinds with controlled sources: 1—(Zhamaletdinov 1990); 2—(Kraev et al. 1947); 3—(Lundholm 1946); 4—(Zijl 1969); 5—(Blohm et al. 1977), 6—(Zhamaletdinov et al. 2011). **b** The soundinds with natural (MT-AMT) sources: 7—(Kovtun et al. 1986); 8—(Vladimirov and Dmitriev 1972); 9—(Vanyan 1997); 10—(Varentsov et al. 2002); 11—section from laboratory data (Cermak and Lastovickova 1987)

in the study of geological structure and composition of the Earth's interior. Two principal concepts of this problem are being developed at present: the electronically conductive and the fluidal one.

The Fluid Conception

Thermal dehydration observed in rocks of low metamorphic grade such as serpentinites and amphibole schists, is commonly accepted as the most probable mechanism of fluid formation at depth. The dehydration phenomenon in interpretation of the origin of crustal electrical conductivity anomalies was first described by Keller (1963) on the example of two samples from a deep sea well in the vicinity of Puerto Rico. During heating of these samples, which had considerable porosity (up to 8%), their electrical resistivity decreased to 10^3 Ω m in the temperature range of 500–600 °C. On further heating to 1000 °C, the resistivity increased to 10^4 Ω m.

The authors explained the obtained trend of temperature curves by the release of bound water and its subsequent evaporation. They supposed that if water evaporation were prevented by confining pressure, the resistivity could decrease to 10 Ω m at 800–900 °C. In the subsequent years, the phenomenon of thermal dehydratation was investigated in greater detail by E. Lyubimova, I. Fel'dman, R. Hyndman, P. Shearer, L. Van'yan and many other researchers.

At any interpretation the fluid conception should assume that in deep layers of the Earth's crust the connected systems of pores filled with brine solutions should exist. However, such an assumption contradicts data of petrology, according to which under deep conditions of granulitic facies metamorphism all free fluid (H_2O, CO_2) is intensely absorbed by the rock and enters into the composition of a crystal lattice (Yardley and Valley 1997). Furthermore, the possibility of free fluid existence in the crystalline Earth's crust of the shields and ancient platforms is doubtful since pores and cracks are closed up at the depth under effect of litho static pressure, and the fluids should move upward under influence of squeezing out effect (Rodkin 1993).

The dehydration can give rise to intermediate conductive layers only in tectonically active zones, where the thermal field is non-stationary, and where above-crystalline solutions can exist in the intergranular space of rocks at the boundaries of temperature fronts. Otherwise, in case of stable lithospheric plates it would be necessary to assume at least a two-layer division of the crust as in the work by Jones (1987).

The first condition is related to the evaluation of the least possible value of electrical resistivity for a fluid-saturated rock (ρ_r). This can be computed with the Archie law $\rho_r = \rho_{fl}/p^n$ where ρ_{fl} is the resistivity of fluid in Ω m, p is the volume porosity of deep matter, n varies from 1.5 to 2 depending on the rocks matrix.

In crystalline rocks we can use $n = 2$ after (Vanyan 1997). The lowest average values of resistivity of brine solutions at depth lie in the range of 0.02–0.04 Ω m, and the possible porosity in the crust according to the data from the Kola Super Deep Well even if seal failure effects are considered, does not exceed 0.015 of the rock volume at a depth of 10 km. Thus, if one assumes $n = 2$, the least permissible value of resistivity of fluid-saturated rocks at the low crust does not exceed 100 Ω m. In (Fel'dman and Zhamaletdinov 2009) the estimate increases up to 1000 Ω m.

The second condition of the fluidal conception is related to the depth of anomalies. The depth is determined by the minimum required temperature of rock dehydration, 500–600 °C. Finally, the third, completely qualitative condition is based on the assumption that the fluidal anomalies should be characterized by smooth boundaries, weak spatial gradients of electric field, small or absent anisotropy, and large sizes comparable with the depth of anomaly occurrence.

If the deep (crustal) anomalies of electrical conductivity doesn't not fit to the framework of the above-mentioned conditions for the fluid mechanism, their nature should be interpreted on the basis of electronically-conductive conception. In Fig. 40.3 is a diagram of the distribution of the largest anomalies of electrical conductivity in all the world.

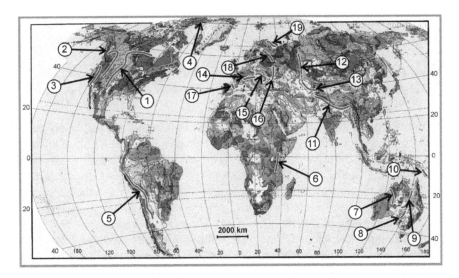

Fig. 40.3 The largest anomalies of electric conductivity in the world: 1—North American; 2—Bitterroot-Cascade Mountains; 3—Sierra Nevada; 4—North Greenland; 5—Andean; 6—Kenya, 7—Carpentaria; 8—Flinders; 9—South-West Queensland; 10—New Guinea; 11—Trans-Himalayan; 12—Urals; 13—South Tien Shan; 14—Alpine-Pannonian; 15—Carpathian; 16—Kirovograd; 17—Pyrenees; 18—Ladoga-Bothnia; 19—Polmak-Pechenga-Imandra-Varzuga

The Electronically-Conductive Conception

Prof. M. N. Berdichevsky, who actively developed the fluid concept and was at the root of its development, was the first to organize the All-Soviet Union program for study the nature of crustal anomalies of electrical conductivity and actually supported the idea of A. S. Semenov (Semenov 1970) and A. Adam (Adam 1974). Figure 40.3 appeared due to result of this activity (Zhamaletdinov 1996, 2014).

At the present time only some conductive anomalies of the continental lithosphere are considered as fluidal, but even then, as a rule, only because they are insufficiently studied. In the overwhelming majority of cases, it has been proved that conductive anomalies in the earth's crust are due to the existence of electron-conducting sulfide-carbon rocks developed in the composition of primary-sedimentary super-crustal complexes of all ages - from the Archaean to the Phanerozoic. They include anomalies of electrical conductivity of up to thousand of kilometers long.

The most outstanding are such anomalies (Fig. 40.3) as Great North American (1), Andean (5), Trans- Himalayan (11), South Tien Shan (13), Alpine-Pannonian (14), Carpathian (15) and many others. Undoubtedly, the transformation of organic residues into sulfide-carbon electron-conducting rocks could not occur without the influence of regional tectonic factors—temperature, pressure and fluids. But they happened millions and billions of years ago. At present time the fronts of high temperature and fluids disappeared and we fix only traces of their activity in the form of sulfide-carbon conductive bands. The area of their manifestation has got the conventional name of "SC-layer" of Semenov.

Fluidal Conception of Dilatancy-Diffusive Nature ("DD-Layer")

Along with this, there is a place in the continental crust for real intermediate conducting layers, which are associated with the presence of fluids. This is so-called layer of dilatancy-diffusive nature(«DD layer»). The notion of "layer" in this case is accepted conditionally, since the «DD layer» is a system of cracks in the form of listrick faults, along which the fluid (meteoric water) penetrates to the depth from the day surface. The fluids at the depths from 2–3 to 7–10 km are preserved in the free state due to the dilatancy-diffusive phenomenon. Dilatancy is an irreversible cracks opening under interaction of tangential and lithostatic pressures described in the geotectonic scheme by Nikolayevsky (1996). Deeper than 10–12 km, the existence of "«DD layer»" is problematic, since the conditions for the existence of free fluids disappear. Rocks at these depths are transformed into a quasi plastic ("ductile") state due to a reduction in the role of tangential pressure.

Especial experiment on the deep sounding has been implemented for to study the structure of the "DD-layer". The scheme of the experiment, named as "Kovdor-2015", is shown on the Fig. 40.4.

Electromagnetic soundings were carried out on the Kovdor–Yensky segment of the White Sea block, which is composed of relatively homogeneous, high resistive rocks of the Archaean basement (Fig. 40.1). The installation includes axial and equatorial multibeam soundings. Mutually orthogonal L-shape grounded lines (dipoles) of about 1.5 km long have been used for that. The distances between the sources and receivers changed around 25 and 50 km. Soundings were carried at two sites (western and eastern) located at 85 km apart from each other (Fig. 40.4). The electromagnetic field was created by an Energia-3 M generator of 29 kW power operating in the frequency range from 4 Hz to 2 kHz. The maximum current was 22 A at an output voltage of 1000 V.

The total results are shown on the Fig. 40.5, where apparent resistivity and inversion data are presented. The intermediate conducting layer, identified with the "DD layer", is clearly visible at all the sounding sites in Fig. 40.5I–III. Along with this, one can notice a regular displacement of the apparent resistivity curves.

The curves for the total magnetic field (Fig. 40.5-III), free from static shift distortions, occupy the highest position on the apparent resistivity scale. The curves, calculated for the total electric field (Fig. 40.5-II) are shifted downward relative to the curves over the total magnetic field (Fig. 40.5-III) by three to five times. Finally, the apparent resistivity curves for the total input impedance (Fig. 40.5-I) are shifted down relative to the curves for the total magnetic field

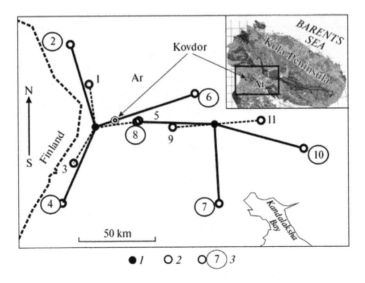

Fig. 40.4 The layout of transmitting and receiving lines in the frequency sounding experiment "Kovdor-2015", implemented for to study the "DD-layer. Legend: 1—centers of L-shape transmitting lines AB, 2—centers and numbers of receiving sites at 25 km distance from AB lines, and 3—the same at 50 km distance from AB lines

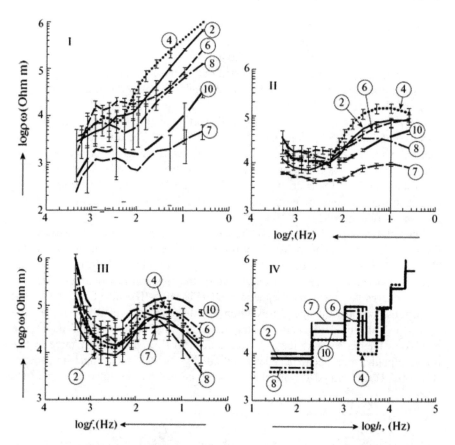

Fig. 40.5 Resulting curves of "DD-layer" study in the Archaean crystalline basement on the Kovdor area. Legend: **I–III** apparent resistivity curves, calculated with the use of total impedance (**I**), total electric field (**II**) and total horizontal magnetic field (**III**). **IV** Inversion data. Arabic digits in circles indicate the numerals of points of sounding with spacing of 50 km between transmitter and receiver. Location of transmitting and receiving sites is shown in Fig. 40.4

(Fig. 40.5-III) by 10–20 times due to the action of static distortions. These offsets are in a good agreement with the theory of static shift distortions (Rokitjansky 1981). Static correction had been taken into account when solution the inversion problem. Results of this inversion are shown in Fig. 40.4-IV. They point out on the existance of intermediate inhomogeneous conductive layer ("DD-layer") in the depth interval from 2–3 to 7–10 km in all observation regions (Fig. 40.4-IV). The longitudinal conductivity of the layer varies from tenths of a share of Siemens to 1–2 S.

Results of the Kovdor 2015 experiment, depicted above, are completely consistent with results of frequency soundings in Central Finland in 1997 (Zhamaletdinov et al. 2002) that gave the first opening of the "DD-layer".

Conclusion

Thus, it should be concluded that the continental Earth's crust for its entire thickness is, in general, "dry". The so-called intermediate conductive layers, erroneously associated with the existence of fluids at depths of the first tens of kilometers, are in fact the result of lateral influence of steeply dipping or inclined sulfide and carbon bearing electronically conducting rocks having longitudinal electrical conductivity up to several thousand Siemens. These rocks are distributed within the upper 10-kilometer stratum of the earth's crust and have received the conventional designation as "SC-layer" of Semenov.

Weak traces of fluids are supposedly detected in the form of an inhomogeneous conductive layer in the uppermost thickness of the continental earth's crust in the depth range from 2–3 to 7–10 km. The longitudinal conductivity of the layer varies from tenths to 1–2 S. Following after the geodynamic model of V.N. Nikolaevsky we explain the nature of the layer by dilatancy-diffusive phenomena in the upper "brittle" part of the Earth's crust. The hypothesis of dilatancy-diffusive phenomena suggests that liquids (water brines) penetrate to the depth through a system of cracks that are flattening with depth under the influence of tangential stresses. The phenomena of dilatancy-diffusive expansion of pores and cracks occur due to rupture of brittle rocks as a result of the interaction of tangential and lithostatic pressures. Due to the dilatancy-diffusive effecting, free liquids fill up the opening cavities. The electrical conductivity of the rocks increases within a fairly narrow interval of depths, forming an intermediate conducting region. The area received the symbol "DD-layer". The "layer" itself should be understood as an inhomogeneous zone of reduced resistivity. This region exists due to a large number of cracks that are known in the geological literature as "listrick" cracks (Goryainov and Davidenko 1979). Thus the nature of the intermediate conducting layers in the continental Earth's crust should be explained either by the presence of electronically conductive sulfide and graphite bearing rocks or by the influence of "DD-layer" having fluidal, dilatancy-diffusion nature of conductivity.

References

Adam A., (1974): Electric crustal anomalies in the Carpathian Basin and their origin in the rock composition. // Acta Geol. Hung. 18. Pp. 13–22.

Blohm E.K., Worzyk P., Scriba H. (1977). Geoelectrical deep soundings in Southern Africa using the Cabora Bassa power line. // Journal of Geophysics, 43. Pp. 665–679.

C˘ermak, V. & Lastovic˘kova, M. (1987). Temperature Profiles in the Earth of Importance to Deep Electrical conductivity Models. // Pageoph, vol. 125. Pp. 255–284.

Fel'dman I.S & Zhamaletdinov AA. (2009). Fluid and thermal conductivity model of the lithosphere. // In: Proc. of Int. Conf. Apatity. Geological Institute of the KSC RAS. Pp. 100–107.

Goryinov P.M.; Davidenko I.V. (1979). Tektonic- kesson effect in rocks and ore deposits – an important new phenomenon ingeodynamics. // Doklady FN USSR, Vol. 247, No 5. Pp. 1212–1215.

Joedicke H. (1992). Water and graphite in the Earth's crust —An approach to interpretation of conductivity models // Surveys in Geophysics, 13. Pp. 381–407.

Jones A.G. (1987). MT and reflection: an assential comb. // Geophys. J.R. astr. Soc., N89. Pp.7–18.

Keller G.V. (1963). Electrical properties in the deep crust. // IEE Trans. Antennas and Propagat., V. 11, N 3. Pp. 615–637.

Kovtun A.A., Moiseev O.N., Vagin C.A., Vardanjants I.L., Kokvina E.L., Saveljev A.A., Uspensky N.I. (1986). MT- and AMT-sounding on the Kola peninsula and in Kareliya. // Deep electrical conductivity of the Baltic shield. Petrozavodsk. Karelian branch of RAS. Pp. 34–48.

Kraev, A.P., Semenov, A.S., Tarkhov, A.G. (1947). Ultra-deep Electrical Sounding. // Razvedka Nedr, 1947, no. 3. Pp. 40–41.

Lundholm R. (1946). The experimental sending of d.c. through the Earth in Sweden. // Proceedings of the Conf. Internat. des Grands Reseaux Electriques a Haute Teusion . Paper No 134

Nikolaevsky V.N. (1996). Cataclastic breaking down of rocks of earth crust and anomaly of geophysical fields. // Izv. Akad. Nauk, Ser. Fiz. Zemlin No. 4, Pp. 41–50.

Rokitjansky I.I. (1981). Inductive soundings of the Earth. // Kiev. Naukova Dumka. 296 p

Rodkin M.F. (1993). The Role of a Deep Fluid Regime in Geodynamics and Seismotectonics // Moscow

Semenov A.S. (1970). The Nature of Electrical Conductivity in the Ancient Basement. // Vestnik SPb Univ. No. 12. Pp. 19–26.

Semenov A.S. & Zhamaletdinov A.A. (1981). Deep Electrical Soundings. // SPb. Vestn. Leningr. Univ., Ser. 7: Geol., Geogr., Vol. 3, No. 18. Pp. 5–11

Vanyan L.L. (1997). Electromagnetic soundings. // Moscow. "Nauchny Mir". 218 p

Varentsov Iv.M., Engels M., Korja T., Smirnov M.Yu. and the BEAR Working Group. (2002). The generalized geoelectric model. // Fizika Zemli, No. 10. Pp. 64–105.

Vladimirov N.P. & Dmitriev V.I. (1972). Geoel. Sect. of crust and upper mantle in the Russian platform acc. to MTS. // Izvestiya Russ. Ac. of Sci. Phys. of the Solid Earth. No 6. Pp. 100–103.

Yardley B.W.D. & Valley J.W. (1997). The petrologic case for a dry lower crust. // Journal of Geophysical Research, V. N B6. Pp. 12173–12185.

Zhamaletdinov A.A. (1996). Graphite in the Earth's Crust and Electrical Conductivity Anomalies. // Izvestiya, Physics of the Solid Earth, Vol. 32, No. 4. Pp. 272–288.

Zhamaletdinov A.A. (1990). Model of electrical conductivity of lithosphere by results of studies with controlled sources (Baltic shield, Russian plateform). // Leningrad. "Nauka". 159 p.

Zhamaletdinov A.A., Shevtsov A.N., Tokarev A.D., Korja T., Pedersen L. Experiment on the Deep Frequency Sounding and DC Measurements in the Central Finland Granitoid Complex. // Electromagnetic Induction in the Earth. 14-th Workshop in Sinaia (Romania), 1998. P. 83

Zhamaletdinov A.A., Shevtsov A.N., Tokarev A.D., and Korja T. (2002). EMFS of the Earth Crust beneath the CFGC // Izvestiya, Physics of the Solid Earth, Vol. 38, No. 11. P. 954–967.

Zhamaletdinov A.A., Shevtsov A.N., T.G. Korotkova et al. (2011). Deep EM Sounding of the Lithosphere (FENICS). // Izvestiya, Physics of the Solid Earth. Vol. 47, No. 1. Pp. 2–22.

Zhamaletdinov A.A, The Largest in the World Anomalies of Electrical Conductivity and their Nature - a review.// Global Journal of Earth Science and Engineering, 2014, 1, 84–96

Zhamaletdinov A. A., E. P. Velikhov, A. N. Shevtsov, V. V. Kolobov, V. E. Kolesnikov, A. A. Skorokhodov, T. G. Korotkova, V. V. Ivonin, P. A. Ryazantsev, and M. A. Biruly. The Kovdor-2015 Experiment: Study of the Parameters of a Conductive Layer of Dilatancy–Diffusion Nature (DD Layer) in the Archaean Crystalline Basement of the Baltic Shield. // Doklady Earth Sciences. 2017. Vol. 474. Part 2. pp. 641–645.

Chapter 41
A Map of the Total Longitudinal Electric Conductivity of the Sedimentary Cover of the Voronezh Crystalline Massif and Its Framing

V. I. Zhavoronkin, V. Gruzdev, I. Antonova and Y. Austova

Abstract In the central part of the region, a digital version of the conductivity map of the sedimentary cover. When constructing the S map, electro-prospecting data was used by the VES method. In general, sedimentary rocks are characterized by a large range of variation in the specific electrical resistance (from 0.1 to several thousand Ω m). The values of S vary from Sm units in the VCM territory to the first thousand Sm in the neighboring valleys. In the arch part of anteclise, the values of the longitudinal conductivity are within the first tens of Sm. The increase in conductivity is observed along the Losevskaya suture zone. In the area of transition of the array to adjacent large-scale tectonic structures, the values of S are from 40 to 120 Sm. In the central part of the DDV and RSP, S is equal to 2000 Sm. However, there is an area with S = 3500–4000 Sm.

Keywords The Voronezh crystalline massif · Sedimentary cover
The total longitudinal conductivity · Geoelectric section

Over the past few years, the Department of Geophysics of the Voronezh State University has been studying the petrophysical characteristics of the sedimentary cover of the Voronezh anteclise and adjacent areas (Muravina et al. 2013). Within the framework of these works, a digital version of the sedimentary cover conductivity map was compiled in the central part of the region. A map of the total

---

V. I. Zhavoronkin (✉) · V. Gruzdev · I. Antonova · Y. Austova
Voronezh State University, Voronezh, Russia
e-mail: vzhavoronkin@yandex.ru

V. Gruzdev
e-mail: grumerr@rambler.ru

I. Antonova
e-mail: mavka_r@mail.ru

Y. Austova
e-mail: walera21111@yandex.ru

© Springer Nature Switzerland AG 2019
D. Nurgaliev and N. Khairullina (eds.), *Practical and Theoretical Aspects of Geological Interpretation of Gravitational, Magnetic and Electric Fields*, Springer Proceedings in Earth and Environmental Sciences, https://doi.org/10.1007/978-3-319-97670-9_41

longitudinal conductivity (S) of the sedimentary rocks of the Voronezh crystalline massif was used. It was compiled in 1982 in the Voronezh geophysical expedition. This map was supplemented by the results of subsequent studies (Muravina and Zhavoronkin 2014).

When constructing the S map, the results of electro-prospecting studies using the VES method were used, performed mainly in the Voronezh geophysical expedition. Electrical exploration using the VES method was carried out in different places on different scales (1:10,000–1:200,000). In some areas (near Voronezh) the data were obtained in 1–2 points at a distance of 10–20 km. The central part of the KMA was unevenly studied, where only a route survey was carried out at certain sites. However, the accuracy of interpretation of the VES, including the determination of parametric conductivity, is high, which is confirmed by drilling.

The value of the electrical conductivity of the sedimentary cover depends on the thickness of the sediments and their material composition. The thickness of the sedimentary cover is determined by the level of the day surface and the relief of the Precambrian base.

The surface in the regional plan is relatively smooth. The absolute level of the watershed spaces is 140–180 m (the Oksko-Don plain) and 220–260 m on the neighboring elevations. At some sites in the valleys of major rivers, there are gutters of the depth of tens of meters.

The relief of the crystalline basement has a much more complex form. In General, the crystalline massif is elongated in a North-westerly direction (azimuth 305°). The surface of crystalline rocks asymmetrical. The greatest dip is noted in the South-Western and South-Eastern slopes. The Northern slope is relatively flat. Crystalline massif articulates with negative structures. The amplitude of the displacements can reach considerable size and be accompanied by a gradient of conductivity zones of the sedimentary cover.

In the central part of anteclise, the relief of the soles of sedimentary formations ranges from 50–100 to 800–1000 m. In the axial part of the Ryazan-Saratov Trench, the thickness of the sediments reaches up to 4000 m. In the Dnipro-Don basin, the thickness of the sediments is 9000 m. This is reflected in the conductivity map.

In general, the relief of the central part of the crystalline massif changes. Against the background of flat areas, individual isometric depressions and ascents of various forms are recorded. This is especially true for sections with linear layers of ferruginous quartzites. These quartzites form crest-like uplifts up to several tens of meters in sedimentary deposits. Linear conductivity anomalies are associated with a decrease in the thickness of the sedimentary cover. The magnitude of such anomalies is the first tens of Cm.

The influence of the lithological composition of sedimentary formations on the regional component of electrical conductivity has not been studied to date. However, it should be expected that the difference in the real composition of locally distributed layers can affect the distribution of the electrical conductivity parameter.

In general, sedimentary rocks are characterized by a large range of variation in the specific electrical resistance (from 0.1 to several thousand Ω m).

The geoelectric section of the sedimentary cover is in the first approximation three-layered. The first layer in the arched part of the VCM is represented by the sediments of the Cenozoic. On the slopes of the crystalline massif and in the valleys it is represented by sediments of the mesocanozoic. Its power varies from 5–10 m to 200–300 m, and the specific electrical resistance varies from 20 to 100 Ω m. The second layer consists of rocks of Cretaceous age (the central part of VCM) or rocks of Carboniferous and Devonian age (slopes of the massif and valleys). Its power is from 10 to 500 m. It has a higher electrical resistivity (from 100 to 1000 Ω m). The third layer is associated with Paleozoic deposits. It has a power from several meters to 2000–4000 m and a specific resistance of 1–20 Ω m. In all layers, the resistivity decreases with depth due to an increase in the mineralization of solutions that saturate the rock.

The values of S vary from Sm units in the VCM territory to the first thousand Sm in the neighboring valleys (Gruzdev 2012). In the arch part of anteclise, the anomalies of the electrical conductivity S form mosaic forms, and the longitudinal conductivity values are within the limits of the first tens of Sm. There are several regions with values greater than 50 Sm. Often the region of conduction coincides with the contours of tectonic structures. For example, against a background of small values (2.5–10 Sm), zones with high conductivity values are observed in the center of the arched part of the VCM. These zones have a north-western extent. They coincide with Tim-Yastrebovskaya and Volotovskaya tectonic structures. A similar increase in conductivity is observed along the Losevskaya suture zone. The Mossic character of the distribution of the total longitudinal electrical conductivity is also observed in the Rossoshan massif. Here, against a background of small values (up to 10 Sm), there are anomalous regions (the first tens of Sm). To the east of the Losevskaya suture zone to the Khopersky megablock, the conductivity values gradually increase to 30 Sm. On the slopes of the massif, the isolines of longitudinal conductivity coincide with the isohypses of the surface of the crystalline basement. In the area of transition of the array to adjacent large-scale tectonic structures, the values of S are from 40 to 120 Sm. In the central part of the DDV and RSP, S is equal to 2000 Sm. However, there is an area with S = 3500–4000 Sm. The conductivity gradient in some places is 100 S/km.

Thus, the constructed digital map of total conductivity is of interest from the point of view of modeling the electrical conductivity of various layers of the earth's crust and upper mantle. If the anomalous features of the conductivity of the sedimentary cover are not taken into account, this can lead to significant errors in the interpretation.

Acknowledgements Work performed under the grant RFBR 16-05-00975a.

References

Gruzdev, V. (2012). Electrical conductivity of the territory of the VKM and adjacent regions. Lithosphere of the Voronezh crystal massif by geophysical and petrophysical data: monograph. Voronezh "Scientific Book", pp. 74–138.

Muravina, O. and V. Zhavoronkin (2014). Magnetic susceptibility of the Phanerozoic deposits of the Voronezh anteclise. Herald of KRAUNTS. Earth sciences. Vyp. No. 23. pp. 79–88.

Muravina, O., V. Zhavoronkin, and V. Glaznev (2013). Petrophysical characterization of the sedimentary cover of the Voronezh anteclise. Herald of the Voronezh state University. Series: Geology. - No. 1. pp. 189–196.

Chapter 42
Geophysical Monitoring for the Preservation of Architectural Monuments

Z. Slepak

Abstract Preservation of buildings and structures always was and remains to be an urgent problem for the mankind. Cracks and deformations of walls and subsidence of foundations eventually resulting in destruction of buildings are very common, and measures taken for their elimination are fairly expensive and often belated. All this makes it necessary to control the course of active geological processes, determine their nature, and take timely measures for prevention and mitigation of possible adverse impacts. Geophysical methods of measurement from the ground surface causing no impact on the state of geological environment should play an important role in preservation of buildings, towns and cities. At the same time, traditional measurements of this kind in urban conditions with limited space and numerous hindrances and interferences are often inefficient and inexpedient. A new methodology of geophysical exploration and studies is put forward and analyzed for investigating the negative influence of geological processes in the upper layer of the geological section on buildings and engineering structures.

Introduction

The importance of a survey of the technical condition of architecture monuments and other structures and the need for restoration and repair work was noted in the publications of many researchers (Aidarov 1978; Bakhireva and Rodina 1992; Vyazkova and Pashkin 1989; Zaitsev 1989 Pashkin 1975; Slepak 1997, 1999a, b, 2007; Slukin 1989).

The service life of buildings depends on their mechanical strength characteristics laid up during construction and, to a considerable extent, on the negative impact of geological processes on their foundations after construction. Such processes

---

Z. Slepak (✉)
Institute of Geology and Petroleum Technologies,
Kazan Federal University, Kazan, Russia
e-mail: Zakhar.Slepak@kpfu.ru

include, first and foremost, changes in hydro-geological conditions of ground waters and modern tectonic movements.

In urban conditions, the influence of these processes is intensified by technogenic impacts associated with water leaks and breakthroughs from underground water pipelines, floods, destruction of dams, etc. Apart from that, the established geological environment is disturbed by construction of facilities with extensive earthwork, archeological excavation, road surface replacement, and other works. Changes in directions of ground water flows and the course of karst, suffusion, and soil slip processes cause more soil subsidence and caving.

This methodology is based on high-precision geophysical measurements and monitoring of physical fields and their variation in time, which reflect the course of geological processes deep under the ground surface. The most efficient methods of geological monitoring are high-precision gravity measurements and transient electromagnetic sounding.

High-sensitivity equipment, specific measuring procedures, special data processing methods, and comprehensive geophysical, geodetic and archeological studies were used for practical implementation of the new geological exploration methodology.

The new methodology, first tested by the author on the territory of the Kazan Kremlin architectural ensemble, made it possible to investigate the course of geological processes both near and immediately underneath buildings and structures, determine their nature, and assess their influence on the foundations of buildings.

Research Results

The Kazan Kremlin standing on a high hill is an architectural complex of unique landscape and buildings. The top of the hill is occupied by architectural monuments, such as the Suumbeki Tower, former Governor Palace, Annunciation Cathedral, Spasskaya Tower, and other structures. The strike of the Kremlin hill is elongated in the northwest direction. From the north, it is bounded by the Kazanka river, which is now somewhat separated from the hill by a rock-fill dam with a road. The Kremlin hill base measures approximately 750×350 m. The highest elevations of the relief equal to 84 m above sea level correspond to the eastern periphery of the hill top (Fig. 42.1).

The Kremlin hill went through most intensive changes in recent years due to construction of the Kul-Sharif Mosque and other buildings, archeological excavations, repair and restoration works, replacement of old and laying of new underground communications, replacement of asphalt with block pavement, extensive excavation works, and operation of heavy construction machinery. All these works and activities disturbed the anthropogenic layer and changed the structure and strength of soil. At the same time, the activity of surface and ground waters intensified and changed. It is clear that one of the most important tasks now is to continue studies of the geological structure and composition of the Kremlin hill and geological processes in its subsoil.

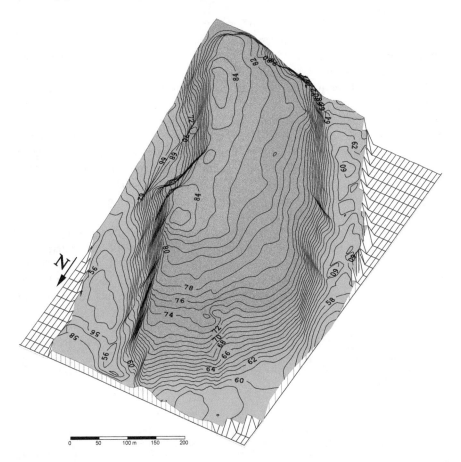

Fig. 42.1 Vertical sections of total electrical conductivity S(H) along profiles on the site of Suyumbeki Tower and the Governor's Palace

The first gravimetric measurements in monitoring mode were carried out in the northern part of the Kremlin hill, on the territory around the Suumbeki Tower and former Governor Palace (now residence of the President of the Republic of Tatarstan). These measurements consisted in two high-precision horizontal gravimetric surveys conducted at the same points of a regular 5 × 5 m grid. The accuracy of measurement of Bouguer anomalies at the survey grid points attained by multiple measurements at every grid point (up to 6–10 measurements in independent series) did not exceed 10 μgal. The results of these surveys were used for construction of maps of gravity anomalies with an isoanomaly interval of 30–40 μgal and detect local field changes caused by active geological processes in the earth crust. Two maps of Bouguer gravity anomalies and anomaly patterns along longitudinal profiles and transverse profiles were constructed according to the results of these gravimetric surveys. The Bouguer anomaly map based on the results of survey 2 is very similar to the map based on the results of survey 1. At the same time, noticeable changes are observed in the strike of certain isoanomaly curves, which points to changes of the gravitational field that took place in the period between the two series of gravimetric surveys.

The map of differences in Bouguer anomalies based on the results of measurements performed at the same points during the two independent surveys conducted in the spring and fall of 1995 makes it possible to state more confidently that the gravitational field changed between the two surveys (Fig. 42.2). In the central part of the surveyed area, a positive anomaly of north-south strike outlined by isoanomaly curve +10 μgal is traced together with several relatively small local anomalies of similar intensity (Fig. 42.2). Gravity changes like those occurred in the period between the two series of surveys are traced in even greater detail along the transverse and longitudinal profiles (Slepak 2007).

The non-tidal gravity anomaly of the positive sign in the central part of the study area (Fig. 42.2) in Fig. 42.3 almost disappeared. Its "remnants" delineated by an isoline of 10 μGal, were preserved only to the northeast of the Syuyumbeki Tower and the gravitational field was "restored". This indicates the loss of water. It follows that the positive anomaly of non-tidal gravity changes can correspond to the natural

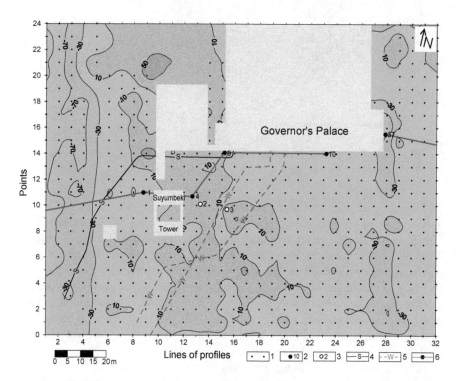

Fig. 42.2 Map of non-tidal gravity changes. Spring 1995–autumn 1995: 1—points of observation, 2—wells drilled before 1999, 3—wells drilled in 1999, 4—sewerage, 5—water conduit, 6—line of geological profile

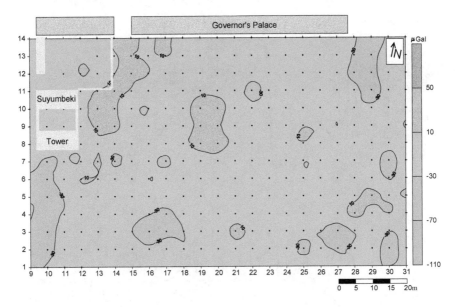

Fig. 42.3 Map of non-tidal gravity changes. Spring 1995–summer 1996

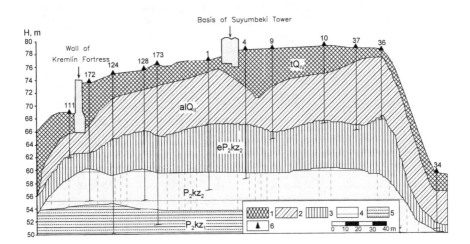

Fig. 42.4 Geological section through the Kremlin Hill: 1—cultural layer, 2—alluvial sediments, 3—siltstone-carbonate dust, 4—dolomite, limestone, 5—sandstone, 6—wells

reservoir of ground and technogenic waters accumulation that could migrate towards the Kazanka River and the lower layers of the Kremlin Hill.

The geological profile that crosses the Kremlin Hill near the Syuyumbeki Tower and the Governor's Palace is characterized by a five-layered structure (Fig. 42.4). The maximum depth of the upper anthropogenic layer at the eastern side of the

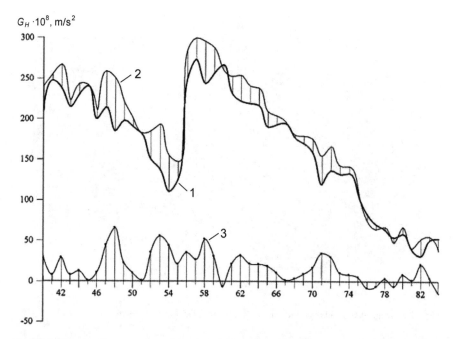

Fig. 42.5 Curves of measured gravity values and non-tidal gravity changes at the eastern wall of Kazan Kremlin along profile 11: curves of gravity values measured in: 1—December1997; 2—December 1998; 3—curve of non-tidal gravity changes between the two series of measurement

Syuyumbeki Tower is 7 m and corresponds to the local anomaly observed in Fig. 42.2.

It is established that ground and technogenic water are the main cause of the tilt (almost 2 m) of the Tower. Also, the negative impact of groundwater on other buildings and fortress walls has been studied and measures for their elimination have been recommended (Fig. 42.5).

The comparison of the measured gravity values graphs shows that gravitational field along the profile 11 in the section of the fortress wall during this period has significantly increased, especially between pickets 47–72, which is due to the accumulation of groundwater at its base in front of the steep slope from the inside walls.

Similar results of gravimetric monitoring were obtained at other sites of the Kazan Kremlin walls.

Also, useful data were obtained on the results of gravimetric measurements in the monitoring mode inside the Taynitskaya Tower, the Governor's Palace and the travel part of the Syuyumbeki Tower.

The presence of a natural water reservoir at the top of the Kremlin Hill (Figs. 42.1 and 42.2) confirmed by the electromagnetic sounding. Changes in the thickness of anthropogenic layer can be seen on the maps of the total electric conductivity S(H), which correspond to the boundaries between different geological strata (Slepak 2007).

The most elevated part of the bottom of anthropogenic layer in absolute depths (in the axonometric projection) is located under the southern pylon of the Syuyumbeki Tower (Fig. 42.6).

The surface is dipping in the northeast direction and forms a deflection, which can be traced near the Governor's Palace, located on the northern slope of the Kremlin Palace. Thus, the results of electromagnetic sounding not only confirmed the existence of an underground reservoir of ground and technogenic waters, identified by the high-precision gravimetric measurements in the monitoring mode, but also allowed to specify its location and direction of strike.

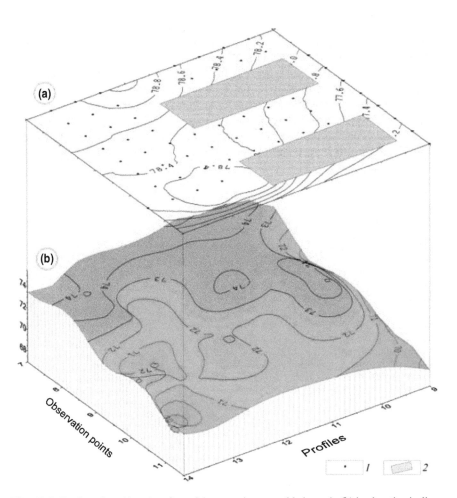

Fig. 42.6 Earth surface (a) and surface of the water-impermeable layer 4a (b) in elevation isolines above sea level according to the results of electromagnetic sounding on the territory of the Suumbeki Tower: 1—electromagnetic sounding points; 2—contours of Suumbeki Tower pylons

Conclusion

The main results of the research are a survey of the technical state of architectural monuments in the city, without violating the foundations of buildings and ecology.

Geophysical monitoring helped us study the variations of physical fields caused by active geological processes. Gravimetric monitoring does not require proper account of the gravitational effects of the relief, buildings and structures. That use of original surveying methods and modern geological data processing technologies makes it possible to study the influence of hydro-geological conditions on architectural monuments and solve various geological and geophysical problems

Geophysical surveys in monitoring mode including high-precision gravimetric surveys and transient electromagnetic sounding turned out very efficient for preservation of architectural monuments of the Kazan Kremlin. The results were demonstrated on an example of the Suumbeki Tower—a unique monument of ancient architecture. The dynamics of long-term negative influence of ground waters and technogenic waters on the foundation of this tower was studied by analyzing the variations of physical fields on the territory of the Suumbeki Tower. The influence of these waters was found to be the main reason behind the tilt of the Suumbeki Tower. The negative impact of ground waters on other structures and fortification walls was studied, the factors of this impact were analyzed, and recommendation for their elimination were worked out.

References

Aidarov S.S. (1978) Architectural Heritage of Kazan, Kazan, 1978. P. 96.

Bakhireva L.V. and Rodina E.E. (1992) Geological engineering surveys for preservation of architectural and historical monuments in urbanized territories (examples of foreign experience) // Inzhenernaya Geologiya, 1992, No.6, pp. 121-127.

Pashkin E.M. (1975) Influence of anthropogenic changes of rocks on integrity of architectural monuments // Monuments of Russia, Moscow, Sovremennik, 1975, pp. 158-163.

Slepak Z. (1997) Complex geophysical investigations for studying the cultural layer and remains of ancient buildings in the territory of Kazan Kremlin, Kazan, Republic of Tatarstan, Russia. // Archaeological prospection, 1997, vol. 4, pp. 207–218. John Wiley & Sons, Ltd.

Slepak Z.M. (1999) Electromagnetic sounding and high-precision gravimeter survey define ancient stone building remains in the territory of Kazan Kremlin (Kazan, Republic of Tatarstan, Russia). // Archeological prospection, 1999, vol. 6, pp. 147–160, John Wiley & Sons, Ltd.

Slepak Z.M. (1999) Geophysical monitoring for preservation of architectural monuments on an example of the Kazan Kremlin, Kazan: Kazan State University Press, 1999, P. 176.

Slepak Z.M. (2007) Urban Geophysics – Tver, GERS Publishing Hous, 2007, 240 pages.

Slukin V.M. (1989) Non-destructive engineering methods for surveys of historical territories and architectural monuments and buildings // Engineering and technical aspects of preservation of historical and architectural monuments, Moscow, Izd-vo NMS MK SSSR, 1989, pp. 132-144.

Vyazkova O.E. and Pashkin E.M. (1989) Engineering geology and preservation of historical and architectural monuments, Moscow, Izd-vo NMS MK SSSR, 1989, pp. 6–15.

Zaitsev A.S. (1989) Application of geophysical methods for assessing the integrity of architectural monuments // Subsoil Exploration and Protection, 1989, No. 10, pp. 58–60.

Chapter 43
Application of Detailed Magnetics in Intensive Industrial Noise Conditions

P. N. Novikova

Abstract The use of detailed magnetics for the localization of underground communications and engineering-geological wells is considered. Special technique aspects of field survey and data processing in the intensive industrial noise conditions are presented. The amplitude-frequency spectrum of technogenic noise is analyzed by means of empirical mode decomposition (EMD) decomposition. The qualitative interpretation results of detailed magnetics are revealed when different types of communications and wells are detected.

Keywords Detailed magnetics · Intensive industrial noise · Empirical mode decomposition (EMD) · Underground pipeline · Well

The study of soils by geophysical methods is an effective approach used in engineering and geological surveys for construction purposes, environmental studies, etc. One of the specific tasks of the urban and industrial areas survey is the current project documentation restoration related to the detection and localization of underground communications and structures. Also, problems of atypical anthropogenic objects searching, such as running and liquidated wells, sunk production equipment, etc., and anthropogenic objects that are have a historical value is appeared. Timely diagnosis of modern cities engineering systems is an important task to prevent dangerous situations and accidents, both at the stage of facilities construction and of their operation process. At the moment, georadiolocation methods, detailed geoelectrics, active and passive electric field recording (locators), acoustic location and infrared thermography are used for these purposes.

In this paper propose an alternative approach to the urban areas geomonitoring—detailed magnetics, tested in several experimental field works.

Performing field measurements of the magnetic field within urban areas is complicated by the presence of intense electromagnetic noise of various nature. A qualitative analysis of the magnetic field variations within the study sites showed

P. N. Novikova (✉)
Mining Institute UB of RAS, Perm, Russia
e-mail: polinagfz@gmail.com

that the measured signal represents a set of rapidly changing low-frequency components (Novikova and Voroshilov 2017; Novikova et al. 2017; Utkin et al. 2010; Fomenko et al. 2016). Thus, for the successful method application in intense industrial noise conditions in the standard processing of magnetic data, it is necessary to place the magnetovariational station as closely as possible or directly at the measurement site with a minimum set of magnetometers during field works (Novikova and Voroshilov 2017).

The modified field survey technique can not fully account for the entire spectrum of industrial noise, so the Empirical Mode Decomposition (EMD) method was used to filter the observed data. EMD is a method of spectral analysis of non-stationary signals using predetermined internal signal modes. In practice, for such a transformation, the function should consist of a series of extrema, along which two signal envelopes are formed—along the maxima and minima. Internal oscillations are defined as the average value for signal envelopes and represent an adaptive basis with variable frequency and amplitude, functionally dependent on the content of the data itself (Fig. 43.1). Ultimately, in an iterative procedure, a complex signal can be represented as the sum of empirical mode functional dependened on the residual trend component (Davydov and Davydov 2010; Dolgal and Hristenko 2017).

Usually the measured magnetic variations contain up to 20 internal close to periodic modes, whose amplitude ranges from 20 to 400 nT, the frequency varies within the range (0.56–133) 10^{-3} Hz (Table 43.1). Figure 43.2 shows the most commonly encountered forms of magnetic field variations within the urban environment, obtained by EMD-decomposition. Comparison of EMD-components of profile and variations measurements makes it possible to obtain adequate data reflecting the magnetic field of near-surface sources.

A qualitative data interpretation of the detailed magnetics is also of special interest. According to the magnetic anomalies morphology, it is possible to separate sources into different technogenic objects types. Below the results of practical application of the described technique for detection of underground engineering infrastructure tasks are offered.

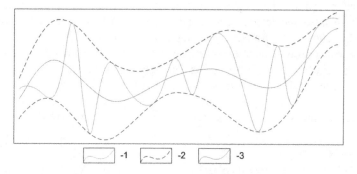

Fig. 43.1 Empirical mode decomposition of nonstationary signal: 1—an analytical signal; 2—signal extremum envelopes; 3—average signal—empirical model

Table 43.1 An example of the internal mode frequency composition of the magnetic variations within the urban environment

Mode	Frequency (10^{-3} Hz)	Maximal amplitude (nT)
1	133.3	±375
2	71.4	±400
3	26.3	±375
4	24.4	±210
5	7.1	±200
6	5.46	±75
7	1.9	±100
8	1.43	±200
9	1.48	±50
10	0.56	±40
11	0.56	±30
12	0.56	±25

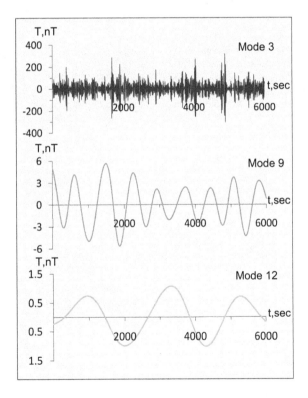

Fig. 43.2 Empirical modes of magnetic variations within the urban environment

In the case of linear underground communications detection, the following types of magnetic anomalies can be classified. High-voltage electric networks are sharply allocated, which are displayed in most cases by intense negative anomalies of several tens of thousands of nT. Low-voltage cable lines for various purposes are displayed in the magnetic field with dynamic, rapidly changing in time both positive and negative anomalies. In this case, the sign of the anomaly ΔTa is usually conserved, and the amplitude can vary by several hundreds and even thousands of nT. Also, along the electrical cable line, there may be separate "emissions" exceeding the total level of the pulsating anomaly (Fig. 43.3a, b).

Water and sewerage pipelines on the map ΔTa are traced by linear positive anomalies with a varying amplitude in the range (200–500) nT. The most intense anomalies are observed over welds and at the pipelines intersection with cable lines. This type of magnetic anomalies is rather difficult to trace by existing methods of recognizing linear structures, therefore at this stage of the investigation the detection of such anomalies was made visually by maps and graphs of an anomalous magnetic field (Fig. 43.3a, c). It is worth noting that most magnetic anomalies are localized in space within 1 m.

As a further example, anomalous magnetic field fragments are presented in the study sites containing the active (Fig. 43.4a) and abandoned (Fig. 43.4c) engineering-geological wells.

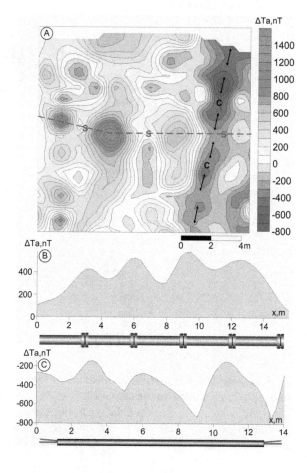

Fig. 43.3 Fragment of anomalous magnetic field containing underground engineering networks effects (**a**) and ΔTa graphics above sewer net (**b**) and low-voltage conductor cable

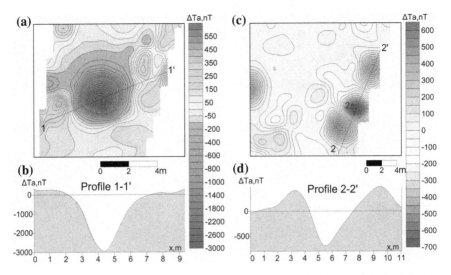

Fig. 43.4 Anomalous magnetic field and ΔTa graphics on running (**a, b**) and abandoned (**c, d**) engineering-geological wells

Magnetic anomalies from wells are extremely localized, isometric, can be both positive and negative and reach several hundred nT. If the bottom hole is at a relatively shallow depth, then there are anomalies accompanying the opposite sign. Sewage wells with cast iron hatches, lampposts, as well as household and construction "magnetic" debris also have a similar form. Therefore, in the case of a search for wells, such "false" anomalies can be observed, which must be eliminated by special filtration methods and direct verification.

The above studies demonstrate the possibility of using detailed magnetic prospecting in conditions of intense technogenic jamming for localization of linear and limited technogenic objects with a special technique of field survey and initial data filtering by means of EMD-analysis. The possibility of recognizing the various linear types of engineering networks based on the magnetic anomalies morphology is shown.

References

Davydov, V.A. and Davydov, A.V. (2010). Purification of geophysical data from noise using the Hilbert-Huang transformation. Electronic scientific publication «Aktual'nye innovacionnye issledovanija: nauka i praktika» . №1.

Dolgal, A.S. and Hristenko, L.A. (2017). Application of empirical mode decomposition in the processing of geophysical data. Izvestija Tomskogo politehnicheskogo universiteta. Inzhiniring georesursov. T. 328. № 1. pp. 100–108.

Fomenko N.E., Zhurbickij B.I., Fomenko L.N. (2016). Prediction of electromagnetic pollution of urban areas using special hardware and software. Geofizicheskie metody pri razvedke nedr: materialy Vserossijskoj nauchno-prakticheskoj konferencii s mezhdunarodnym uchastiem, posvjashhennoj 70-letiju osnovanija v Tomskom politehnicheskom institute kafedry «Geofizicheskie metody poiskov i razvedki mestorozhdenij poleznyh iskopaemyh». Tomskij politehnicheskij universitet. Tomsk : Izd-vo Tomskogo politehnicheskogo universiteta. pp. 283–288.

Novikova, P.N. and Voroshilov, V.A. (2017). Detection of underground infrastructure in conditions of intensive industrial noise on magnetic field data. Geophysics. № 5. pp. 4–9.

Novikova, P.N., Voroshilov, V.A., Kopytin, V.V., Subbotin, P.A., Kalashnikova, M.M., Temirov, P.A. (2017). Detection of underground communications by engineering magnetic prospecting in conditions of anthropogenic origin noise. Vosemnadcataja ural'skaja molodezhnaja nauchnaja shkola po geofizike: Sbornik nauch. Materialov. Perm: GI UrO RAN. pp. 147–151.

Utkin, V.I., Tjagunov, D.S., Sokol-Kutylovskij, O.L., Senina, T.E. (2010). Distortion of magnetic field by electromagnetic noise of low frequencies of technogenic origin. Vestnik KRAUNC. Serija: Nauki o Zemle. Vyp. 15. № 1. pp. 216–222.

Chapter 44
The Results of Numerical Simulation of the Electromagnetic Field Within the Voronezh Crystalline Massif and its Framing

V. Gruzdev and I. Antonova

Abstract On the territory of the Voronezh crystalline massif the results of numerical simulation of the electromagnetic field are obtained. The resistivity curves contain information on regional changes in the conductivity of the sedimentary cover of the VKM and adjacent structures. Areas of minimum and maximum values are identified, as well as the character of the change in the apparent resistance along different profiles, secant structures. The estimation of galvanic and induction distortions of apparent resistance curves in different VCM zones is given. Attention should be paid to the large errors in the interpretation of the experimental MTZ curves within the framework of one-dimensional models, especially in the central part of the BKM. Therefore, it is necessary to take into account the two-dimensional distribution of conductivity in the structures of VCM, RSP and DDV in interpreting the MTZ curves.

Keywords Voronezh crystalline massif (VCM) · Numerical simulation of electromagnetic fields · Apparent resistivity curves · The total conductivity of sedimentary cover

As a result of numerical modeling of the electromagnetic field, curves of apparent resistivity ρ_a^M were obtained in the range of periods from 900 to 43,200 s for various points on the territory of the Voronezh crystalline massif (VKM), the Ryazan-Saratov trough (RSP) and the Dneprovo-Donetsk depression (DDV) (Gruzdev 2012). Model curves ρ_a^M contain information about the regional changes of conductance of the sedimentary cover of all considered structures. Model curves ρ_a^{XYM}, ρ_a^{YXM} of apparent resistivity along a profile intersecting structures of VCM,

---

V. Gruzdev (✉) · I. Antonova
Voronezh State University, Voronezh, Russia
e-mail: grumerr@rambler.ru

I. Antonova
e-mail: mavka_r@mail.ru

© Springer Nature Switzerland AG 2019
D. Nurgaliev and N. Khairullina (eds.), *Practical and Theoretical Aspects of Geological Interpretation of Gravitational, Magnetic and Electric Fields*, Springer Proceedings in Earth and Environmental Sciences, https://doi.org/10.1007/978-3-319-97670-9_44

RSP and DDB in their Central parts are shifted on the y-axis under the influence of the conductive sedimentary cover of 4 orders of magnitude. At the central part of the VCM the resistivity curves ρ_a^{XYM}, ρ_a^{YXM} are maximally displaced along the ordinate axis to values of 10,000 Ω m. The values of the longitudinal apparent resistivity ρ_a^{XYM} 10 times higher than the values of the transverse apparent resistivity ρ_a^{YXM}. Within the central part and slopes of the Voronezh crystal massif the resistivity curves ρ_a^M are shifted along the vertical axis by 2 orders of magnitude. The minimum values of the apparent resistivity ρ_a^M are observed in the central part of the РСП and DDV (up to 0.2 Ω m). In this case, the values of the longitudinal resistivity curves ρ_a^{XYM} are 10 times smaller than the values of the transverse resistivity curves ρ_a^{YXM}. For periods of 900–432,00 s in the central part and on the slopes of the VCM there is a weak dependence on the period of the magnitude of the displacement of the resistivity curves ρ_a^M along the ordinate. This indicates a slight effect on the apparent resistivity curves of regional induction effects. Whereas in the Central part of the Ryazan-Saratov trough and Dniprove-Donets basin, these distortions become significant.

When carrying out numerical simulation, a number of simplifications of the model were made, which naturally affect the quality of the interpretation of the experimental data.

First, the calculations did not take into account the leakage of electric currents from the sedimentary cover to the crystalline part of the crust. Thus, the conductivity of the upper part of the crystalline basement was not taken into account. What for the territory of the VCM is essential. As a consequence, the value of the calculated resistivity curves ρ_a^M does not reflect their actual position, but only the displacement along the ordinate axis as a result of the presence of a regional galvanic effect due to the sedimentary cover.

Secondly, averaging of the values of S was carried out over a rather large cell (35 × 35 km) and local variations in the longitudinal conductivity of the sedimentary cover in the calculations were not taken into account. The values of S differ significantly in the arched part of the VCM and changed in the range from 2.5 up to 20–30 Sm. Therefore, local galvanic and inductive distortions of the electromagnetic field in the examined structures, this modeling does not take into account.

The values ρ_a^{XYM} for T = 1800 s along the profile vary from fractions of units of Ω m (RSP, DDD) to over 10000 Ω m. There is also strong differentiation and large gradients ρ_a^{XYM} in VCM junction areas with RSP and DDB, as well as in the central part of the VCM. For T = 1800 s the values ρ_a^{XYM} and ρ_a^{YXM} change more smoothly along the profile, which crosses the tectonic structures (from tens of Ω m to hundreds of Ω m). It should be noted a significant difference between values ρ_a^L for one-dimensional local models and ρ_a^{XYM}, ρ_a^{YXM} for the two-dimensional model. This fact indicates large errors in the interpretation of the experimental MTZ curves in the framework of one-dimensional models, especially in the central part of the VCM. Two-dimensional distribution of conductivity within the structures of VCM, RSP and DDV in the interpretation of the MTZ curves must be considered.

The results of numerical simulation of film allowed us to assess the effect of sedimentary cover on the low-frequency branch of the experimental curves ρ_a. Most of the observed curves ρ_a within the VCM and its slopes in varying degrees are lower level compared to the model curves (Zakutsky 1984). This can be explained by local galvanic effects due to the influence of well-conducting sediments of the upper part of crystalline basement. The individual observed curves ρ_a in the low frequency region is similar in level to model curves. Consequently, they do not contain local galvanic distortion. It is of great importance for interpretation of experimental curves ρ_a. This allows them to be divided into distorted and not distorted by local galvanic effects.

Acknowledgements Work performed under the grant RFBR 16-05-00975a.

References

Gruzdev, V. (2012). Electrical conductivity of the territory of the VKM and adjacent regions. Lithosphere of the Voronezh crystal massif by geophysical and petrophysical data: monograph. Voronezh "Scientific Book", pp. 74–138.

Zakutsky, S. (1984). On the results of magnetotelluric sounding on the Voronezh crystalline massif. Korovye anomalies of electrical conductivity. - L., pp. 90–100.

Chapter 45
Predural Depression Structures in the Arctic Urals Magnetic Field

V. A. Pyankov and A. L. Rublev

Abstract We have investigated the morphology of the negative regional magnetic anomaly of Arctic Urals. This magnetic anomaly spatially coincides with structures of Predurals depression. In this paper we applied the method of construction 3-D surfaces using magnetic data. We suggest modified iterative local correction method for solving structure magnetic inverse problem. The 3D-magnetic model of the Arctic segment of the Earth crust has been constructed. The main elements of subsurface structure at southern latitudes are preserved.

Keywords Magnetic data · Depression · Local correction method

Introduction

Crustal structure of the Arctic sector of Urals fold system is of great interest due to the complex nature of the positional relationship of geological structures of the first order. The greatest interest, in our opinion, is the study of the deep structure of Predurals depression. This structure is the reference for the entire length of the Urals fold system. In a magnetic field, it is manifested in the form of regional submeridional negative magnetic anomaly rather complex morphology. This is apparently due to the fact that there are certain differences in the geological structure of the different segments of the Urals fold system. The Arctic Ural segment is characterized by a complex mosaic geological structure, which is reflected in the magnetic field. In the deep structure of arctic part of the Urals, there are certain similarities and differences. Similarity is in the first place, in the presence of a single sequence general structural elements. The most important sources of complex geological and geophysical data on the deep structure of the Urals fold system are the Arctic Ural transects (Rybalka et al. 2011), and magnetic data intersecting the Northern Urals at different altitudes profiles aeromagnetic survey (Shapiro et al. 1993). The need to pass several refer-

---

V. A. Pyankov · A. L. Rublev (✉)
Institute of Geophysics, Ural Branch of RAS, Ekaterinburg, Russia
e-mail: a.roublev@list.ru

ence profiles was called as a desire to create a three-dimensional model of the Urals fold system, and the existence of certain differences in the geological structure of the main fragments of the Urals.

Restoration of the Layer Surface on the Magnetic Data

We introduce the rectangular Cartesian coordinates with axis Z pointing downwards and plane XOY coinciding with the surface of the observations. We consider the two-layer model of the earth's crust. Surface separating the upper and lower layers at a sufficient distance from the center of area goes to the asymptote. The vertical component of the magnetic induction Z at the point (x, y) on the ground surface of the contact surface, separating the layers with vertical magnetization, I1 and I2 is calculated by the formula:

$$Z(x,y) = \Delta I \int_{-\infty}^{+\infty} \int_{-\infty}^{+\infty} \left(\frac{z(x,y)}{\left[(x-x')^2 + (y-y')^2 + z^2(x,y)\right]^{3/2}} - \frac{H}{\left[(x-x')^2 + (y-y')^2 + H^2\right]^{3/2}} \right) dxdy, \quad (45.1)$$

where $z(x, y)$—the equation of the surface separating the upper and lower layers, the ΔI—magnetization jump at the boundary layers, H—horizontal asymptote.

To solve this equation and find the function $z(x, y)$ programs have been developed based on the modified method of local corrections (Martyshko and Prutkin 2003). Method of local corrections were proposed for the approximate solution of nonlinear inverse problems of gravimetry (Prutkin 1986) and is based on the assumption that a change in the value of the field at some point most affected by the change in the nearest to a given point of the surface S, which is the boundary between the two layers with different physical properties.

We have developed an iterative method of finding the borders of magnetized layers defined by the equation $z = z(x, y)$. At each step, an attempt is made to reduce the difference between the given and the approximate values of the field at a given point is only due to changes in the values of the required function at the same point. Discretization of Eq. (45.1) leads to the following system of nonlinear equations:

$$c \sum_i \sum_j K_{i_0 j_0}(z_{ij}) = U_{i_0 j_0}, \quad (45.2)$$

where c—a weighting factor of the cubature formula, $U_{i_0 j_0} = \Delta Z(x_{i_0}, y_{j_0}, 0)$—left-hand side of Eq. (45.1), $z_{ij} = z(x_i, y_j)$, $K_{i_0 j_0} = K(x_{i_0}, y_{j_0}, x_i, y_j, z_{ij})$—the integrand.

As a result, we received an iterative formula for finding z_{ij}^{n+1}: $\left(z_{ij}^{n+1}\right)^2 = \frac{\left(z_{ij}^{n}\right)^2}{1+\alpha\left(z_{ij}^{n}\right)^2\cdot\left(U_{ij}-U_{ij}^{n}\right)}$, where α—regularization parameter, $\{z_{ij}^{n}\}$—the values of the unknown function $z(x, y)$, n—the number of iteration.

Simulation of the Magnetization Distribution in the Crust of the Arctic Segments of the Urals Fold System

To construct surfaces of separation necessary to know the parameters of the asymptotic surface and the estimated magnetization of layers. It is believed that the first layer is practically non-magnetic $I_1 = 0.1$ A/m (Shapiro et al. 1993), and the magnetization of the next layer is produced from a study of data interpretation of the magnetic observations profile. To this end, in the Urals within the trapezoid 52–72°E * 52–68°N special precision aeromagnetic survey on 8 latitude geotraverses and two submeridional profiles at altitudes of 150–4000 m was carried out (Shapiro et al. 1993). The main purpose of this survey was to study the regional structure of the magnetoactive layer of the lithosphere of the Urals and adjacent regions. The final stage of this research was the interpretation of the anomalous magnetic field.

Interpretation of magnetic anomalies was carried out according to the method developed at the Institute of Geophysics, UB RAS (Martyshko and Prutkin 2003). According to this technique, anomalous magnetic field of each profiles approximated by singular sources fields that allowed to allocate the sources that generate the regional component of the magnetic field. As a result of research 2D-magnetic models for all profiles have been constructed. For the western margin of the Urals intensive depression under the folded Urals extended object with magnetization $I_2 = 0.7–1$ A/m was noted. Carried out researches of the features of the magnetization distribution of rocks with depth show that the magnetization much larger 1–2 A/m for large blocks improbably To define the parameters of the asymptotic surface seismic data of Arctic Urals transects are used. Arctic Urals transect has a length of 300 km and crosses the main structure of the Urals fold system. The western boundary of the transect is located in Predurals depression, further east it crosses the plate packs of Lemva allochthon (Fig. 45.1) (Rybalka et al. 2011).

To divide the long—and short-wave components of the amplitude spectrum of anomalies, geophysicists utilize numerical methods of field simulation at various altitudes (Martyshko et al. 2015a). In order to eliminate the influence of surface sources, we used the method of upward and downward magnetic data continuation. In our paper we investigate the magnetic field for a rectangular area of 62–64°E * 65–67°N. The software implementing parallel algorithms of altitude transformations are developed using the MVAPICH2 library in Fortran programming language. The parallel algorithms for recalculation were implemented in the

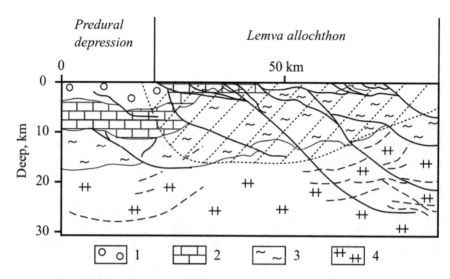

Fig. 45.1 Schematic geological-geophysical model of the western segment of the Polar Urals transect. Legend: 1—Permian and early Mesozoic deposits of Predurals depression, 2—shallow terrigenous-carbonate formation, 3—Late Proterozoic volcanic and sedimentary formations, 4—gneisses and amphibolites of pre-Riphean basement projections

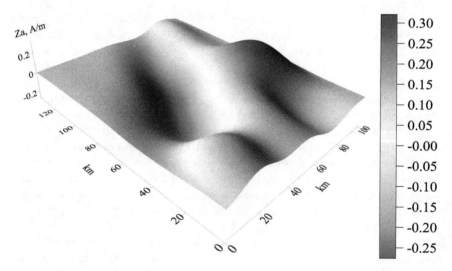

Fig. 45.2 Magnetic field obtained by upward and downward continuation to 10 km

multiprocessor computing machine for 512 × 512 points of the mesh using 512 cores (processors) (Martyshko et al. 2015b).

In Fig. 45.2 a map of Z component of the magnetic field obtained by upward continuation to 10 km, and then downward continuation to the zero level is shown.

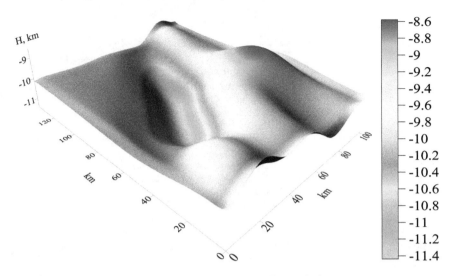

Fig. 45.3 The contact surface, calculated by local correction method

For recalculated field Z 3D-magnetic inverse problem for of the contact surface with the asymptote $H = 10$ km and jump in the magnetization of 1 A/m by local corrections method is solved. The depth to the asymptote is selected based on the analysis of geological and geophysical structure of the crust of the Arctic Transect.

By analyzing of the contact surface morphology (Fig. 45.3), we conclude that the depth to the surface increases to the south-east by a few kilometers, i.e., dipping of the contact surface under the Ural is observed. Apparently, thus subduction zone manifests in a magnetic field.

Conclusions

Our investigations have shown:

1. Analysis of the morphology of the regional negative magnetic anomaly for the Arctic Urals characterizes mainly a single structure Predurals depression.
2. As a result of solving the magnetic inverse problem by local corrections method 3D-model of the distribution of the magnetization of the Arctic Urals crust was obtained. As a consequence, it is shown that the contact surface separating nonmagnetic and magnetic rock is inclined to the south-east under the Urals. Thus, the dipping of the Timan-Pechersk plate to the south-east is observed (subduction zone).

Acknowledgements The study is funded by the RAS Institute of Geophysics, UB RAS (project 18-5-5-23) and carried out in the Institute of Geophysics, UB RAS.

References

Rybalka A.V. et.al. (2011) The deep structure of the Urals based on data of Polar-Urals transect, Regionalnaya geologiya I metellogeniya, Russia, pp 25–36.

Shapiro V.A. et. al. (1993) The structure of the magnitoactive layer of the Northern Urals based on the geomagnetic data, Dokladi Akademii nauk, Geofizika, Russia, vol.330/issue 6, pp 771–777.

Martyshko P.S. and Prutkin I.L. (2003) Method of separation of gravity field sources at depth, Geophys. Zh, (In Russian), pp 159–170.

Prutkin I.L. (1986) The solution of three-dimensional gravimetric problem in the class of contact surfaces by the method of local correction, Physics of the Solid Earth, vol. 22, issue 1, pp 49–55.

Martyshko P., Pyankov V., Rublev A. (2015a) Manifestation of the predurals depression structures in the magnetic field of the Arctic Urals // 15th International Multidisciplinary Scientific GeoConference SGEM 2015, Conference Proceedings, Albena. Bulgaria. Book1 Vol. 3, 887–894 pp. https://doi.org/10.5593/sgem2015/b13/s5.115.

Martyshko P.S., Pyankov V.A., Rublev A.L. (2015b) The new technique of solving the inversemagnetic problem for Denezhkin Kamen dunite-gabbro massif // XIVth International Conference – Geoinformatics: Theoretical and Applied Aspects. Kiev, Ukraine. https://doi.org/10.3997/2214-4609.201412427.

Chapter 46
Results of the Complex Airborne Geophysical Survey in the Central African Ridge Area

Yu. G. Podmogov, J. Moilanen and V. M. Kertsman

Abstract We review the results of different-scale airborne geophysical survey performed in the territory of the Republic of Rwanda. Such parameters as the modulus of the magnetic field vector, electromagnetic sounding data in time and frequency domain, as well as data from a gamma-ray spectrometer with a total scintillator volume of 32 L were continuously recorded during the flight. The software automatically performed the necessary navigation management on a real-time basis. Automated execution of flight tasks allowed to exclude an operator from the surveying process. It took only 6 months of field works to perform the survey at a scale of 1:50,000 over the area of 26,000 km^2, as well as detailing works. The total workscope amounted to 57,718 line km. Maps of anomaly magnetic field and its local part, apparent resistivity maps for the most informative frequency and time channels, Th, U, K concentration maps and Th-K ratio maps have been prepared for a short period of time. Based on the results of map interpretation we selected prospective sites for further study. Detailed airborne survey was performed for 5 sites, and ground geological and geophysical survey was performed for 3 sites. Based on the detailed survey results we selected new local objects associated with quartz-magnetite veined mineralization, dyke bodies, and mafic rock intrusions. Multiple pegmatite intrusions were identified. The information about magnetic, electrical and radiometric characteristics of rocks allowed to reliably classify the identified anomaly objects.

Keywords Electromagnetics · Magnetometry · Gamma-ray spectrometry
EQUATOR

Yu. G. Podmogov
LLC«Geotechnologies», Moscow, Russia

J. Moilanen (✉)
ICS RAS, Moscow, Russia
e-mail: info@geotechnologies-rus.com

V. M. Kertsman
Geological Faculty of Lomonosov MSU, Moscow, Russia

Introduction

Airborne geophysical survey began in Rwanda using EQUATOR (Fig. 46.1) technology (Felix et al. 2014, Karshakov et al. 2017) by 12th of October 2016. It was fully completed by 12th of April 2017. A survey at a scale 1:50,000 and infill works was completed on a square of 26,000 km^2. Total survey volume is 57,718 line km. About half of the survey time was in the rain season. And half of the survey area are situated in the rough terrain conditions. Average productivity of airborne geophysical survey is 9620 line km per month. Detailed characteristics of EQUATOR system are described by Karshakov et al. (2017).

Mapping of Granite Intrusions and Their Material Composition Definition

The local component of the magnetic field (Babayants and Tararuhina 2009) is weakly differentiated for granite massifs of different composition in Rwanda. Its intensity does not exceed the first units of nT (Fig. 46.2a). The outer contours of granite intrusions are determined reliably.

The apparent resistivity of granites, depending on their types, varies from 300 to 2500 Ω m (Fig. 46.2b). The most significant granites differ in their radiogeochemical specialization (Fig. 46.2d). The map of classes of radiogeochemical specialization (Babayants et al. 2015) was built on the basis of a two-dimensional correlation analysis. Originally maps of K and Th values were considered as

Fig. 46.1 Original construction of towed platform of EQUATOR technology

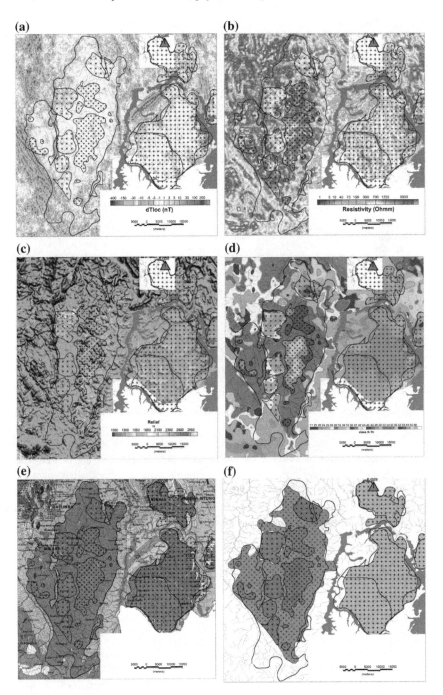

Fig. 46.2 Mapping of granite intrusions according to the technology of EQUATOR: **a** local component of the magnetic field; **b** apparent resistivity at a time gate 5 mcs; **c** digital elevation model; **d** map of radiogeochemical specialization classes for K-Th; **e** geological map; **f** varieties of granites from the interpretation of airborne geophysical data

Table 46.1 Physical properties of different types of granites

N	Pattern	K_Th class	dTloc (nT)	ρk (Ω m)	K (%)	Th (ppm)	Characteristic of granite massive
1		22–24	0–4	500–700	0.2–0.4	3.6–17.1	Low-radioactive granites, characterized by low potassium content, relatively low resistivity.
2		53–54	0–4	700–2500	2.0–2.15	11.2–16.4	Radioactive granites, high potassium content and medium thorium content, high resistivity.
3		33–35	0–4	300–600	0.86–0.91	10.9–24.2	Leucogranites, the content of potassium is below average, the content of thorium is medium, conductive.
4		63–65	0–4	1300–2500	3.6–3.9	10.4–24.3	Highly radioactive potassium granites, high content of potassium and thorium, high resistivity.

independent random variables. Each map was represented in the form of "stable-homogeneous" areas. The belonging of regions to different general populations was determined by the known rules of mathematical statistics. After this, algebraic intersections of maps of the domains K and Th were formed. As a result, stable areas of joint distribution of these elements were obtained. In general, the outer boundaries of granite massifs are in good agreement with the geological map. According to airborne geophysical data, it was possible to clarify their internal structure and distinguish four different types of granitoids (Table 46.1).

Results of Airborne Geophysical Survey at a Scale of 1:10,000

The work was carried out with the aim of detailing high-intensity linear magnetic anomalies and accompanying high-conductivity zones, revealed by the results of airborne geophysical survey at a scale of 1: 50,000. Detailed studies allowed to determine the shape, dimensions, and epicenters of magnetic anomalies (Fig. 46.3). A rapid assessment of the depths to the upper edge of the magnetic objects was performed. Objects with a minimum depth were selected for further study.

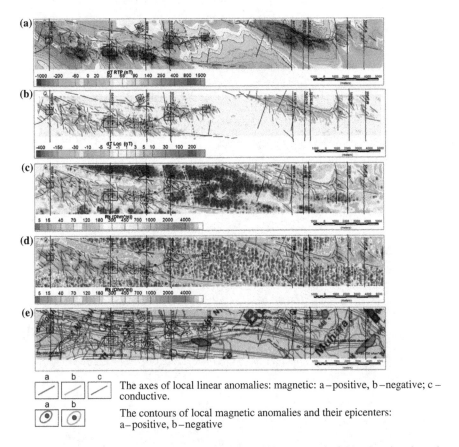

Fig. 46.3 Example of infill survey at a scale of 1: 10,000: **a** magnetic field reduced to the pole; **b** local component of the anomaly magnetic field; **c** apparent resistivity map at a gate of 5 μs; **d** map of apparent resistances for a window of 1000 μs; **e** fragment of the geological map

All picked up magnetic anomalies are situated in linear zones. They are accompanied by an increased density of lineaments which were picked up along the axes of linear magnetic and electrical anomalies, as well as relief forms.

Apparent resistivity maps show rocks with anomalously high resistivity (1000–10,000 Ω m). They correspond to quartzites. Rocks with medium resistivity are typical for metamorphic shales. Anomalously conductive zone (20–250 Ω m) with substantial vertical thickness (more than 100 m) is very interesting. Its conductivity sharply increases with depth. There is a magnetic anomaly in that zone. Such low-resistivity zones do not always coincide with depressive forms of relief and are most likely associated with the intrusion of igneous rocks of a younger age. Within the infill area, low-resistivity anomalies with limited vertical thickness (up to 30 m) are also fixed. We associate them with the weathering crust.

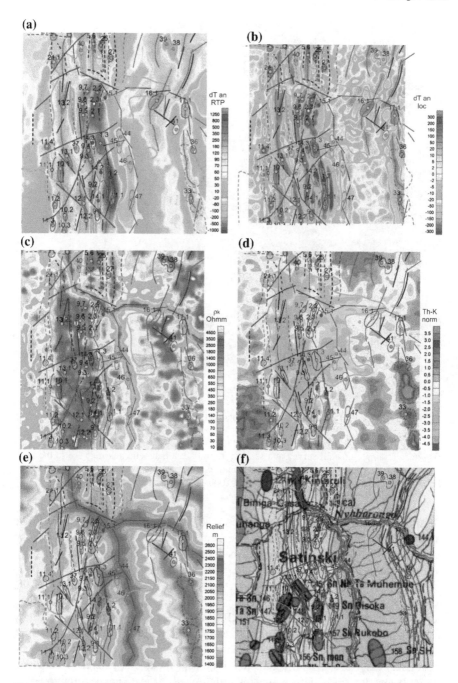

Fig. 46.4 Results of airborne geophysical survey at a scale of 1: 25,000. **a** an anomaly magnetic field map reduced to the pole; **b** local part of anomaly magnetic field map; **c** apparent resistivity map for depth interval 0–30 m; **d** a map of the difference of the normalized concentrations of Th-K; **e** a map of the digital elevation model; **f** a fragment of the geological map

Results of Airborne Geophysical Survey at a Scale of 1:25,000

The area (Fig. 46.4) is situated in the field of shale development and granite-gneiss rock formation. The basic tectonic faults have a meridional strike, and orthogonal to them—sublatitudinal. This is also emphasized by modern hydro-network and relief. A large number of ore occurrences of cassiterite, wolframite and coltan are concentrated to the south of the infill block.

We have identified a number of intense anomalous objects in various geophysical fields. They are not reflected in the modern geological map. They represent a significant search interest.

Intrusions of ore pegmatites are distinguished by weakly-medium intensive (20–100 nT) isometric magnetic anomalies with dimensions of 300–500 m. It has been established on known objects. Pegmatite bodies are usually characterized by high resistivity of 900–2000 Ω m. Unfortunately, they do not create contrast anomalies with host rocks.

It is interesting that the revealed magnetic anomalies have different radiogeochemical specialization. There are individual objects with a dominant potassium or thorium, as well as a general increase in radioactivity.

A number of intense linear magnetic anomalies of the meridional orientation are revealed. They are accompanied by linear conduction zones. Probably, magnetic anomalies are associated with dikes of amphibolites, and conduction zones with sulphide mineralization. These bodies are absent on the geological map and represent a certain search interest.

Conclusions

Survey at a scale of 1:50,000 allowed:

- to significantly clarify information about the geological structure of the territory;
- to classify granite massifs;
- to pick up linear and isometric intrusions of the basic and ultrabasic composition;
- to pick up laterite weathering crusts of bauxite and (or) rare-earth types.

Different geochemical specialization of young volcanogenic formations in the north-west of the territory was revealed. 35 perspective targets perspective for various minerals have been identified.

Infill survey at a scale of 1:10,000–1:25,000 allowed to identify contours, shape, size and position of the epicenters of local geophysical anomalies and their radiogeochemical specialization. Local anomalies are also highlighted which associated with intrusions of mafic rocks, promising for sulfide, copper-nickel mineralization.

Infill survey results allow to reduce volume of ground geophysical and drilling operations.

References

Babayants P.S., Tararuhina N.M., (2009). Principles of the modern interpretation technology of complex airborne geophysics data in a broader development of trap rocks. Modern airborne geophysical methods and technologies. vol. 1 issue 1, p. 71–110 (in Russian).

Babayants P.S., Kertzman V.M., Levin F.D., Trusov A.A. (2015). Principles of modern airborne gamma-ray spectrometry. Exploration and conservation of mineral resources. p. 11–16 (in Russian).

Felix, J.T., Karshakov, E.V., Melnikov, P.V., and Vanchugov, V.A. (2014). Data comparison results for airborne and ground electromagnetic systems used for kimberlites exploration in the Republic of Angola: Geophysika, 4, 17–22 (in Russian).

Karshakov, E.V., Podmogov, Yu. G., Kertsman, V.M., Moilanen J. (2017). Combined Frequency Domain and Time Domain Airborne Data for Environmental and Engineering Challenges. Journal of Environmental & Engineering Geophysics, Allen Press, Inc. 22(1):1.

Chapter 47
The Forecast of the Structural Surfaces Along the Top of the Pre-Jurassic Base on the Gravitational Field and the Evaluation of Productivity in the Poorly Studied Regions of Western Siberia at Various Stages of Work

N. N. Yaitskii, I. I. Khaliulin and M. V. Melnikova

Abstract The principal possibility and necessity of using the method of correlation separation of the gravitational field for the prediction of the structural plan of the foundation roof in poorly studied territories. Based on the results of the studies conducted in the poorly studied areas of Western Siberia, promising areas were identified, within which the subsequent drilling operations opened hydrocarbon deposits and proved the prospects of new areas. The further use of the similar studies allows us to hope for the discovery of new oil and gas fields in poorly studied areas of the AP.

Keywords Gravity prospecting · Magnetic prospecting · Integration of geophysical methods · Seismic survey

The Method of Predict Structural Plans for Potential Fields

A full cycle of the geological research, with the aim of forecasting and searching the hydrocarbon fields, is conventionally divided into two main stages in any territory: regional—the study of the general features of the structure of the study area in order to elucidate its potential oil and gas content and detailed—the research and study of specific hydrocarbon deposits).

The structural bases within Western Siberia in recent years are mainly based on a seismic data based on drilling results, but in the poorly explored areas other

---

N. N. Yaitskii (✉) · I. I. Khaliulin · M. V. Melnikova
Gazprom Geologorazvedka", Tyumen, Russia
e-mail: yaickiy@ggr.gazprom.ru

approaches should be used to forecast the structural plan. The best way out in such conditions for the construction of the structural surfaces is to use the gravitational and magnetic fields in combination with the available data of exploratory drilling and regional seismic exploration.

There are various methods for forecasting structural plans for potential fields (Berezkin 2002; Kosarev and Gerasimova 2000; Segal et al. 1996, Smirnov 1970, Shraybman et al. 1977, 1980, 1984; Shchekin et al. 1998). The use of data on potential fields for forecasting the structural plan has been practiced since the middle of the last century using various methods. In general, this is the research of Berezkin (2002), (Shraybman et al. 1977, 1980, 1984) etc. For Western Siberia—the work of Smirnov(1970) (60s), (Kosarev and Gerasimova 2000), S.N. Pianov (Segal et al. 1996); (Shchekin et al. 1998) (in recent decades).

The authors used the correlation field separation method (KOMR) to solve the structural geological problems, using gravimetric data. This method makes it possible to extract a local component in the gravitational field, that is most correlated with a given geological surface and then use it to construct the predictive structural maps. The direct calculations of the structural surfaces were performed by using the program "Correlation method of prediction" of various versions and modifications (author—S. N. Pianov), where the algorithms of correlation and correlation interpolation methods for separating fields within a sliding window of variable size are realized. These algorithms are a further development of the method COMR (Shraybman et al. 1977, 1980, 1984).

The authors of this paper have a positive experience of applying the KOMR method for constructing predictive structural maps in Western Siberia (Segal et al. 1996; Yaitskiy and Segal 2006; Yaitskiy 2010). The authors developed a scheme for predicting the structural surfaces of the roof of the pre-Jurassic basement and the bottoms of the platform cover using this technique, and also using them to identify promising areas for industrial hydrocarbon deposits (Fig. 47.1).

From the end of the last century to the present time, a number of predictive structural maps for the roof of the pre-Jurassic basement in poorly studied regions of Western Siberia (the Urals part of the AP, the Priobskaya zone of the Yamal-Nenets Autonomous District, the Yamal Peninsula region, the eastern part of the AP, etc.) were built by the authors on this method. In some areas, the subsequent seismic and the exploratory drilling operations in general, the structural trend and the prospects of these sites on industrial hydrocarbon deposits were confirmed (Gemini and Menshikov 2002; Yaitskiy and Segal 2006; Yaitskiy 2010). At one of the sites identified by the authors, as a promising openly commercial oil and gas field. Earlier (Yaitskiy 2010), the authors predicted a promising site in the region of this hydrocarbon field based on the results of the regional structural construction. At present, in the area of the open field of hydrocarbons, a forecast was made for the productivity of sediments in two stages. From the beginning, the structural plan was refined by the method of correlation separation of the gravitational field. Taking into account the new borehole data and using new parameters of the sliding window, more detailed structural constructions of the pre-Jurassic foundation were carried out (Fig. 47.2).

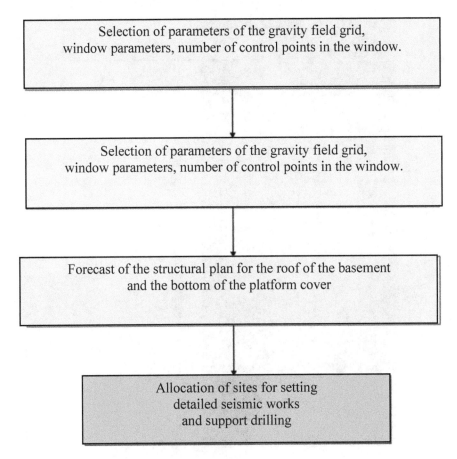

Fig. 47.1 The scheme for predicting the structural surfaces of the roof of the pre-Jurassic basement and the bottoms of the platform cover using the KOMR

The Method Predict Areas Promising for the Accumulation of Hydrocarbon Deposits

In the second stage, the "Pangea® IP" was used to identify the areas promising for the accumulation of hydrocarbon deposits. For the potential fields, as well as using the new structural constructions on the foundation, forecasting of the areas of the most probable presence of hydrocarbon raw materials was made (Fig. 47.3).

The forecast was based on the theory of regression analysis of predicted parameters from wells-standards to the adjacent area. The more detailed structural constructions and the forecast of productivity made on its basis, allow to count on similar industrial deposits of oil and gas in the given area.

The forecasting of oil and gas potential at the regional stage of works for the purpose of zoning of the poorly studied lands according to the degree of their

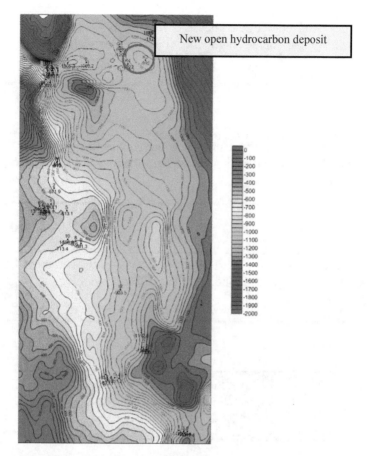

Fig. 47.2 The predict structural surfaces of the roof of the pre-Jurassic basement in the area of the open field of hydrocarbons on the gravitational field using the KOMR

prospects allows concentrating further work in the most important areas, significantly shortening the time for identifying priority sites for setting prospecting operations, and improving the reliability of the forecast of oil and gas potential.

The Conclusions

In the course of the research carried out by the authors of many years of research:

1. The principal possibility and necessity of using the method of correlation separation of the gravitational field for the prediction of the structural plan of the foundation roof in poorly studied territories.

Fig. 47.3 Forecasting of the areas of the most probable presence of hydrocarbon raw materials was made

2. Based on the results of the studies conducted in the poorly studied areas of Western Siberia, promising areas were identified, within which the subsequent drilling operations opened hydrocarbon deposits and proved the prospects of new areas.
3. The further use of the similar studies allows us to hope for the discovery of new oil and gas fields in poorly studied areas of the AP.

References

Berezkin V.M. (2002) The full gradient for geophizical work – M: Nedra, 1988. – 188 c.

Gemini MT, Menshikov Yu.P. "Bright spot' in the Lower Cretaceous deposits of the Shaim oil and gas bearing region of Western Siberia "Geophysics No. 4, 2002 M.

Kosarev I.V., Gerasimova E.V. (2000) Integration of geological and geophysical information and data on the morphology of the modern relief with the purpose of forecasting buried uplifts // Sb. "Geology, drilling, development and operation of gas and gas condensate fields". The gas industry, No. 5, M., 2000. p. 3-6.

Segal Yu.Z., Yaitskiy N.N., Pyanov S.N. (1996) "Local forecast of hydrocarbon deposits in poorly studied regions" Geophysics No. 2 for 1996.

Smirnov V.G. (1970) Anomalies of gravitational and magnetic fields as search criteria for the structures of the cover of the West Siberian plate within the boundaries of the Tyumen region). - Tyumen, ZapSibNIGNI, 1970. P. 3–10.

Shraybman V.I., Zhdanov M.S., Vitvitsky O.V. (1977) Correlation methods of transformation and interpretation of geophysical anomalies. M., Nedra, 1977.

Shraybman V.I., Zhdanov M.S., Vitvitsky O.V. (1980) Complex interpretation of field geophysics data, based on correlation transformations. - Geology of oil and gas, 1980, №7.

Shraybman V.I., Fuchs I.B., Vitvitsky O.V., Titkova N.G. (1984) Study of the geological structure of the southern part of the Siberian platform by correlation methods. - Geology of oil and gas, 1984, №3.

Shchekin S.N., Nezhdanov A.A., Turenkov N.A., Mikolaevsky E.Yu. (1998) "New methods for integrating geophysical methods in the Pangea system for forecasting oil and gas potential" Geophysics No. 6, 1998.

Yaitskiy N.N., Segal Yu.Z. (2006) "Efficiency of joint interpretation of geological and geophysical information in poorly explored areas" Geophysics No. 1, 2006.

Yaitskiy N.N. (2010) Forecast of the structural plan of the roof of the pre-Jurassic base in the Urals part of the AP on the gravitational field and the assessment of the prospects of oil and gas bearing on its basis // Tyumen, Gornye Vedomosti, 2010. №7. Pp. 52–59.

Chapter 48
Well Logging During the Processes of Field Development of Native Bitumen and Super-Viscous Oil Deposits

S. I. Petrov, R. Z. Mukhametshin, A. S. Borisov and M. Y. Borovsky

Abstract The article proposed to use the cross-well tomography to improve the efficiency of the bitumen deposits development and to obtain reliable information about the features of the geological structure and, consequently, for the optimal choice of sites for well testing and operations. Also proposed to use well logging techniques for investigating of the Permian hydrocarbons deposits in the deep wells. We consider the complex of well logs applied to the specific features of these wells. Noted that the design used in the Permian deposits has a critical diameter (0.4 m) for well logging and it is recommended to drill the first part of the well with a smaller diameter (not exceed 8.5″). More intensive introduction of "new" well logs, tested and successfully applied in the study of traditional oils (gamma spectrometry, C/O logging, VIKIZ, dielectric scanner and others) is needed.

Keywords Well logging · Native bitumen · Super-viscous oils

The main objects of bitumen-field development in Tatarstan region are reservoirs of the bedding of ufimian sand pack occurring at a depth of 150–200 m. The standard suite of well logging methods is currently being used to study native bitumen (NB) or super-viscous oils (SVO) deposits in terrigenous sediments of permian system, that is to solve the following common tasks of well logs (Khisamov et al. 2007; Petrov 2014; Abdullin and Rakhmatullina 2012): (a) separation of layers and estimation of reservoir properties—porosity factor (PF) and shale volume (Vsh); (b) evaluation of bitumen saturation factor.

As the analysis of similar investigations shows, there is no clear-cut solution of this problem. Consequently, the problem of increase of the efficiency of well logging for native bitumen and high-viscosity oil arises. It should be noted, unlike

---

S. I. Petrov (✉) · R. Z. Mukhametshin · A. S. Borisov
Kazan Federal University, Kazan, Russia
e-mail: sergey.petrov@kpfu.ru

M. Y. Borovsky
Geophysservic Ltd, Kazan, Russia

conventional hydrocarbon accumulations, NB and HVO deposits have some specific characteristics:

- the boundary surface between bitumen and water-saturated sandstones is rough and heterogeneous (Borovskiy et al. 2000);
- the content in the porous space of free water capable to migration, along with bitumen and bound water;
- the presence of water-saturated lenses as well as free of bitumen interlayers in the productive part of the profile which, according to V. N. Napalkov, I. M. Klimushin et al. mainly accounts for the water encroachment in the bitumen well stream.

Low salinity and some variability in chemical composition (Table 48.1) and various resistivity of intracounter formation waters of NB (HVO) deposits introduce complementary errors in evaluation of saturation in the productive layer.

Geophysical service companies are currently bringing into use new techniques of well-logging and integrating modern equipment. As it is known, in early stages of choice of rational well-logging complex for native bitumen electromagnetic propagation log (EMPL) was tested to identify oil-bitumen saturation factor. The results of field trials on the territory of the republic of Tatarstan are indicative of rather high efficiency of the given technique. Afterwards, due to missing of production equipment, EMPL was no longer used in standard complex. At the same time, many well-log analysts (Kozhevnikov et al. 2001) find the application of EMPL desirable for

Table 48.1 Salinity (S) and electrical resistivity of intracounter water (R_w) within the sandstone pack P_{1n}

Deposit, area	S, g/dm^3	R_w, Ω m		Composition of water NB (HVO) deposits
		from—to	average	
Averyanovskoye	4.4–4.96	No data		Sodium bicarbonate (sodic) and calcium bicarbonate type
Ashalchinskoye	3.1–5.9	No data		Sodium bicarbonate (sodic)
Chumachkinskoye	5.3–6.7	No data		Sodium bicarbonate (sodic)
Kamenskoye	2.5–3.7	No data		Sodium bicarbonate (sodic)
Karmalinskoye	2.7–5.0	No data		Sodium sulfate
Melnichnoye	7.02	1.1	1.1	No data
Mordovo-Karmalskoye	1.02–4.9	2.2–7.0	1.9	Sodium bicarbonate (sodic)
Nizhne-Karmalskoye	2.04–7.8	1.05–3.6	2.25	No data
Sarabikulovskaya	3.8	2.1	2.1	No data
Severo-Ashalchinskoye	2.5–4.1	No data		Sodium bicarbonate (sodic)
Vostochno-Sheshminskoye	2.3–5.9	No data		Sodium bicarbonate (sodic), sodium sulfate
Yuzhno-Ashalchinskoye	2.9–7.4	1.1–2.6	1.4	No data

thorough watercut interval-boundary resolution and enhancement of accuracy of bitumen saturation assessment for well-logging (under low bitumen saturation); but as far as thin water-bearing interbed resolution in the productive layer of the reservoir concerns, it is required to use microzonation data. The application of nuclear-physical methods in modification of gamma-ray spectrometry GRL-S (Kozhevnikov et al. 2001; Borovskiy et al. 2007; Borovskiy et al. 2000) and carbon-oxygen logging C/O (Akhmetov et al. 2013) also appears to be effective. The use of gamma-rays spectral log makes it possible (even in limited scope in key wells) to enhance the accuracy of quantitative identification of petrophysical properties of formations by specifying the content of polymictic sandstones (Kozhevnikov et al. 2001).

The error of measuring of bitumen content value in the layer depends on (Petrov and Abdullin 2016) the validity of porosity factor determined from well-logging data. The porosity histogram according to well-logging data is often directly opposed to the porosity histogram determined using core test. Logging data indicates porosity index increase in low and medium porous layers and, on the contrary, decrease in high porosity reservoirs is noted. The similar pattern can be seen while comparing the distribution of the values of bitumen content ratio. It has been shown (Abdullin and Rakhmatullina 2012, Petrov and Abdullin 2016) that the essential part of the reservoir according to well-logging data has more than 56% bitumen content, but according to laboratory data the main masses of the same layers are characterized by less than 44% bitumen content ratio. Decrease of porosity value by WL data is likely to affect the calculations of bitumen content ratio value by WL. Inaccuracy in assessment of porosity factor (PF) and resistivity of formation water (R_w) lead to obtaining inadequate values of bitumen content ratio in the reservoir. Gamma-ray spectrometry (GRL-S) data make it possible to do reliable calculations of effective porosity of the layer that estimate thorium (Th) and potassium (P) content; this enables to measure the quantity of hydromica concentration in matrix of reservoir rock (Table 48.1).

Foreign service companies often use Pickette plott to evaluate the resistivity of formation water (Rw) which is based on Archei algorithm. In this case we need porosity and electrical resistivity data for estimation of resistivity of brine water. Using the plot for water-bearing interval (if there is any) electrometry readings are taken and porosity is calculated. Pickette plot is also needed to evaluate readings of R of formation water.

Figure 48.1 presents the results of well-logging interpretation from well No.433 of Bolshekamensky upheaval of native bitumen deposits of RT. The left illustration shows the standard suite data (SPL, GRL, CALI, RL) within the interval of Sheshminsky horizon. The next track indicates core and neutron log-derived reservoir parameters with the account of clay content. In the figure we can observe the coincidence of estimated value on log and core based porosity. The fourth track gives the evaluation of bitumen saturation ratio (S_{b_ZAK}) by R_w 2.0 Ω m (as established by interpretation model) and bitumen saturation ratio (S_b_cor) calculated by R_w equal to 3.9 determined using Pickette plot. The given bitumen saturation is somewhat lower than S_b stand, but practically coincides with core samples saturation. Consequently, we can come to the conclusion that undercount of the degree of porous water salinity substantially distorts the results of standard WL interpretation of bitumen content estimation.

Salinity assessment of intracounter waters in different deposits or in some of their parts has a significance while determining their conservation (degree of disintegration). In this connection, occurrence of low-salinity water of sodic composition with the degree of 1.02 g/dm^3 salinity in the region of well 4 of Mordovo-Karmalskoye field is accounted for by the destruction of the deposit, which also validates high content of hydrogen sulphide (Khisamov 2016). Hydro-chemical anomalies can reflect fractured caprock, that is necessary to take into account in choosing thermo-chemical methods of impact (Mukhametshin and Punanova 2012).

In order to enhance the efficiency of bitumen deposits development works in the RT and to obtain the relevant information on the characteristics of the geological feature, and, as a result, to make an optimal choice of the objects to be tested and operated we suggest that interwell tomography seismic data be used. Sample data in conditions of oil and gas field in Bashkortostan and abroad indicate a high potential of method of Tatarstan super-viscous oil field structure investigation and the reserves recovery control. The work (Borovskiy et al. 2007) presents a program of

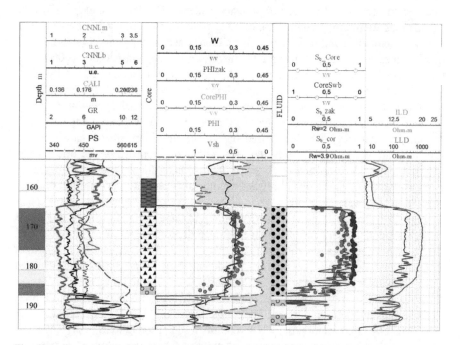

Fig. 48.1 Comparison of S_b determined by standard technologies of well-logging interpretation using Pickette plot with the results of the laboratory core research (well 433, Bolshekamensky upheaval) (according to figures from TNG-Group Co)

assessment of cross-hole tomography capability to reveal "delicate" geological aspects of native bitumen and super-viscous oil deposits. Investigation methodology includes:

- carrying out of field interwell observation on the field test site;
- processing of experimental investigation materials using interwell seismic tomography method;
- geological interpretation of data received (interpretation in conjunction with well log data);
- formulation of recommendations of cross-hole tomography method application for revealing of structural features of native bitumen accumulations in natural conditions and in case of thermal bed stimulation (applicability of cross-hole tomography is essentially increased when SAGD technology is applied while using super-viscous oil deposits in operation).

The problem of recovery of hydrocarbon feedstock reserves includes the search of additional sources of information about availability of super-viscous oil accumulations in the near-surface geological section of the perspective territories. An option of hydrocarbons search in the upper stages of geological cross section was proposed by Akhmadishin et al.(2014) during the drilling of deep wells. A suite 7of well logging methods has been reviewed with respect to specific features of the design of such wells. It is noted that the design applied in the intervals of Permian deposits provides for use of a drill steel guide, which has a critical diameter of 0.4 m for geophysical well logging. According to company TNG-Group data the standard suite of well logging methods ensures the radial depth of investigation less than 0.5 m (Akhmadishin et al.2014).

An example of geophysical well logging of Permian deposits at the raising "Odinochnoye" of Sokolkinskiy field is given in the report. In the exploratory well 1057, in the interval of 0–310 m, the logging of well was performed at the scale of 1:200 before running in hole of 13″ drill steel guide, using the following methods: inclinometry, GRL, NNL, SPL, LL, IEL, ASL, CL, RL, resistivimetry and NML. Based on well logging data terrigenous and carbonate reservoirs have been identified and their petrophysical characterstics have been determined. It is indicated that the above set of methods is insufficient for identification of hydrocarbon (HC) saturated intervals. To determine oil and gas saturation factor it is necessary to complement the well logging suite with the following methods: gamma ray spectrometry logging (GRL-S), gamma-gamma density logging (DL), carbon-oxygen logging (COL). At the same time it is noted that the proposed geophysical methods have small radial depth of logging. In the interval of Permian deposits it is suggested to drill with the drill bit of 8.5″, to apply a complemented suite of well logging with the subsequent reaming of the diameter of the well bore with the drill bit of 15.5″ and case the well with a drill guide (12.75″) for continuation of drilling in the sublayers. Thus, it will be possible to obtain important information about saturation of Permian deposits with hydrocarbons and determine reservoir properties of identified intervals.

Monitoring of development of native bitumen and super-viscous oil deposits is of importance currently. It is essential to determine the character of the current and residual bitumen saturation of productive reservoirs. In order to solve this problem, importance is being increasingly attached to PNGL-S method and its modification—carbon/oxygen logging (C/O logging) (Akhmetov et al. 2013). Separation of developed terrigenous reservoirs into oil saturated and water-encroached types is the main common task of carbon/oxygen logging. To specify the possibility of quantitative determination of terrigenous reservoir bitumen saturation factor, a test survey was carried out by OOO TNG Group for pilot testing of suite of nuclear-physical methods, including C/O and GRL-S logging at the well of Mordovo-Karmalskiy region. Experience of application and the results of C/O logging, new for bitumen saturated terrigenous reservoirs, show a principal possibility of quantitative determination of current bitumen content.

According to geophysical well logging data an algorithm of isolation of gas bearing formation can also be referred to modern geophysical technologies. During well construction identification of gas accumulations in Permian deposits, as well as their predicting, help to anticipate and prevent unforeseen consequences from the point of view of both negative impact on environment and safety engineering. Critical values of NL and GGL readings at which gas bearing rocks are identified have been determined for Permian deposits of Tatarstan. Gas saturated intervals in the terrigenous part of Permian deposits are clearly identified at NL readings of higher than 1.4 relative units and in the carbonate part—of higher than 2.4 relative units, while GGL readings should be higher than 5 relative units. These digital criteria are correct at full dissipation of mud-filtrate penetration area. The analysis of geophysical well logging data, based on a wide range selection of wells of Novo-Yelhovskiy, Arkhangelskiy and other fields, allowed to identify the presence of multiple gas bearing beds in the productive interval of Ufimian age Permian deposits.

Conclusions

The following three ways of increasing the efficiency of geophysical well logging for native bitumen (super-viscous oil) on the territory of Tatarstan Republic have been specified:

- successful approbation and large scale implementation of the modern technology of interwell seismic tomography will allow to increase the efficiency of geological exploration works both at preparation stage and during the processes of development of native bitumen (super-viscous oil) deposits;
- development of rational suite of well logging methods for obtaining important information under different mining technological conditions (deep drilling wells);

- implementation of geophysical well logging methods appraised at a small number of prototype objects and successfully applied for study of conventional oils (gamma spectrometry, C/O logging, high frequency induction logging with isoparametric sounding (HFL), electromagnetic propagation log (EMPL) etc.)

References

Abdullin R.N., Rakhmatullina A.R. (2012). New technology of determining reservoir properties and bitumen saturated factor of Ufimian age terrigeneous deposits // High viscosity oils and native bitumen: problems and increase of efficiency of exploration and development of fields. International Research and Practice Conference. – Kazan: Fen Publ., pp. 35–37.

Akhmadishin F.F., Suleymanov A.Y., Musayev G.L. (2014). Prospecting for high viscosity oils in the Permian deposits at deep well drilling // The book of scientific works of TatNIPIneft, publication, No. 82. – Moscow : OAO VNIIOENG, pp. 240–242.

Detailed elaboration of geological structure of bitumen deposits (2007) // Current problems of petroleum geology: International Research and Practice Conference / M.Y. Borovskiy, A.G. Bolgarov, I.N. Faizullin et.al. – St. Petersburg : VNIGRI, pp. 301–304.

Gamma spectrometry in the suite of well logging methods for study of bitumen deposits of Tatarstan (2001) / D.A. Kozhevnikov, N.E. Lazutkina, G.A. Petrov et al. // Geofisika, No. 4, pp. 82–86.

Geophysical methods of preparation and monitoring of operation processes of native bitumen deposits (2000) / M.Y. Borovskiy, E.K. Shvydkin, R.Z. Mukhametshin et al.; under the editorship of R.Z. Mukhametshin. – Moscow : Geos, 170 p.

Hydrogeological conditions of heavy high viscosity oils and bitumen fields (2016) / R.S. Khisamov, R.N. Gatiyatullin, R.L. Ibragimov et.al.; under the editorship of R.S. Khisamov. – Kazan : Ihlas, 176 p.

Khisamov R.S., Borovskiy M.Y., Gatiyatullin N.S. (2007). Geophysical methods of exploration and prospecting for native bitumen deposits in the Republic of Tatartsan. – Kazan : Fen, 247 p.

Mukhametshin R.Z., Punanova S.A. (2012). Non-traditional sources of hydrocarbon raw material: geochemical features and aspects of development // Neftyanoe khozyaystvo – Oil Industry, No. 3, pp. 28–32.

Petrov S.I. (2014). Current status and perspectives of studying Permian bitumen in Tatartstan applying well logging techniques // Hard to recover and non-conventional hydrocarbon reserves: experience and prediction: International Research and Practice Conference. – Kazan : Fen Publ., pp. 313–316.

Petrov S.I., Abdullin R.N. (2016). Determination of oil/bitumen saturation of Ufimian age sanstone series based on well logging data under the conditions of variable salinity of brine water. Innovations in exploration and development of oil and gas fieds: proceedings of V.D. Shahin Int. Research and Practice Conference. – Kazan : Ihlas, vol. 2. pp. 208–211.

The specifics of interpretation and perspectives of application of nuclear-physical suite of well logging methods for native bitumen and high viscosity oil deposits of the republic of Tatarstan (2013) / B.F. Akhmetov, V.V. Bazhenov, L.I. Limonova, D.R. Abdullina // Geophysical, geochemical, and petrophysical investigations and geologic modelling for exploration and production monitoring of oil and gas fields: reports of Int. Research and Practice Conference (1–4 of October, 2013, Bugulma). – Moscow : VNIIgeosystem, pp. 29–36.

Chapter 49
Preliminary Results of a Satellite Image Frequency-Resonance Processing of the Gas Hydrate Location Area in the South China Sea

S. Levashov, N. Yakymchuk, I. Korchagin and D. Bozhezha

Abstract The experimental studies with the frequency-resonance method of satellite images processing using were carried out on a local site in the South China Sea, studied by a 3D seismic survey and drilling, where gas hydrates were successfully extracted by Chinese oilmen. In the area of the drilled wells that opened the gas hydrates, the resonant frequencies of the gas hydrates were refined, with which further investigation was conducted. At the specified resonant frequency, two anomalous zones "Gas-hydrate-1" and "Gas-hydrate-2" were detected and mapped on the entire image area. These anomalies are located within the BSR zones, identified by 3D seismic data, and are significantly smaller in area. At resonance frequencies of gas, four anomalous zones of the "gas reservoir" type have been mapped over the entire area of the image: Gas-1, Gas-2, Gas-3 and Gas-4. Within all these anomalies the reservoir pressure intervals were estimated: (1) 19.8–21.0 MPa; (2) 21.0–21.5 MPa; (3) 21.0–21.5 MPa; (4) 21.2–21.5 MPa. In the contours of "Gas-hydrate-1" and "Gas-hydrate-2" anomalies, anomalous responses at the resonance frequencies of gas were not fixed. It can be concluded that there are no sub-hydrate deposits of gas here. Within the surveyed area, anomalous zones at resonant frequencies of oil were not detected. In the Gas-2 anomaly contour a channel for deep fluids vertical migration was revealed and localized—a small local area with very high reservoir pressure values of 67 MPa. There are good reasons to state that the obtained results are a significant addition to the data of earlier studies on this local site.

S. Levashov · N. Yakymchuk · D. Bozhezha
Institute of Applied Problems of Ecology, Geophysics and Geochemistry,
Kiev, Ukraine
e-mail: geoprom@ukr.net

I. Korchagin (✉)
Institute of Geophysics, NAS Ukraine, Kiev, Ukraine
e-mail: korchagin.i.n@gmail.com

Introduction

In May 2017, the information (data) about the successful extraction of gas hydrates in the South China Sea by Chinese oilman's appeared in the mass media (Internet sites including) (On the eve…, 2017). This event caused a wide resonance in the world. It was estimated by some sources as a revolution in the energy sector of the world economy.

This event was not ignored by the authors of this paper. Our interest in this event is due to the fact that for many years we have purposefully carried out experimental studies, aimed at introducing mobile, low-cost and direct-prospecting methods into oil and gas exploration (and gas hydrate, including). These methods were also used to search for accumulations of gas hydrates in the Antarctic region during seasonal work in Ukrainian Antarctic expeditions (Soloviev et al. 2017; Yakymchuk et al. 2015).

In this regard, the authors have the opportunity to promptly (operatively) conduct additional testing of the frequency-resonance method of satellite images processing on a sufficiently well studied by 3D seismic survey and drilling site (area) of a successful gas-hydrate production.

Object and Purposes of Research

Experimental studies using mobile and direct-prospecting technology were carried out on a local site in the South China Sea, studied by a 3D seismic survey and drilling, where gas hydrates were successfully extracted with technology using, developed by Chinese specialists.

Additional approbation of the frequency-resonance method of processing and interpretation (decoding) of remote sensing data of the Earth (satellite images) for the purpose of detection and localization of hydrocarbon accumulation areas (oil, gas, gas-condensate, gas hydrates) on onshore and on the offshore. Demonstration on a specific example of the potential of this mobile, low-cost and direct-prospecting technology in a thoroughly studied local area.

Method of Research

Mobile technology of frequency-resonance processing of remote sensing data (Levashov et al. 2010; 2011; 2012) is widely used during the experimental studies conducting of various kinds. The individual components of this technology have been developed on the principles of "substance" ("matter") paradigm of geophysical research (Levashov et al. 2012), the essence of which is searching for a specific (desired in each case) substances—oil, gas, gas condensate, gold, iron, water, etc.

Distinctive features of the methods are described in many publications, including those listed in the list of references (Levashov et al. 2010, 2011, 2012, 2016, 2017a; 2017b; Yakymchuk et al. 2015). Some information about this technology may be found also at the site [http://www.geoprom.com.ua/index.php/ru/].

The search work by mobile methods can be carried out in three main stages (Levashov et al. 2017a): (1) frequency-resonance analysis of satellite images of the major search areas (blocks) in a relatively small scale (*the study of reconnaissance character*); (2) a detailed frequency-resonance analysis of satellite images of certain areas (site) of anomalous zones, allocated at the first stage (*detailed work*); (3) geoelectric field work on the most promising local sites, allocated at the second phase of the work (*ground-based studies*).

Processing of satellite images of search areas, taken from open access sources, is carried out operatively in the laboratory. In this regard, this technology can be considered as a super-operative.

Preliminary Research Results

In the course of the research, the materials, published in journal articles, were used to study the area of 3D seismic prospecting and wells drilling (Su et al. 2016; Zheng et al. 2011).

Thus, when preparing a satellite image of a site for processing, the information (data) from (Zheng et al. 2011) were used: site coordinates, contours of the 3D seismic survey area, location of the drilled wells, contours of the gas hydrate reservoir (BSR zone), detected by seismic data.

The satellite image of the survey site was placed on a sheet of A3 format (an element of the technological processing scheme) on a scale of 1: 70,000 (Fig. 49.1). Some data from (Zheng et al. 2011) are shown on satellite image.

When carrying out the frequency-resonance processing of the prepared image, materials from paper (Su et al. 2016) were used, including Fig. 49.2, which presents more complete information (data) on the site structure, based on the results of the studies. In particular, it shows another zone of gas hydrate location (BSR zone), as well as the outlines of gas columns (pipes), detected by seismic study.

At the initial stage of the image processing, in the area of the drilled wells that opened the gas hydrates, the resonant frequencies of the gas hydrates were determined (or, more accurately, refined), with which further detection and mapping of anomalous zones of the "gas hydrate deposit" type was conducted. We note here that the refined values of the resonant frequencies differ slightly from the frequencies, which were used during the gas hydrate accumulation searching in the Antarctic region (Soloviev et al. 2017; Yakymchuk et al. 2015).

At the specified resonant frequency, two anomalous zones of the "gas hydrate deposit" type ("Gas-hydrate-1" and "Gas-hydrate-2") were detected and mapped on

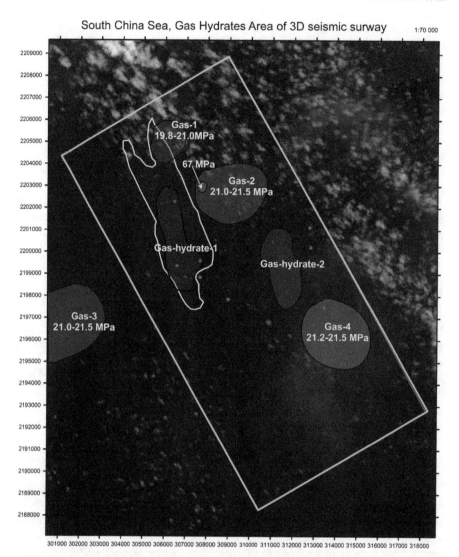

Fig. 49.1 Sketch-map of anomalous zones of the "gas" and "gas-hydrates" type on a satellite image of the 3D seismic survey area in the South China Sea (based on the results of frequency-resonance processing of a satellite image). Yellow circuit—gas hydrate zone according to seismic data (BSRs zone); red dots—drilled wells; a small area (red color) in the Gas-2 anomalous zone is the vertical channel of deep fluids migration with a reservoir pressure of 67 MPa. Gas hydrates are found in drilled wells that enter the contour of the anomalous zone "Gas-hydrate-1"

the entire area, shown in the image (including outside the 3D seismic study contour) (Fig. 49.1 and Fig. 49.2). These anomalies are located within the BSR zones, identified by 3D seismic data, and are significantly smaller in area.

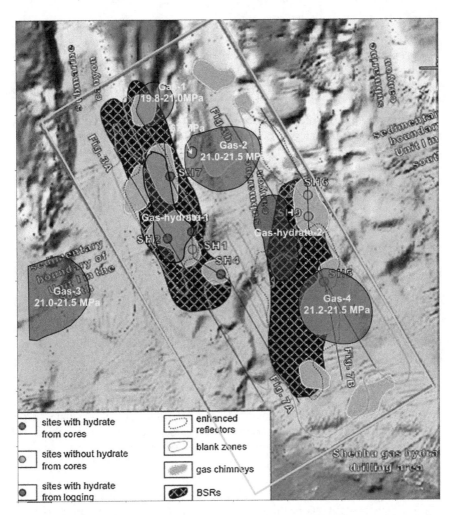

Fig. 49.2 Sketch-map of anomalous zones of the "gas" and "gas-hydrates" type superimposed on figure with the results of the 3D seismic survey at the site (Su et al. 2016) in the South China Sea (based on the results of frequency-resonance processing of a satellite image). Red dots—drilled wells; a small area (red color) in the Gas-2 anomalous zone is the vertical channel of deep fluids migration with a reservoir pressure of 67 MPa. Gas hydrates are found in drilled wells that enter the contour of the anomalous zone "Gas-hydrate-1"

Let's pay attention to the fact that wells, drilled on this local area, do not get into the contours of the anomaly of "Gas-hydrate-2".

In the next stage of studies on resonance frequencies of gas, four anomalous zones of the "gas reservoir" type have been detected and mapped over the entire area of the image: Gas-1, Gas-2, Gas-3 and Gas-4 (Fig. 49.1 and Fig. 49.2). In this case, the Gas-3 anomalous zone was detected outside the 3D seismic survey area. Within all the anomalous zones, the formation (reservoir) pressure intervals were estimated: (1) 19.8–21.0 MPa; (2) 21.0–21.5 MPa; (3) 21.0–21.5 MPa; (4) 21.2–21.5 MPa.

In the contours of the "Gas-hydrate-1" and "Gas-hydrate-2" anomalous zones, anomalous responses at the resonance frequencies of the gas were not fixed. In this connection, **it can be concluded that there are no sub-hydrate deposits of gas here**.

Within the surveyed area, anomalous zones at resonant frequencies of oil were not detected (satellite image processing was also performed with resonant frequencies of oil using).

Additional studies in the contour of the Gas-2 anomalous zone revealed and localized a channel for deep fluids vertical migration—a small local area with very high reservoir pressure values of 67 MPa (Fig. 49.1 and Fig. 49.2).

Possible Additional Studies

A small volume of research using the frequency-resonance method of satellite images processing was operatively (quickly) performed at the site of the survey. The carried out researches can be classified as reconnaissance works of a demonstration character. The mobile method used allows the following additional work to be carried out operatively.

1. In the contours of the detected "Gas-hydrate-1" and "Gas-hydrate-2" anomalous zones, a geological cross-section can be scanned to estimate the depths and thicknesses of anomalous polarized layers (APLs) of "gas-hydrate" type. Scanning in a sufficient number of points along the area of anomalous zones will allow to estimate the volumes of gas hydrate deposits, as well as the forecasted gas resources in them.
2. Scanning of the geological cross-section within the anomalous zones of the "gas" type will provide an opportunity to determine the volumes of the APLs of "gas"type in each anomalous zone, and also to estimate the predicted gas resources in the contours of the anomalous zones. In the process of scanning by a detailed technique, the reservoir pressures in individual APLs of "gas" type as well as porosity of reservoir can also be assessed.
3. In order to improve the accuracy of the results obtained, satellite images of the locations of individual anomalous zones can be processed on a larger, detailed scale. Other channels of deep fluids vertical migration can also be detected. To this we should add that in Fig. 49.2, the contours of 13 gas columns (pipes), detected by seismic data, are shown in the survey area.

When processing satellite images of individual sites of the survey area, the anomalous zones of small dimensions can be detected also.

4. The ones shown in Fig. 49.2 gas columns (pipes), as well as the detected channel of deep fluids vertical migration indicate at a deep source of gas. In the

limits of the detected anomalous zones of the "gas" type, the reservoir pressures in one interval of the cross-section were estimated. However, during the vertical migration of gas, its accumulations can be formed in reservoirs, located in different intervals of the cross-section. In this regard, in the contours of gas-type anomalies, it is advisable to perform a procedure for estimating reservoir pressures to a prior predetermined depth.

Conclusions

1. Investigations of reconnaissance character within the surveyed area in the South China Sea were carried out quite operatively (in a very short time period). In comparison with the time, spent on seismic studies conducting, these are disparate values (quantities).
2. Taking into account the considerable volume of earlier conducted experimental studies in different regions of the world, the authors have good reasons to state that the obtained results are a significant addition to the materials of earlier studies on this local site. Analysis of Fig. 49.2 allows us to conclude the following:
(a) the contours of the anomalous zones of the "gas hydrate" type are refined (localized); wells in which gas hydrates are found fall into the contours of detected anomalies;
(b) four anomalous zones of the "gas" type were additionally identified and mapped; by the total area, these zones are even larger than the area of anomalies of the "gas hydrates" type; one anomalous zone of the "gas" type was detected outside the site of seismic studies;
(c) the materials (data) of conducted research can be used to select the location of production wells;
(d) the channel of deep fluids vertical migration is detected, which can be considered an important factor (element) for understanding the mechanisms of formation of gas and gas hydrate accumulations.
3. The volume and quality (reliability) of the received information can be significantly increased if more detailed studies with using the technology of frequency-resonance processing of satellite images will be carried out within the survey area.
4. In paper (Wu et al. 2008) there is a map of the location of BSR zones and sites of wells drilling in the northern and western parts of the South China Sea. It is possible to operatively assess the prospects for the detection of industrial accumulations of gas hydrates and gas within these zones using the frequency-resonance method of satellite images processing and decoding.
5. We also note that the authors carried out a large volume of experimental research in the areas of the crystalline rocks distribution (the Ukrainian and Baltic shields), in coal basins (Donbass, Kuzbass, England), in the areas of the

distribution of shales, dense sandstones and rocks of the Bazhenov suite. In paper (Zou et al. 2013) there is a map of the location of unconventional oil and gas resources on China onshore. The technology of frequency-resonance processing of remote sensing data of the Earth (satellite images) can also be used for operative detection and localization of "Sweet spots" zones in the areas of distribution of unconventional collectors (hydrocarbons) in China.

6. The results of long-term application of direct-prospecting methods to search for oil and gas accumulations have led the authors to the position of proponents of deep (abiogenic) synthesis of hydrocarbons. This was repeatedly noted by us in many published works. In articles and presentations, we also many times referred to publications (Krayushkin, 1986 p. 582; Kutcherov and Krayushkin 2010, p. 5), in which the mechanism of hydrocarbon accumulations formation is formulated in the following form: "… The formation of oil and gas deposits occurs differently. Rising **from sub-crustal layers abiogenically synthesized oil and gas along the fault and its feathering fractures** are **"injected" under tremendous pressure of mantle fireplace in any porous and permeable environment**, extending within it from the fault like a mushroom cloud. They remain relatively fixed, not float in any anticline or syncline or in an inclined or horizontal formation till the new portions of oil and gas not promoted their deposit. This is indicated by experiments and practice of construction of underground gas storage facilities in the horizontal and inclined water-saturated layers of sand or sandstone".

In the above reference, we draw attention to a fragment of the text **"oil and gas along the fault and its feathering fractures"** are injected **"under tremendous pressure of mantle fireplace in any porous and permeable environment."** On the area of the survey, the following facts may be indicative for the described mechanism: (a) the presence of a significant number of faults, pockmarks and gas columns (pipes), established by seismic studies; (b) channel of deep fluids vertical migration, local area with very high reservoir pressure—67 MPa; c) anomalous zones of the "gas" and "gas hydrates" type.

It follows from the described mechanism of hydrocarbon accumulations formation that deposits of gas, oil, gas hydrates can be formed only in collectors (porous and permeable rocks). The absence of a sub-hydrate gas in the contours of anomalies of the "gas hydrates" type may be due to the fact that in this part of the cross-section there are no reservoir (collectors) rocks. In this situation, gas into the sub-hydrate rocks can not be "pumped" even under great pressure.

The research, operatively carried out by authors in 2016–2017 within many areas (sites), located in different regions of the globe, have provided a considerable amount of additional and independent information as of petroleum prospects of the surveyed areas, so and of methodological features of practical application of mobile and direct-prospecting technology for the specific prospecting tasks solving. To carry out such volume of research in such a short time allows only the super-mobile and super-operative frequency-resonance technology of RS data processing and decoding. The above experimental results once again demonstrate the feasibility of this technology using in the search and exploration process. More active and purposeful use of different technologies components to solve specific practical problems will significantly speed up, streamline and reduce the cost of exploration process for industrial (commercial) oil and gas accumulations prospecting and exploration in reservoirs of traditional and non-traditional type. During the sharp drop of oil prices in the world, this problem is extremely urgent.

Testing of advanced techniques and methodological procedures of remote sensing data processing and interpretation (decoding) on the fields and promising areas in different regions (onshore and offshore), and the received results provide additional evidence (arguments) for the understanding of the oil and gas genesis and the nature of their industrial accumulations formation. Thus, numerous data on the existence within the surveyed areas of anomalous zones with multiple intervals of reservoir pressure and of the vertical channels of deep fluid migration can be considered as powerful arguments in favor of the endogenous (deep) origin of hydrocarbons.

The results of numerous experimental studies in various regions indicate that the use of mobile and operative methods of "direct" searching for hydrocarbon accumulations in areas of non-conventional and conventional reservoirs spreading will significantly increase drilling success rate (an increase in the number of wells with commercial hydrocarbon inflows). The wells laying within areas of vertical channels of fluids migration location may lead to an increase in hydrocarbon inflows.

Proven direct-prospecting technology of remote sensing data frequency-resonance processing is recommended to be used for the preliminary assessment of petroleum potential of large by area (remote and inaccessible) poorly studied blocks. Application of this technology can bring significant impact during the search for commercial hydrocarbon accumulations in unconventional reservoirs (including the areas of shale, rocks of the Bazhenov formation, coal-bearing formations and crystalline rocks spreading). Mobile technology can also be successfully used during studies within the poorly studied areas and blocks in the known oil and gas-bearing basins.

References

Krayushkin V.A. (1986) Mestorozhdenija nefti i gaza glubinnogo genezisa. Zhurnal Vsesoyuznogo khimicheskogo obshchestva im. D.I. Mendeleeva, vol. 31, no. 5, pp. 581–586 (in Russian)

Kutcherov, V. G., and V. A. Krayushkin (2010), Deep-seated abiogenic origin of petroleum: From geological assessment to physical theory, Rev. Geophys., 48, RG1001, https://doi.org/10.1029/2008rg000270. http://onlinelibrary.wiley.com/doi/10.1029/2008RG000270/pdf

Levashov S.P., Yakymchuk N.A., Korchagin I.N. (2010) New possibilities of the oil-and-gas prospects operative estimation of exploratory areas, difficult of access and remote territories, license blocks. Geoinformatics, 3, 22–43. (in Russian)

Levashov S.P., Yakymchuk N.A., Korchagin I.N. (2011) Assessment of relative value of the reservoir pressure of fluids: results of the experiments and prospects of practical applications. Geoinformatics, 2, 19–35. (in Russian)

Levashov S.P., Yakymchuk N.A., Korchagin I.N. (2012) Frequency-resonance principle, mobile geoelectric technology: a new paradigm of Geophysical Investigation. Geophysical Journal, 34, 4, 167–176. (in Russian)

Levashov S.P., Yakymchuk N.A., Korchagin I.N., Bozhezha D.N., Prylukov V.V. (2016) Mobile direct–prospecting technology: facts of the channels detection and localization of the fluids vertical migration - additional evidence for deep hydrocarbon synthesis. Geoinformatics, 2, 5–23 (in Russian)

Levashov, S.P., Yakymchuk, N.A., Korchagin, I.N. and Bozhezha, D.N., (2017a), Application of mobile and direct-prospecting technology of remote sensing data frequency-resonance processing for the vertical channels of deep fluids migration detection. NCGT Journal, v. 5, no. 1, March 2017, p. 48–91. www.ncgt.org

Levashov Sergey, Yakymchuk Nikolay, and Korchagin Ignat. (2017b), On the Possibility of Using Mobile and Direct -Prospecting Geophysical Technologies to Assess the Prospects of Oil -Gas Content in Deep Horizons. Oil and Gas Exploration: Methods and Application. Said Gaci and Olga Hachay Editors. April 2017, American Geophysical Union. p. 209–236.

On the eve of an energy revolution: China was able to produce "combustible ice" (На пороге энергетической революции: Китай смог добыть « горючий лед ») [in Russian]. http://toomth.livejournal.com/6016584.html

Soloviev V.D., Levashov S.P., Yakymchuk N.A., Korchagin I.N., Bozhezha D.N. (2017) The experience of integrated mobile technologies used for deep hydrocarbon accumulation prospecting and geophysical mapping at the Western Antarctic bottom structures. Geophysical Journal, 39, 1, 123–143. (in Russian)

Su M., Yang R., Wang H., Sha Z., Liang J., Wu N., Qiao S., Cong X. (2016) Gas hydrates distribution in the Shenhu area, northern South China Sea: comparisons between the eight drilling sites with gashydrate petroleum system. Geologica Acta, Vol. 14, N° 2, June 2016, 79–100. https://doi.org/10.1344/geologicaacta2016.14.2.1

Wu, N.; Yang, S.; Zhang, H.; Liang, J.; Wang, H.; Su, X; Fu, S. (2008) Preliminary discussion on gas hydrate reservoir system of Shenhu Area, North Slope of South China Sea. In Proceedings of the 6th International Conference on Gas Hydrates (ICGH 2008), Vancouver, Canada, 6–10 July 2008.

Yakymchuk, N. A., Levashov, S. P., Korchagin, I. N., & Bozhezha, D. N. (2015, March 23). Mobile Technology of Frequency-Resonance Processing and Interpretation of Remote Sensing Data: The Results of Application in Different Region of Barents Sea. Offshore Technology Conference. https://doi.org/10.4043/25578-ms. https://www.onepetro.org/conference-paper/OTC-25578-MS

Zheng Su, Yuncheng Cao, Nengyou Wu and Yong He (2011) Numerical Analysis on Gas Production Efficiency from Hydrate Deposits by Thermal Stimulation: Application to the Shenhu Area, South China Sea. Energies 2011, 4, 294–313; https://doi.org/10.3390/en4020294

Zou Caineng, Zhang Guosheng, Yang Zhi, Tao Shizhen, Hou Lianhua, Zhu Rukai, Yuan Xuanjun, Ran Qiquan, Li Denghua, Wang Zhiping. (2013) Concepts, characteristics, potential and technology of unconventional hydrocarbons: On unconventional petroleum geology. PETROL. EXPLOR. DEVELOP., 2013, 40(4): 413–428. http://www.sciencedirect.com/science/article/pii/S1876380413600531

CPSIA information can be obtained
at www.ICGtesting.com
Printed in the USA
LVHW082359100319
610164LV00006B/109/P